European Simplified Methods
for Active Solar System Design

A publication of the Solar Energy Applications to Buildings R&D Programme of the Commission of the European Communities

The design tool software for personal computers referred to in the Handbook is distributed on behalf of the Commission of the European Communities by University College Dublin.

Further information may be obtained from

Energy Research Group
University College Dublin
Richview, Clonskeagh
IRL-Dublin 14

Telefax: +353-1-838908

EUROPEAN SIMPLIFIED METHODS FOR ACTIVE SOLAR SYSTEM DESIGN

by

BERNARD BOURGES

Consultant, Rennes, France

in collaboration with the Centre d'Energétique,
ARMINES – Ecole des Mines de Paris, France

KLUWER ACADEMIC PUBLISHERS

DORDRECHT / BOSTON / LONDON

ISBN 0–7923–1230–9

Publication arrangements by
Commission of the European Communities
Directorate-General Telecommunications, Information Industries and Innovation, Scientific and Technical Communication Unit, Luxembourg

EUR 13355 EN

Published by Kluwer Academic Publishers,
P.O. Box 17, 3300 AA Dordrecht, The Netherlands.

Kluwer Academic Publishers incorporates the publishing programmes of
D. Reidel, Martinus Nijhoff, Dr W. Junk and MTP Press.

Sold and distributed in the U.S.A. and Canada
by Kluwer Academic Publishers,
101 Philip Drive, Norwell, MA 02061, U.S.A.

In all other countries, sold and distributed
by Kluwer Academic Publishers Group,
P.O. Box 322, 3300 AH Dordrecht, The Netherlands.

Printed on acid-free paper

Printed in the Netherlands

FOREWORD

The events leading up to the publication of this book started effectively in 1976 with the exchange of information between those modelling teams in Europe which were involved in the Commissions R&D-programme on Solar Energy.

When it became clear that the availability of experimental data for model validation was next to nothing, the Commission took the initiative to support in the frame of the Commissions' Solar Energy R&D-programme the construction of Solar Pilot Test Facilities on 8 sites in Europe. Each experimental facility consisted of two real solar heating systems with collectors, storage, controls, and associated piping, but with the dwelling thermal distribution system replaced by a physical load simulator. One of the two systems on each site was a reference system identical for the 8 participating teams. The simulator was capable of producing a typical thermal load for a house, interactive with the actual weather and taking account of the effects of the occupants.

With data from these facilities not only national simulation programs were validated, but also the meanwhile commonly accepted modular structured European program EMGP2.
EMGP2, which only could be run on a mainframe computer or under special conditions on a mini-computer, formed in turn the basis for the development of EMGP3, EURSOL and ESM 1+2.

ESM 1+2 are the first fully validated, user friendly, simplified methods for active solar system design.

Although ESM1+2 are the products of efforts of a large group of European researchers, guided by their coordinator Professor Willy Dutré, the author of ESM1+2, Bernard Bourges, and his co-workers from the Ecole des Mines de Paris, have to be particularly thanked for the profound scientific work they have invested in the development of these simplified methods.

Unless all signs fail an upturn in the interest in and application of active solar heating systems seems forthcoming. This upturn is carried by the waves of a more stable and steady environmental current than the energy price issue was in the seventies and the eighties.
ESM1+2 will, therefore, be ready in time to serve as effective and reliable tools for the engineers to design or to assess the performance of environment-friendly solar heating systems.

Theo C. Steemers,

Commission of the European Communities.

IN MEMORIAM

All those who have worked together for the accomplishment of this book, i.e. all the participants in the European research projects SPTF (Solar Pilot Test Facilities) and OPSYS (active solar model development and model validation) would like to honour the memory of

Ir Ed van Galen,

who died during the course of these projects, to which he contributed so much with great enthusiasm. Ed was recognized by all as a scientist and engineer of great international stature; he paired engineering professionalism with sound organisational skill.

The European research community has lost a young scientist, who played already such a promising role in international research.

ACKNOWLEDGEMENTS

This handbook was prepared within the framework of the concerted research action OPSYS (of the European Solar Energy Research and Development programme), co-ordinated by W.DUTRE (K.U.LEUVEN).

The author would like to thank the many experts who participated in the preparation of this handbook. In particular the following persons should be thanked:

Jerome ADNOT, Lamine KADI, Bruno PEUPORTIER, and Philippe CALVET (Centre d'Energetique, Ecole des Mines de Paris), for providing the major technical information needed to produce this handbook.

Soren Ostergaard JENSEN (Thermal Insulation Laboratory, Lyngby, Denmark), Aart DE GEUS and Pieter BERGMEIJER (TPD, TNO/TH, The Netherlands), Willy DUTRE and Arnold DEBOSSCHER (K.U. Leuven, Belgium), and Cees DEN OUDEN, for their important technical contributions and useful comments.

Stephen GROVE for a careful and complete review of the first draft of this handbook.

Laurence CROSSBY for his editorial assistance.

The research which led to this book was integrally funded by the Commission of the European Communities (DG XII). The project was supervised by Theo C.STEEMERS.

The development of the methods and programmes described in this document was made possible by the continuous support of the Commission of the European Communities (DG XII) and the French Energy Management Agency (Agence Française pour la Maitrise de l'Energie, AFME).

Inquiries concerning the availability of the computer programmes described in this document (ESM1.4, ESM2.0) should be addressed to:

Centre d'Energetique
Ecole des Mines de Paris
60, Bd St-Michel
75272 PARIS CEDEX 06
France

TABLE OF CONTENTS

APPENDICES

PARTICIPANTS IN THE CEC OPSYS PROJECT

Co-ordinator: Willy DUTRE (Katholieke Universiteit Leuven, Belgium)

Katholieke Universiteit Leuven (Belgium):
Arnold DEBOSSCHER

Thermal Insulation Laboratory, Technical University of Denmark
Soren Ostergaard JENSEN

TPD-TNO-TH, Delft (The Netherlands)
Aart de GEUS
Pieter BERGMEIJER

Centre d'Energetique ARMINES-Ecole des Mines de Paris
Jerome ADNOT, Lamine KADI, Philippe CALVET
Bernard BOURGES

Commission of the European Communities
Theo STEEMERS, Cees DEN OUDEN

ILLUSTRATIONS

Thanks are due to the following persons, institutions and companies for providing figures or photographs and granting permission to reproduce them:

Pieter W. BERGMEIJER (TNO Institute of Applied Physics, Technisch Physische Dienst), Delft, The Netherlands
Figures 1.1.2, 1.1.4, 1.3.4.

CIAT, Culoz, Ain, France
Figures 2.2.4, 2.4.2.

DE DIETRICH, Niederbronn, Bas-Rhin, France
Figure 2.4.1.

Cees DEN OUDEN, The Netherlands
Figure 1.3.2.

Pr. Willy DUTRE (Catholic University of Leuven, KUL) Belgium
Figures 1.1.1, 2.2.1, 8.1.1, 8.1.2, 8.1.3, 8.1.4.

Daniel MANDINEAU (Ecole Supérieure d'Ingénieurs de Marseille), France
Figures 1.4.3, 1.4.11.

Michel ORSET, Arzano , Finistère, France
Figure 1.1.5.

Soren OSTERGAARD JENSEN (Technical University of Denmark, Lyngby)
Figure 1.3.8.

T. PAFELIAS, Greece
Figure 1.3.6.

Gérard SAVATIER (Association Savoyarde por le Développement des Energies Renouvelables, ASDER), St-Alban Leysse, Savoie, France
Figure 1.1.3.

SOLECO, Castelnau de Médoc, Gironde, France
Figures 2.1.3a, 2.1.3b, 2.1.12

Editions TECHNIQUES et DOCUMENTATION (Tec et Doc - Lavoisier), Paris, France
Figure 4.1.2 (from the book "Le gisement solaire", by Ch. PERRIN de BRICHAMBAUT and Ch. VAUGE, 1982).

TECSO, Nazaré, Portugal
Figure 2.1.5.

VICARB, St-Martin d'Hères, Isère, France
Figure 2.2.5.

WIELAND, Ulm, F.R.Germany
Figure 2.2.6.

The following figures are reproduced from documents published by the Commission of the European Communities:

R.GODOY and R.FERRARO. Solar space heating: an analysis of design and performance data. EUR 8004 EN (1982)
Figures 1.3.1, 1.4.7, 1.4.8, 1.4.9.

R.FERRARO and R.GODOY. Solar heating: Performance of recent systems. EUR 8947 EN (1983).
Figures 1.3.3, 1.4.10, 8.1.5.

R.FERRARO and R.GODOY. Solar heating: performance and cost improvement by design. EUR 8948 (1983).
Figure 1.4.12.

W.GILLETT. Solar collectors: Test methods and design guidelines. EUR 9778 EN (1985)
Figures 2.1.2, 2.1.9, 2.1.10.

J.MOON. Results and analysis of a round robin test series using an evacuated tubular solar collector. EUR 8757 (1983)
Figure 2.1.8.

W.DUTRE. EMGP2: a European transient simulation model for thermal solar systems. EUR 9777 (1985)
Figures 2.2.8, 2.2.9.

Abstract

This Handbook deals with the design of thermal active solar systems for space heating and hot water. Emphasis is placed on the European Simplified Methods, ESM1 and ESM2. These are design tools developed within the framework of the European project OPSYS. Active solar space heating (with possible combined hot water) is taken up in ESM1; domestic water heating (DHW) is taken up in ESM2. Computer programmes are available for both these methods. ESM1 and ESM2 make it possible to estimate the longterm performance of a given system as well as compare several alternatives and thus, optimize the system design.

This Handbook is both a practical guide for solar system designers, and the scientific reference manual for the ESM1 and ESM2 methods. Active solar systems are described from several points of view: schematics; components; operation and control. The components of the solar loop and other basic items required for solar system design (SH and DHW loads; meteorological data) are analysed. ESM1 and ESM2 are presented in detail with their complete algorithms, input parameters, worked examples, and software user's guide. Results of validation in comparison with the detailed simulation programme EMGP2 are also given. A special chapter is devoted to the principles of economic comparison. Additional information is provided in seven appendices (meteorological data; properties of materials; aide-mémoire for computer programmes; and "manual" version of ESM1).

INTRODUCTION:

This Handbook deals with the design of thermal active solar systems for space heating and hot water. Emphasis is placed on the European Simplified Methods, ESM1 and ESM2. These are design tools developed within the framework of the European Modelling Group and, more recently, the project OPSYS. Active solar space heating (with possible combined hot water) is taken up in ESM1; domestic water heating (DHW) is taken up in ESM2. Computer programmes are available for both these methods. ESM1 and ESM2 make it possible to estimate the long-term performance of a given system as well as compare several alternatives and thus optimize the system design.

The principal objectives of this Handbook are first, to serve as a practical guide for solar system designers and secondly, to serve as the scientific reference manual for the ESM1 and ESM2 methods. Furthermore, to meet the needs of a wide variety of readers, the fundamentals of active solar systems are explained and a general introduction is given in the first two chapters. Thus the Handbook can be used in several ways, depending upon the interest and need of the reader.

CONTENTS OF THE HANDBOOK

Chapter 1 describes a variety of active solar systems, both space heating and hot water systems, and how they work.

Chapter 2 analyses the basic components of active solar systems, including various types of solar collectors, storage tanks, and heat emitters. Similarly, characteristic parameters for computation are reviewed.

Chapter 3 presents the theoretical principles of the European Simplified Methods, ESM1 and ESM2: threshold irradiance, base temperature, collectable solar energy, and closed-form system modelling. Readers who would be daunted by equations may "by-pass" sections 3.3 and 3.4.

Chapter 4 provides detailed information about solar radiation data required by ESM's: cumulative frequency curves (CFC's) of solar irradiance. Data sources and useful simplified correlations are also indicated in this chapter.

Chapter 5 deals with the calculation of monthly space heating and hot water loads, which is a basic input for any calculation method. Conventional buildings are taken up in this chapter as well as partially "passive" solar buildings with direct solar gains.

In Chapter 6, the method ESM1 (Active solar space heating with or without combined domestic hot water) is described in detail. The calculation principles and algorithms, and a description of input parameters are outlined. Likewise, a detailed description of the method is given. Results of the validation in comparison with the detailed simulation programme EMGP2 are presented.

In Chapter 7, the method ESM2 (Solar hot water systems) is described in detail. As in Chapter 6, the calculation principles and algorithms, and a description of input parameters are outlined. Likewise, a detailed description of the method is given. Results of the validation in comparison with the detailed simulation programme EMGP2 are presented.

Chapter 8 presents guidelines for the optimisation and economic analysis of solar systems.

Finally, Chapter 9 contains the "User's guide to computer programmes". This chapter explains the installation procedure for ESM1.4 and ESM2.1 (or higher versions) and how to use these programmes on one's own compatible computer. Some practical advice for ESM users is also given in this chapter.

Additional information is provided in seven appendices. Appendix A (Solar radiation data) and Appendix B (Average temperature and degree-Days) provide sample information for twelve representative European locations. Appendix C provides information on properties of materials and collector fluids (e.g., water, and water + antifreeze). A manual version of the method ESM1 for space heating systems, including calculation worksheets, is described in Appendix D (Procedure for manual calculation ESM1). Appendix E (Aide-memoire of the ESM1.4 programme - Space heating) and Appendix F (Aide-memoire of the ESM2.1 programme - Hot water systems) provide practical memory aids, and input data sheets, to facilitate data input and output. Useful processing routines for solar radiation data are presented in Appendix G (CFC computation and fitting: Fortran sub-routines).

HOW TO USE THIS HANDBOOK

This Handbook was written in order to answer a variety of questions from a variety of readers, and it was assumed that not all readers would need the same type of information. Hence, the Handbook was designed so that each chapter may be viewed as a small handbook within a larger handbook and may be read independently of the other chapters. Therefore, readers interested in one particular aspect or aspects (e.g., solar collectors - section 2.1 or solar radiation data - chapter 4) can consult the relevant chapter or section directly without worrying about the material given elsewhere in the Handbook.

However, it is recommended that Chapter 1 be read first: this describes the design and operation of the systems considered in this Handbook. After reading Chapter 1, readers are free to work their way through the Handbook as they please. The following table was designed to help readers to find the sections or chapters they need or are interested in. For example, after viewing the table, the hurried reader who wants to know how to use the ESM1 method (space heating, SH) will see that he or she may go directly to Chapter 6 in which the basic information about the method is given. Section 6.5 (Validation results) may be skipped and one can move directly to Chapter 9 to receive some practical information about the software. Similarly, the hurried reader desiring information in regard to ESM2 (hot water) will see that he or she may go directly to Chapters 7 and 9. In order to facilitate chapter or section selection, a subject index is provided at the end of this Handbook.

WHICH CHAPTER OR SECTION TO READ?

First, read Chapter 1. Then choose the sections to read according to your particular interests. Guidance is given below.

	Sections
1 - DESIGNING AN ACTIVE SOLAR SYSTEM FOR SPACE HEATING	
1.1 - Is ESM1 adapted to your system?	6.1
1.2 - Do you want to use the software ESM 1.4 directly?	6.1 - 6.2 - 6.4 9.1 - 9.2 - 9.4 Appendix E
1.3 - Do you want to use the manual version of ESM1 ?	6.1 - 6.2 - 6.4 Appendices D, A, B
1.4 - Do you need more information about one particular component or input parameter required by the method?	
System parameters	Chapter 2
Solar radiation data	Chapter 4
Loads (Space Heating; Hot Water)	Chapter 5
1.5 - Are you looking for a method of economic optimization ?	Chap. 8
1.6 - Would you like to know the accuracy and limitations of the method?	6.5
1.7 - Do you want to know the theoretical principles of the ESM1 method and how it was developed?	Chapter 3 Chapter 6 (6.5)
2 - DESIGNING AN ACTIVE SOLAR SYSTEM FOR HOT WATER	
2.1 - Is ESM2 adapted to your system?	7.1
2.2 - Do you want to use the software ESM 2.1 directly?	7.1 - 7.2 - 7.4 9.1 - 9.3 - 9.4 Appendix F
2.3 - Do you need more information about one particular component or input parameter required by the method?	
System parameters	Chapter 2
Solar radiation data	Chapter 4
Loads (Hot Water)	Chapter 5

2.4 - Are you looking for a method of economic optimization ?	Chap. 8
2.5 - Would you like to know the accuracy and limitations of the method?	7.5
2.6 - Do you want to know the theoretical principles of the ESM2 method and how it was developed?	Chapter 3 Chapter 7 (7.5)

3 - LEARNING ABOUT ACTIVE SOLAR SYSTEMS

3.1 - What is an "active" solar system? How does it work? How should we control it?	Chapter 1
3.2 - What are the main components of an active solar system? How are they characterized from the physical point of view?	Chapter 2
3.3 - Are there simple mathematical models of active solar systems?	Chapter 3
3.4 - How can we characterize and estimate the climatic solar resource for a solar system?	
Solar radiation	Chapter 4
Temperature	5.2
3.5 - How to estimate the other basic constraint for a solar system: the load?	
Space heating load	5.1
Degree-days	5.2
Hot water load	5.3
3.6 - How to estimate the thermal long-term performance of an active solar system for space heating (and combined hot water)?	Chapter 6 Appendix D
3.7 - How to estimate the thermal long-term performance of an active solar system for hot water (individual or collective)	Chapter 7
3.8 - How to optimize a system from an economic point of view?	Chapter 8
3.9 - How to use the ESM software	Chapter 9 Appendices E and F

NOMENCLATURE

This list contains the symbols most commonly used in this Handbook. A short definition of each symbol is given, as well as its dimensions. Sections in which the symbol is defined are indicated as well as other chapters in which it is used. Symbols used infrequently in this Handbook are not listed here; they are defined locally within the Handbook.

$a, a_1 \ldots$	Non-dimensional parameters of the analytical model of solar systems (Sect. 3.3; Chap. 7)	
a_i	Fitting parameter of the CFC (Cumulative Frequency Curve of solar irradiance) (Sect. 4.2; Chap.4)	
A	Collector surface area (Sect. 2.1)	(m^2)
A_aU_a	Rate of heat loss for DHW back-up tank (Sect. 5.3)	(W/K)
A_cU_c	Rate of heat loss for DHW recirculation loop (Sect.5.3)	(W/K)
A_dU_d	Rate of heat loss for DHW distribution pipes (Sect.5.3)	(W/K)
$A_{dh}U_{dh}$	Heat transfer coefficient of the DHW pre-heating heat-exchanger (submerged in the storage tank) (Sect.2.3)	(W/K)
A_eU_e	Heat transfer coefficient of the solar heat-exchanger (2.2)	(W/K)
$A_{in}U_{in}$	Heat loss for external piping (collector inlet) (Sect.2.2)	(W/K)
$A_{ou}U_{ou}$	Heat loss for external piping (collector outlet) (Sect.2.2)	(W/K)
A_pU_p	Heat loss for indoor solar loop connecting pipes (Sect.6.2)	(W/K)
A_sU_s	Heat loss coefficient of the solar storage tank (Sect.2.3)	(W/K)
B	Direct (beam) irradiance at normal incidence	$(W.m^{-2})$
b_0	Coefficient of the ASHRAE incidence angle modifier formula (2.1)	
B_0	Solar constant (Sect.4.1)	$(W.m^{-2})$
$B_0(\alpha,\beta)$	Extraterrestrial irradiance on a tilted surface (Sect.4.1)	$(W.m^{-2})$
B_{0d}	Daily extraterrestrial irradiation on a horizontal plane (Sect.4.1)	$(MJ.m^{-2})$
B_c	Direct normal irradiance on clear days (Sect.4.1)	$(W.m^{-2})$
c	Water heat capacity (Sect.2.3)	$(kJ.K^{-1}.kg^{-1})$
C_{min}	Minimum capacitance rate of a heat-exchanger (Sect.2.2)	(W/K)
C_s	ESM1 storage correction factor (Sect.6.3)	
d_{on}	Starting control differential of solar loop (Sect.2.2)	(K)
d_{off}	Stopping control differential of solar loop (Sect.2.2)	(K)
D	Diffuse irradiance on a horizontal plane (Sect.4.1)	$(W.m^{-2})$
DD(T)	Degree-days at base temperature T (Sect.5.2)	
e_1	Effectiveness of the solar loop heat-exchanger (Sect.2.2)	
e_2	Heat emitter effectiveness (space heating) (Sect.2.4; 6.2)	
e_3	DHW pre-heating exchanger effectiveness (Sect.2.3)	
e_r	Air recuperator effectiveness (Sect.6.2)	
E_c	Collected energy (Chap.3)	(MJ)
E_d	Energy delivered to the load (Chap.3)	(MJ)
E_s	Stored energy (Chap.3)	(MJ)

f	Monthly solar fraction (fraction of monthly load provided by solar)	
f_a	Annual solar fraction	
$f_i(y)$	CFC fitting orthogonal polynomial (i^{th}-order) (Sect. 4.2)	
F'	Collector heat transfer factor (Sect. 2.1)	
F_r	Collector flow-rate correction factor (Sect. 2.1)	
F_r^*	Collector circuit (collector+piping) correction factor (Sect.2.2)	
F_r'	Solar loop correction factor (Sect. 2.2)	
G	Global irradiance on a horizontal plane (Sect. 4.1)	$(W.m^{-2})$
$G(\alpha,\beta)$	Global irradance on a tilted plane (Sect. 4.1)	$(W.m^{-2})$
G_c	Global irradiance on a horizontal plane on a clear day (4.1)	$(W.m^{-2})$
G_{cst}	System heat storage factor (or daily system efficiency) (constant irradiance) (Sect. 3.3)	
$G_m(\alpha,\beta)$	Monthly average daily irradiation on a tilted plane (4.1)	$(MJ.m^{-2})$
G_{sin}	System heat storage factor (sinusoidal irradiance) (Sect. 3.3)	
G_{str}	System heat storage factor (constant irradiance, stratified store) (Sect. 3.4)	
GV	Rate of heat loss from building (Sect. 5.1)	(W/K)
H_t	Daily sum of solar radiation on the collector surface (4.2)	$(MJ.m^{-2})$
H_{av}	Available or collectable solar energy above a given threshold or a given base temperature (Sect.3.2; 4.2)	$(MJ.m^{-2})$
I	Solar irradiance on the collector surface (2.1; 3.1; 4.2)	$(W.m^{-2})$
I_c	Critical irradiance (or threshold irradiance) (Sect.3.2)	$(W.m^{-2})$
I_{max}	Maximal solar irradiance on the collector surface (Sect.4.2)	$(W.m^{-2})$
k	Thermal conductivity	$(W.m^{-1}.K^{-1})$
$K(\theta)$	Collector incidence angle modifier (Sect. 2.1)	
$\bar{K}$	Monthly average incidence angle modifier (Sect. 4.4)	
K_s	Phillips' coefficient of stratification (Sect. 3.3)	
K_t	Hourly or instantaneous clearness index (Sect. 4.1)	
K_{tm}	Monthly clearness index (Sect. 4.1)	
K_1	Collector stratification coefficient (Sect. 3.4)	
K_2	Storage loss stratification coefficient (Sect. 3.4)	
L	Monthly total (SH + DHW) load (Sect. 6.2)	(MJ)
L_0	Monthly space heating load (Sect. 5.1; 6.2)	(MJ)
L_0'	Corrected monthly space heating load (Sect. 6.3)	(MJ)
L_1	Monthly domestic hot water load (Sect. 5.3)	(MJ)
m	Daily domestic hot water demand (Sect. 5.3)	(kg)
m_e	Effective daily domestic hot water demand (Sect. 5.3)	(kg)
$\dot{m}_0c_0$	Collector fluid capacitance rate (mass flow-rate $\dot{m}_0$ multiplied by heat capacity c_0) (Sect. 2.1)	(W/K)
$\dot{m}_1c$	Solar loop heat-exchanger fluid capacitance rate (between heat exchanger and storage tank) (Sect. 2.2)	(W/K)
$\dot{m}_2c$	Heat emitter fluid capacitance rate (Sect.2.4)	(W/K)
$\dot{m}_3c$	Hot water draw-off capacitance rate	(W/K)
M	Solar storage tank capacity (Sect. 2.3)	(kg)
M_a	DHW back-up tank capacity (Sect. 5.3)	(kg)
n	Number of cycles (stratified storage) (Sect. 3.3)	

n_h	Duration of critical intensity overstepping (CFC abscissa) (4.2)	
N	Number of days in month	
NTU	Number of transfer units (heat exchanger) (Sect. 2.2)	
P	Collected power	(W)
P_{fg}	Rate of building free heat gains (Sect. 5.1)	(W)
P_s	Rate of storage heat loss (Sect. 3.1)	(W)
P_{sh}	Rate of heat demand for space heating (Sect. 5.1)	(W)
q_f	Fresh air flow-rate (Sect. 5.1; 6.2)	(m^3/h)
q_r	Recycled air flow-rate (Sect. 6.2)	(m^3/h)
r	Ratio of capacitance rates of a heat exchanger (Sect. 2.2)	
r	Sun-Earth distance corrective factor (Sect. 4.1)	
r	Interest rate (Sect. 8.1)	
S	Standard deviation of monthly average temperature (Sect.5.2)	(K)
S	Sunshine duration (Sect. 4.1)	(s)
S_0	Time between sunrise and sunset (solar day length) (Sect.4.1)	(s)
t	Time	(s)
t_a	DHW back-up operating time (off-peak energy) (Sect.5.3)	(s)
t_c	Duration of one cycle (stratified storage) (Sect.3.3)	(s)
t_d	Daily radiation period (generally time between sunrise and sunset) (Sect. 3.3; 4.3)	(s)
t_e	Monthly heat emitter operating time (Sect. 3.2)	(s)
t_h	Daily period (24 hours or 86400 s)	(s)
t_L	Daily operating time of the DHW recirculation loop (Sect.5.3)	(s)
t_{mix}	System time constant (mixed storage) (Sect. 3.3)	(s)
t_{op}	Daily collector operating time (Sect. 3.3)	(s)
t_r	ESM1 reference system time constant (75850 s) (Sect.6.3)	(s)
t_s	DHW daily draw-off time (Sect. 2.3; 5.3)	(s)
t_{str}	System time constant (stratified storage) (Sect.3.3)	(s)
T_a	External ambient temperature (instantaneous or monthly)	(°C)
T_{abs}	Absorber plate temperature (Sect. 2.1)	(°C)
T_{av}	Collector fluid average temperature (Sect. 2.1)	(°C)
T_b	Base temperature of the system (Sect. 3.2)	(°C)
T_{bot}	Storage bottom temperature (Sect. 3.4)	(°C)
T_c	Demand temperature for space heating (Sect. 5.1)	(°C)
T_e	Temperature of the storage tank environment (Sect. 2.3; 3.1)	(°C)
T_f	Flow temperature to heat emitters (Sect. 2.4)	(°C)
T_h	Lowest acceptable temperature of domestic hot water (5.3)	(°C)
T_{he}	Effective required temperature of hot water (Sect. 5.3)	(°C)
T_{in}	Temperature of the fluid at the collector inlet (Sect.2.1)	(°C)
T_L	Collector limiting temperature (Sect. 3.3)	(°C)
T_L	Linke turbidity factor (Sect. 4.1; Chap.4)	
T_m	Mains cold water temperature (Sect. 5.3)	(°C)
T_{max}	Maximal air blowing temperature (SH air distribution) (6.2)	(°C)
T_{nh}	No-heating temperature (for degree-days calculation) (5.1)	(°C)
T_{out}	Collector outlet temperature (Sect. 2.1)	(°C)
T_r	Heat emitter return temperature (Sect. 2.4)	(°C)

T_s	Time-averaged storage temperature (Sect. 3.3)	(°C)
T_0	Initial temperature of the storage (Sect. 3.3)	(°C)
T_1	Storage temperature at the end of one cycle (stratified)	(°C)
u	Linear heat loss coefficient (e.g. for thermal bridges or piping losses)	$(W.m^{-1}.K^{-1})$
U	Absorber plate heat loss coefficient (Sect. 2.1)	$(W.m^{-2}.K^{-1})$
U_L	Collector heat loss coefficient (Sect. 2.1)	$(W.m^{-2}.K^{-1})$
U_L^*	Collector circuit heat loss coefficient (collector+pipe) (Sect. 2.2)	$(W.m^{-2}.K^{-1})$
U_s	Storage tank heat loss coefficient (Sect. 2.3)	$(W.m^{-2}.K^{-1})$
U_1	Collector heat loss coefficient: constant term (2.1)	$(W.m^{-2}.K^{-1})$
U_2	Collector heat loss coefficient: proportional term (2.1)	$(W.m^{-2}.K^{-2})$
v	CFC reduced dimensionless integral (Sect. 4.2)	
x	CFC reduced time (probability) (CFC dimensionless abscissa) (4.2)	
y	CFC reduced irradiance (CFC dimensionless ordinate) (Sect.4.2)	
y_c	Reduced critical intensity (Sect. 6.3)	
z	Storage / heat-exchanger flow-rate ratio (Sect. 3.4)	
α	Azimuth angle of the collector (Orientation) (0°= South, positive : West; negative: East)	(°)
α	Absorptance of the collector absorber plate (Sect.2.1; chap.2)	
β	Inclination angle of collector with respect to the horizontal	(°)
γ	Solar elevation (altitude angle of the sun above horizon) (4.1)	(°)
δ	Solar declination (Sect. 4.1)	(°)
η	Collector efficiency (Sect. 2.1)	
η_0	Optical efficiency of the collectors (normal incidence) (2.1)	
η_0^*	Equivalent optical efficiency of solar loop (corrected for piping losses) (Sect. 2.2)	
η_0'	Monthly average optical efficiency of the collector (corrected for pipe losses + Incidence angle modifier) (Sect. 6.3; 7.3)	
θ	Angle of incidence of the sun's rays on the collector (4.1)	(°)
μ	Absolute viscosity (Sect. 2.2)	(Pa.s)
ρ_g	Ground albedo (Sect. 4.1)	
σ	Stefan-Boltzmann constant (Sect. 2.2)	$(W.m^{-2}.K^{-4})$
τ	Transmittance of the collector transparent cover (Sect.2.1)	
$\emptyset$	Monthly or daily utilizability of solar radiation (Sect.3.2; 4.2)	
φ	Latitude (Sect. 4.1)	(°)
ψ	Solar azimuth (Sect. 4.1)	(°)
ω	Solar hour angle (Sect. 4.1)	(°)
ω_0	Sunset hour angle (Sect. 4.1)	(°)

CHAPTER 1

REVIEW OF ACTIVE SOLAR SYSTEMS FOR SPACE HEATING AND DOMESTIC HOT WATER

This handbook discusses thermal applications of solar energy working at temperatures ranging from 20° to 80°C. The range up to 100°C is referred to as low-temperature heat. Typical low-temperature applications of solar energy are in dwellings, either for space heating or domestic water heating. A few other related applications, such as industrial process heating (IPH), are also briefly analysed. These industrial applications do not differ essentially from a domestic application, even when operating at higher temperatures (100°-200°C).

The discussion is limited to a particular type of solar thermal system known as an "active system". Systems of this type as well as their common features and variants, are described in this chapter.

The complexity of all these systems, either from an operational or a computational point of view, results from the interaction of the various components, which we call "system effects". Therefore, the first step in the design process is the selection of a suitable configuration. In particular, this concerns the system schematics and control.

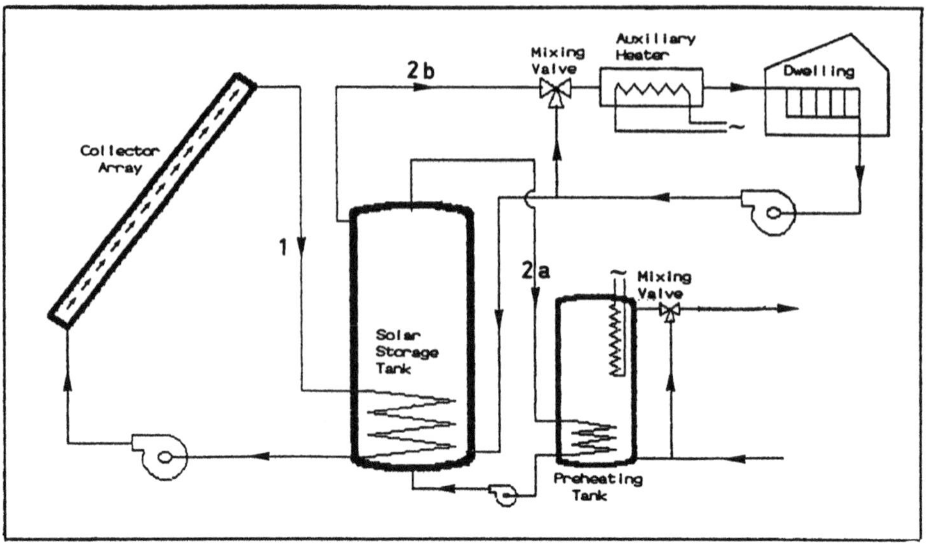

Fig. 1.1.1: Simplified schematic of a typical active solar system for space heating and hot water production.
(1) Solar loop (2a, 2b) Distribution loops.
(Document from Catholic University of Leuven, KUL).

Fig. 1.1.2: Flat-plate collector in a test loop (Doc. TNO/TPD).

1.1 - ACTIVE SOLAR SYSTEMS: A TYPICAL EXAMPLE

1.1.1 - Typical schematic and components

A typical active solar system for the provision of space heating (SH) and domestic hot water (DHW) is shown in simplified form in Fig. 1.1.1. As seen below, a number of variations are possible, however the basic functions always remain the same.

The solar loop (1) also known as the collector loop, comprises the components which collect solar energy (electromagnetic radiation from the sun), convert it to heat and transfer the heat to the storage. The basic component of the solar loop is a set of solar collectors through which a heat-transfer fluid circulates. This fluid is generally water or an anti-freeze solution, such as mono-ethylene glycol + water, but other fluids (e.g. air or an organic thermal fluid) can also be used.

Several types of collectors exist on the market. The most common for low-temperature applications are flat-plate solar collectors (Fig. 1.1.2), however other types can also be used, such as evacuated tubular collectors (ETC). Solar collectors (as well as all the components introduced here) are described in detail in Chapter 2.

The solar loop is completed by a piping circuit, a circulation pump, and several safety and maintenance devices (e.g., an expansion tank, a filter, an air bleed, a safety valve, and a non-return valve).

The heat from the solar loop is usually transferred to a heat storage device, either directly or through a heat-exchanger when different fluids are used in the collectors and in the storage. The solar loop is controlled by a differential thermostat which switches the pump on and off to ensure that fluid circulates only if a net gain of energy is possible.

A storage is required because energy demand and solar availability rarely coincide. Storage is generally as sensible heat (sometimes latent heat) and the most common form is an insulated water tank. A thermal stratification is observed in storage tanks: hot water, lighter than cold water, rises to the top of the tank. In well designed systems, this physical phenomenon is enhanced to improve the performance.

The connection with the load in the distribution loop varies according to the application. In the example shown in Figure 1.1.1, there are two different heat distribution loops. Loop

Fig. 1.1.3: Active solar house with direct solar underfloor heating. (Document from ASDER, Savoie, France).

Fig. 1.1.4: Active solar houses in Zoetermeer (Netherlands). (Document from Technisch Physische Dienst TNO).

(2a) is for preheating the domestic hot water (DHW) supply, and loop (2b) is for preheating the space heating (SH) via the heat emitters (for instance radiators) in the dwelling.

1.1.2 -Active and passive systems

From the example described in section 1.1.1, it can be seen that whatever its particular application, any active thermal solar system must perform the following functions: i) collection of electromagnetic radiation from the sun and conversion into heat; ii) transfer of the heat; iii) storage; and iv) distribution.

Furthermore, these four basic functions will be encountered in all thermal solar systems, even in those systems which are considered as non-active, such as passive solar buildings.

When an "active" solar system is compared with a "passive" system, an important difference can be noted. In an active system, each function (conversion, transfer, storage, and distribution) is carried out by specific, separate components, whereas a purely passive system has no specific components.

This distinction is brought out in the definition given in the "European Passive Solar Handbook" /1.1/:

> "The essential quality of a passive solar design is that it makes use of the conventional elements of a building to perform the functions of collection, storage and distribution of solar energy... In its simplest definition a passive solar building is one in which the necessary elements of a building -its walls, floors, roofs and windows- are arranged in a manner that maximises the benefits of solar heat in the winter months, whilst avoiding risks of overheating in the summer."

The distinction between active and passive systems is sometimes made according to other criteria, as follows. Auxiliary energy is generally required to make the transfer fluid circulate in active systems(forced circulation), whereas natural circulation (of air) takes place in passive buildings. System operation is managed by a specific control device in active solutions, whereas passive systems are more or less self-regulating. However, these distinctions are much less systematic and therefore these two criteria do not permit an acceptable distinction to be made between active and passive systems; for example, mechanical ventilation is often encountered in passive houses.

The first major distinction made above (according to components) will suffice, and it has major architectural implications (Figs. 1.1.3 to 1.1.5).

Passive solar system technology is sometimes preferred because of its architectural flexibility. However, this flexibility is offset by a number of constraints on building orientation. In particular, main rooms must be oriented towards the south and winter shading must be avoided. For active systems, orientation imposes fewer constraints. Southerly orientation concerns the collectors, but not necessarily the building, and roof-integrated collectors are less likely to be subject to shading.

The current trend is towards an integration of active and passive technologies. Forced circulation and control is becoming more and more usual in passive applications; certain passive components are even appearing on the market. On the other hand, better building and architectural integration of active systems is being achieved.

This integration is typified by "hybrid" systems which incorporate both active and passive concepts. Direct solar under-floor-heating is a good example in space heating (SH) applications, while integrated collector-storage (ICS) is another for domestic hot water (DHW) (see section 1.3.1).

Fig. 1.1.5: Passive solar house (Document M.ORSET, Arzano, France).

1.2 -THE SOLAR LOOP: SCHEMATICS, OPERATION AND CONTROL

Although it is possible to imagine a large variety of active solar systems (even with a simple application such as space or water heating), there are far fewer possibilities for the solar loop itself. A small number of basic designs exist, with only a few variants, and these are independent of the application. A distinction can be made between a solar loop without a heat exchanger (direct loop), and one with a heat-exchanger (indirect loop).

1.2.1 - Direct loop

In this simple configuration (Fig. 1.2.1a), solar heat from the collector is transferred directly to the storage (or to the distribution loop when there is no storage in the system). The same fluid circulates throughout, consequently the direct loop is unsuitable when anti-freeze fluid is required in the collector.

The direct loop is usually controlled by a differential thermostat (between the collector outlet and the bottom of the storage tank). This thermostat controls the circulation pump and lets it operate only if the temperature difference between the two sensors is greater than some set value.

In a variant of this system (Fig. 1.2.1b) the pump operates continuously during daylight hours and the differential thermostat controls a three-way valve.

These two types of control are roughly equivalent in terms of thermal performance /1.2/, therefore the former type (the simplest) is often preferred. Nevertheless, it is preferable to avoid starting and stopping the pump too frequently.

Guidelines for setting the on and off temperature differentials are provided in section 2.2.4.

Fluid flow rates of the order of 40-80 l/h per square metre of collector surface are generally used, but much lower values have been recently recommended for DHW applications (see section 2.3.2).

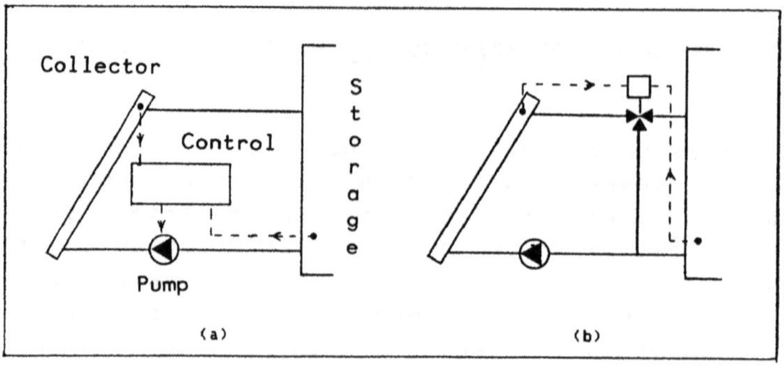

Fig. 1.2.1: Direct collector loop: (a) pump control.
(b) three-way valve control.

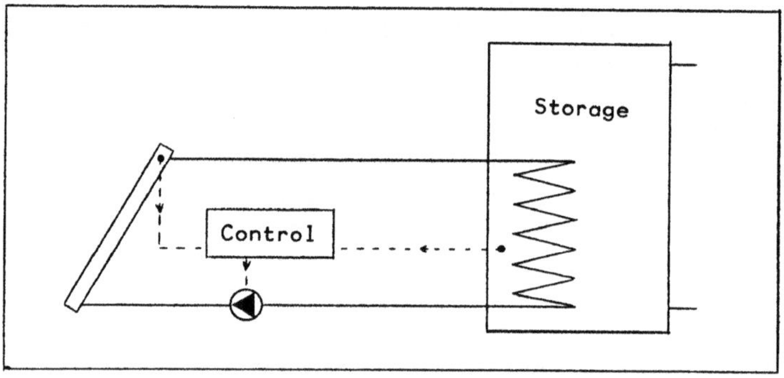

Fig. 1.2.2: Indirect collector loop with internal heat-exchanger.

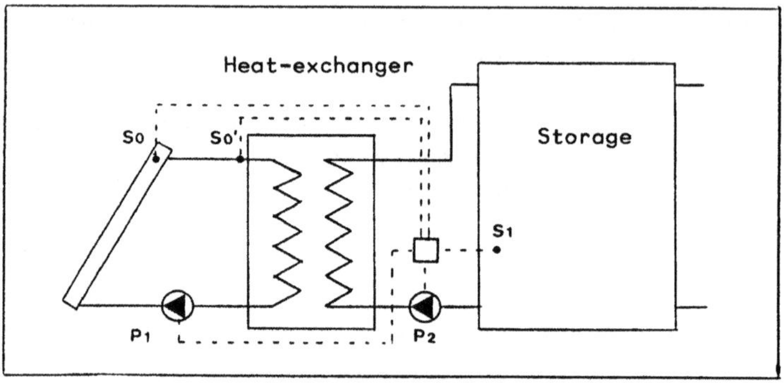

Fig. 1.2.3: Indirect collector loop with external heat-exchanger.

1.2.2 -Indirect loop with internal heat exchanger

Frost protection of a collector loop is usually achieved by the use of an anti-freeze solution, but this is rarely used as a storage fluid owing to its cost. A heat-exchanger is therefore required between the collector loop and the storage. The simplest case is that of a heat-exchanger submerged within the storage tank, most often in the lower part (Fig. 1.2.2). This design is often used for single-family solar water heaters. The control strategies mentioned above still apply here with similar sensor locations. As a general rule, this configuration does not allow the storage to remain stratified when the collector operates, thus low flow-rates are not suitable.

Nevertheless, a certain degree of stratification can be maintained with another device, namely the wrap-around heat exchanger (a steel mantle is welded around part of the inside surface of the tank).

Heat transfer in the tank itself takes place by free convection, and heat transfer coefficients are therefore more difficult to estimate (see Chapter 2).

The main advantage of this configuration is its simplicity (a simple heat-exchanger is used; only one pump is required and control is as straightforward as for a direct loop); this often compensates the decrease in performance caused by the introduction of a heat-exchanger.

1.2.3 - Indirect loop with external heat-exchanger

In this loop, the primary side of a tubular or plate heat-exchanger is inserted in the collector loop, and the secondary side is connected to the storage (Fig. 1.2.3). A secondary pump is needed and the control is more sophisticated.

One simple configuration has the primary pump P1 working continuously (during daylight hours), while a differential thermostat between the collector outlet and the bottom of the storage controls the secondary pump P2.

Another possible solution for large collector arrays has been proposed /1.6/. This solution incorporates three temperature sensors, one at the collector outlet (S0), one at the heat exchanger inlet (collector loop, S0'), and one at the bottom of the storage (S1). The control operates the primary pump P1 as soon as TS0 > TS1; after a delay, the pump is then controlled by the sensor S0' which replaces S0. The primary pump goes on working only if TS0' > TS1; in this case, the secondary pump P2 is also switched on.

For large plants, many sophisticated control strategies have been proposed in the past, including attempts at optimum control of several storage tanks at different temperatures. In most cases however, the practical operating problems outweighed the improvement in performance.

1.2.4 - Practical problems

Control devices are often the weak element of the solar loop and are not as reliable as they should be owing to the lack of electrical protection (breakdown may occur due to overvoltage or lightning). Sensor location must be carefully determined and checked after installation. Periodic checking and adjustment is required.

Heat losses from pipes are sometimes high and cause a considerable decrease in performance, particularly if high performance collectors are used. Reverse thermosiphonic flow sometimes occurs during the night; this can be prevented by the use of a suitable non-return valve.

Jet effects inside the storage tank tend to break down the thermal stratification and thus cause mixing of cold and hot water. Jet effects can be avoided by fitting an efficient nozzle device.

In large collector arrays, thermo-hydraulic problems often cause poor flow distribution, making a large fraction of the collector area useless /1.7-8/.

Some basic measuring equipment is necessary (even in small systems) for periodic checking of performance.

This list is of course not complete, and detailed additional information can be obtained from a survey of actual operations /1.9-10/.

1.3 - HOT WATER PRODUCTION

Some examples of solar systems for hot water production are presented here. First, individual water heaters will be discussed, then collective hot water systems for a block of flats. Finally, industrial process heating will be analysed.

For each type, both a theoretical basic design and practical examples are given.

1.3.1 - Individual water heaters

These systems produce domestic hot water (DHW) for a single family. It is the favoured field for the development of completely manufactured systems, with a large number of manufacturers in the market-place /1.11/. Most of these systems follow the general concepts indicated above, but a number of different designs exist.

1.3.1.1 - Forced circulation systems

The general schematic of a common type of individual solar water heater with forced circulation includes a single storage tank with an internal heat-exchanger (wrap-around or submerged coil). Auxiliary energy is provided at the top of the tank, either by an electric immersion heater (Fig. 1.3.1) or by another heat-exchanger connected to the space heating gas- or oil-fired boiler (Fig. 1.3.2). The auxiliary energy could also be provided in a second tank, the first one being used only as a solar pre-heating tank.

A differential temperature controller between the collector outlet and the bottom of the storage operates the pump. An anti-freeze solution protects the collector from frost.

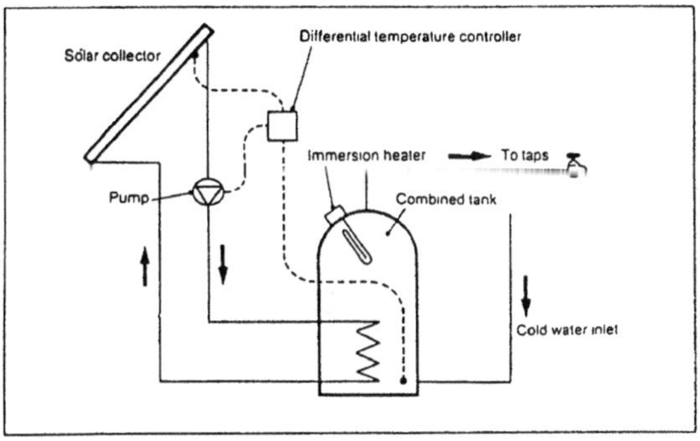

Fig. 1.3.1: Forced-circulation individual water heater: Combined tank system (electric back-up) (from Godoy /1.10/).

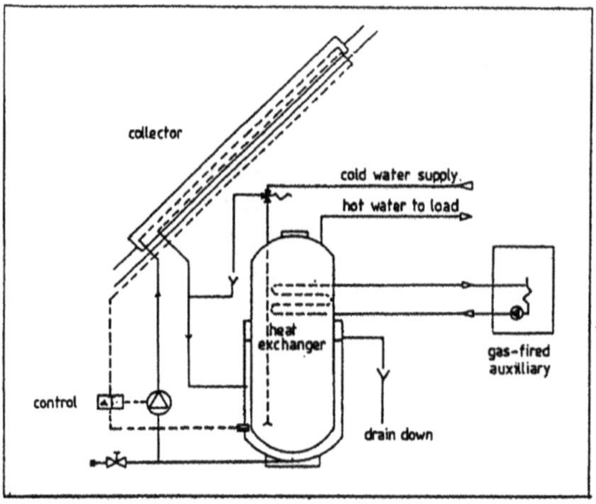

Fig. 1.3.2: Forced-circulation individual water heater: Combined tank system (gas fired auxiliary) (Doc. C.DenOuden).

Alternatively, the system can be protected by draining the collector when freezing is likely to occur. An example of such a configuration is shown in Fig. 1.3.3. Collector fluid (water) is stored in an open tank and a heat-exchanger is required in the distribution circuit. The collectors must be placed above the top of the storage tank; when the pump is switched off, the water in the collector drains down to the tank, hence water freezing cannot occur in the collector /1.6/.

Another possible design of such a "drain-back" system is shown in Fig. 1.3.4: the mantle heat-exchanger serves as a storage vessel for the water from the collector loop.

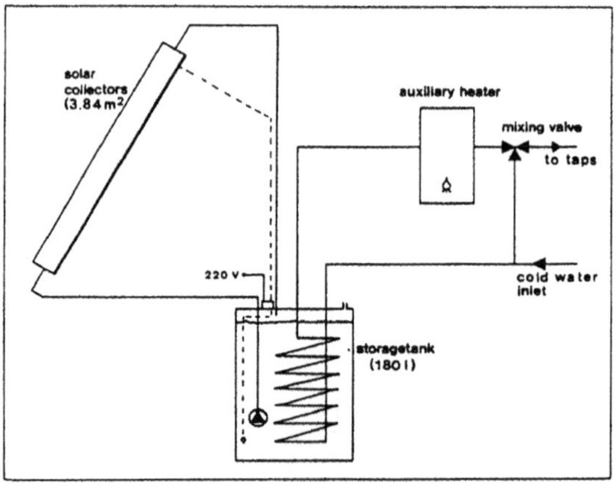

Fig. 1.3.3: Forced-circulation individual water heater: Drain-down system (from Ferraro /1.9/).

Typical sizing is based on a 4-person family consuming about 150 litres of water per day at 50°C. The storage volume is about equal to the daily demand, and the collector area ranges from 2 to 5 m2.

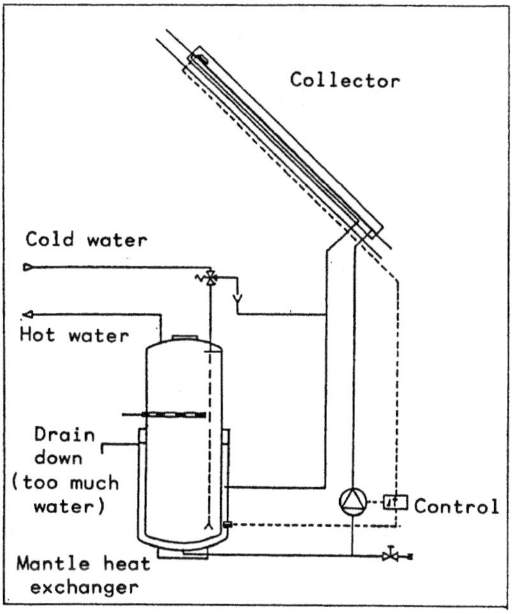

Fig. 1.3.4: Forced-circulation individual water heater: Drain back system (Doc. TNO/TPD).

1.3.1.2 - <u>Other types of solar water heater</u>

Besides the forced circulation water heaters following the general concepts defined in section 1.1 and 1.2, there are a number of other commercial systems.

Thermosiphon systems have been available for many years (Fig. 1.3.5). The water is circulated naturally by gravity convection (warm water in the collector rises and is replaced by cold water from the circuit) /1.12/. This requires careful design, but it leads to compact systems and is well adapted to sunny climates (Fig. 1.3.6). Thermosiphon systems with heat-exchangers also exist for northern climates /1.3/.

Integrated collector-storage (ICS) systems are another example of a recent design in which a single "box" simultaneously collects and stores the solar heat; these systems are simple and their performance rather low, but they are relatively inexpensive /1.14-15/.

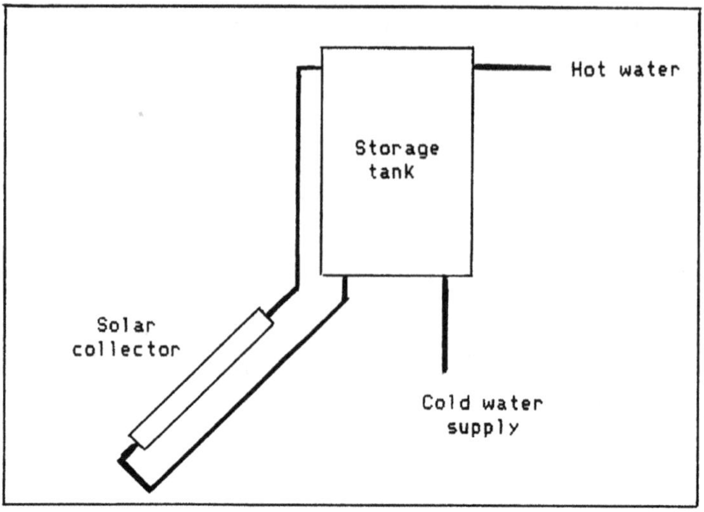

Fig. 1.3.5: Diagram of a thermosiphonic individual water heater.

Fig. 1.3.6: Typical thermosiphonic water heater
(Document from T.Pafelias, Greece).

Boiling-condensing systems are an example of more sophisticated ICS; the collector fluid boils in the collector and condenses in a heat exchanger situated in the water tank. The fluid is circulated by gravity (the tank is located above the top of the collectors) /1.16/.

The above examples give an idea of the wide variety of possible designs.

1.3.2 - Collective domestic hot water systems

Collective DHW solar systems (for a block of flats, a hospital or any "large" system), are in some respects simpler than individual ones: only the forced-circulation schematics can be used, since special designs such as thermosiphon or ICS are not suitable for large collector areas. Nevertheless, economies of scale permit a number of sophistications to be introduced. This applies both to layout and control. In large systems, the relative cost of an additional component (e.g. micro-processor control) is small, and it may seem attractive to complicate the system. This is often a dangerous temptation which can lead to many practical operating problems.

The schematic of collective hot water systems(Fig.1.3.7) brings out some particular features of large (in contrast to individual) systems. Auxiliary energy (either from electricity or a fossil fuel-fired boiler) is generally provided in a special tank. A recirculation loop is often used to ensure that the water is hot at the point of use, no matter how far from the storage. The recirculation loop is represented in Fig. 1.3.7 by a dashed line. It involves an additional pump. All the possibilities for the solar loop described in section 1.2 are design options. Occasionally a direct loop is used, but more usual is an indirect loop with an internal or external heat exchanger.

Fig. 1.3.8 shows a variation for the solar loop; the cold water supply is connected to the inlet of the heat-exchanger. This maintains a lower fluid temperature (and a better collector efficiency), particularly if flow rates are high and the storage is fully mixed /1.6/.

Another type of modification concerns the storage. Large systems require high storage capacities (typically a few cubic metres) and the storage is often divided into two or more tanks. It is possible to optimize the control of the system through three-way valves as in the Freiburg plant /1.9/.

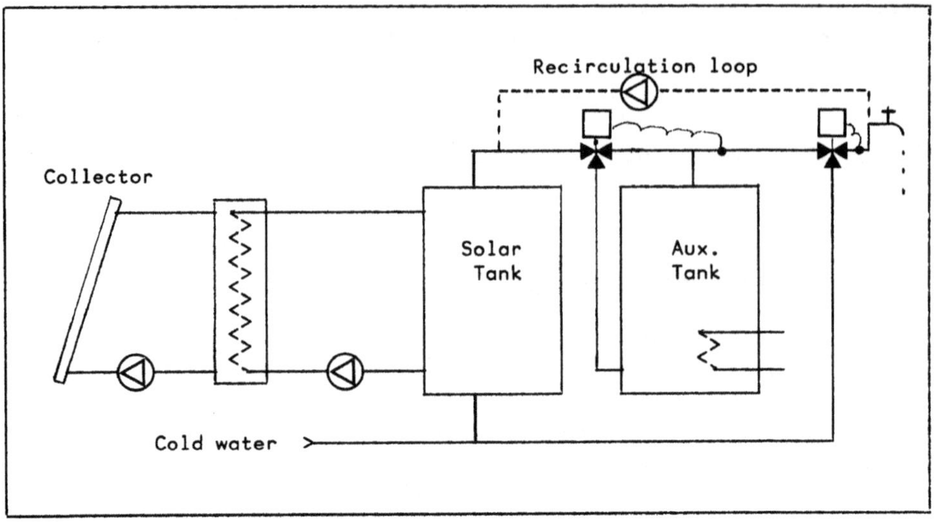

Fig. 1.3.7: Collective solar water heating: Typical schematic with recirculation loop.

A final variation presented here concerns the distribution circuit. In the schematic of Fig. 1.3.7, heat losses from the recirculation loop can be considered as part of the hot water load, but in such a configuration the solar energy cannot make even a partial contribution to this additional load. A solution is shown in Fig. 1.3.8; depending on the temperature level of the storage tank, the recirculation loop is directed towards either the solar tank or the auxiliary tank.

1.3.3 - Industrial process heating

Industrial process heating (IPH) differs from domestic hot water (DHW) systems in that the heat demand is constant during working hours, whereas the DHW load is highly time-dependent during the day. This applies to energy supplied as hot water or as water vapour.

The constant demand makes it possible to use particularly simple systems without storage (Fig.1.3.9) in which the process fluid goes directly through the collectors or through an intermediate heat-exchanger. This type of system must be sized to

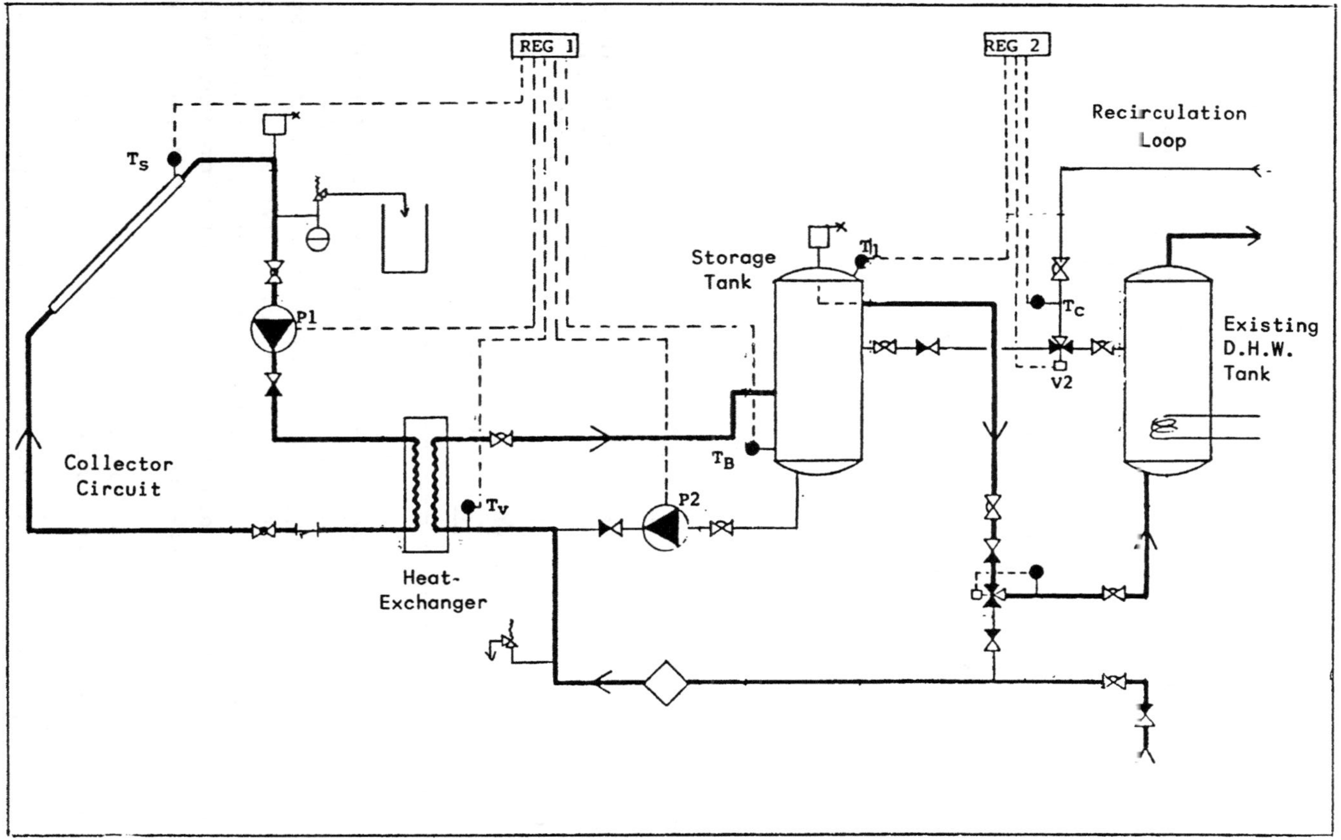

Fig. 1.3.8: Collective solar water heating system: Recirculation loop coupled both to solar tank and auxiliary tank, with control (Document from Technical University of Denmark).

avoid energy dumping; the collector outlet temperature must never exceed the required process temperature. The fraction of the load met by solar energy (which is referred to as the "solar fraction") is therefore low but these systems are cheap and the energy output per square metre of collector is maximised.

A variation of this simple configuration, the so-called "single pass open-loop design" incorporates a variable volume tank. This makes it possible to store heat for use in cloudy periods without increasing the collector fluid temperature (and thus, decreasing collector efficiency) (Fig. 1.3.10).

This type of system has been analysed extensively by several authors and reviewed by A.RABL /1.17/.

Fig. 1.3.9: Industrial process heating: Solar system without storage.

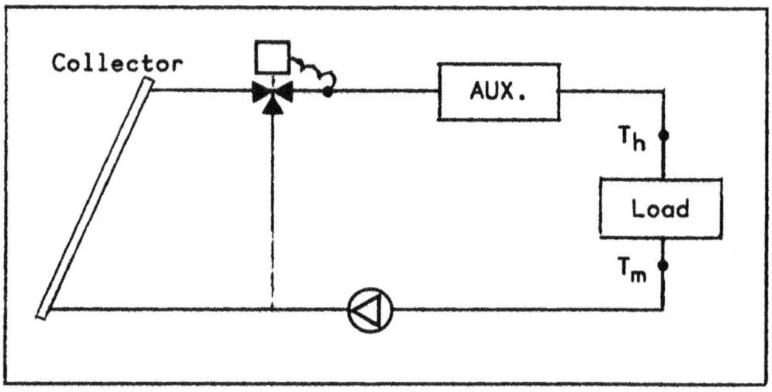

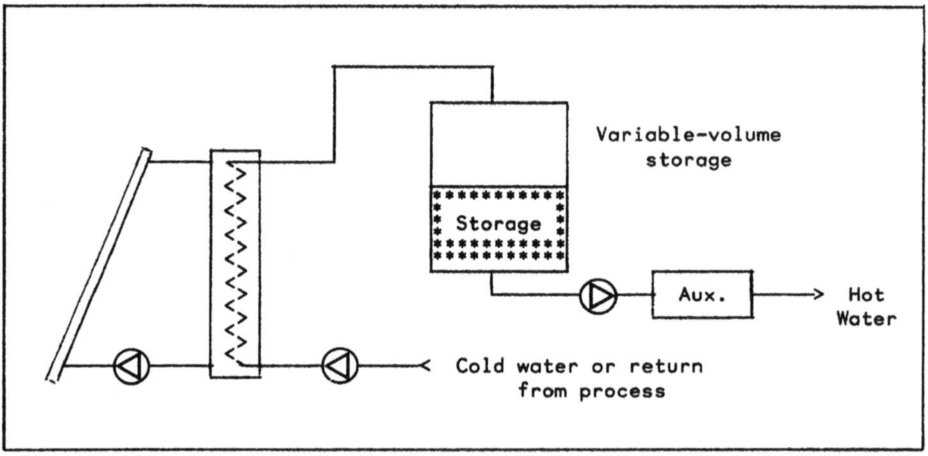

Fig.1.3.10: Industrial process heating: Solar single pass open-loop system.

1.4 - SPACE HEATING SYSTEMS

Space heating (SH) is another typical application for active solar systems. At mid-latitudes, space heating constitutes a much higher yearly load than domestic hot water (DHW), but is much more season-dependent.

1.4.1 - Space heating only

Active solar systems contributing only to space heating are rare, because they are uneconomic, except in regions where space heating is required all year round. Existing systems are generally demonstration plants. Nevertheless, it is interesting to consider them, because they facilitate an understanding of the space heating operation.

The first example is shown in Fig. 1.4.1. The solar loop does not differ essentially from the basic design (see section 1.2). This system is known as "parallel auxiliary"; the auxiliary energy is provided independently (e.g, by electric convectors). Solar energy can be distributed by conventional heat emitters (such as radiators), but low temperature emitters such as underfloor heating or low-temperature convectors are highly preferable. A three-way valve in the distribution circuit limits the distribution temperature (sometimes matching it to the heat demand). The circulation pump is generally controlled by a thermostat situated in the house.

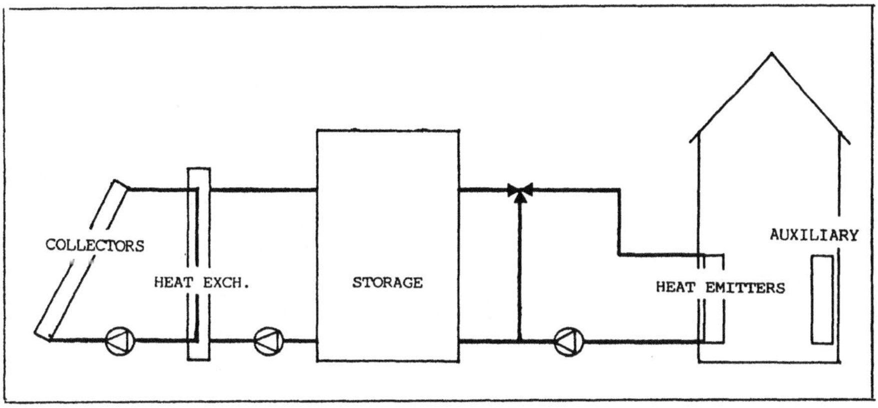

Fig.1.4.1: Schematic of an active solar space heating system with liquid circulation. Parallel auxiliary.

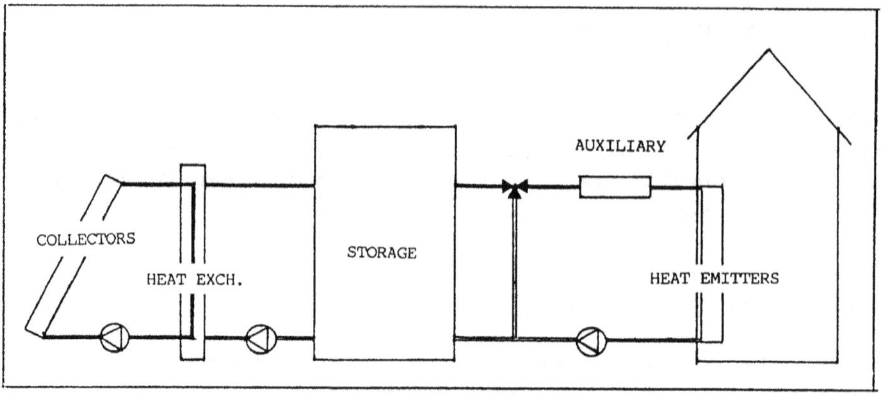

Fig.1.4.2: Schematic of an active solar space heating system with liquid circulation. Series auxiliary.

This configuration follows a basic principle of solar systems; "do not mix energies". However, it requires two distribution networks, it is also more expensive to install and the "series auxiliary" configuration (Fig. 1.4.2) is sometimes preferred.

In the latter case, the three-way valve has the additional function of preventing the fluid entering the storage tank (and making it warmer) when the heat emitter return temperature is higher than the storage tank temperature.

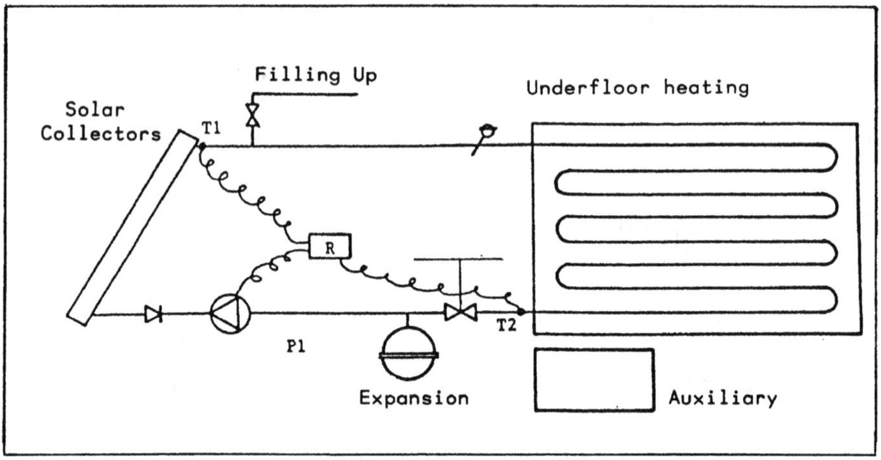

Fig. 1.4.3: Direct solar underfloor heating (Doc. ESIM, Marseille, France).

"Direct solar underfloor heating" is a worthwhile variation in which energy is stored directly in the concrete floor. Thus, the floor serves for both storage and distribution, and no separate storage tank is required. Typical collector areas of 0.2 square metres per square metre of floor can be used with 25 cm thickness of concrete (Fig. 1.4.3) /1.18/.

1.4.2 - Domestic hot water coupling

The efficiency of active solar space heating systems can be greatly improved by coupling them with domestic hot water (DHW) which in most parts of Europe is the only load during the summer. Most practical systems provide DHW as well as space heating (SH). Coupling with DHW can be done in several ways.

The simplest solution is an internal pre-heating heat-exchanger, generally a simple pipe, connected to the usual DHW auxiliary (Fig 1.4.4). The solar energy extracted is very sensitive to the draw-off flow rate, and is highest at low values. This simple method of installing a hot water system can give remarkably poor results, especially in summer when there is usually more than enough solar energy (see section 2.3.4).

Fig.1.4.5 shows another system, in which a DHW pre-heating tank is connected to the main tank through a heat-exchanger and a circulation loop. The circulation pump is controlled by a differential thermostat between the two tanks.

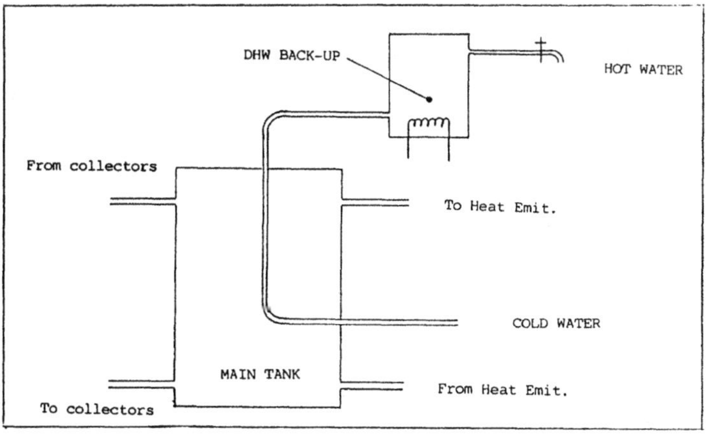

Fig.1.4.4: Active solar heating system incorporating preheating of DHW (by simple pipe or internal heat-exchanger, without preheating tank).

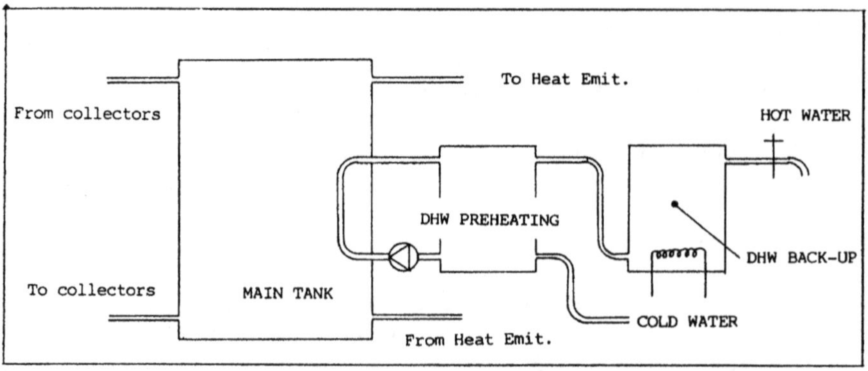

Fig.1.4.5: Active solar heating system incorporating preheating of domestic hot water (external pre-heating tank with a heat-exchanger).

Another possibility is a DHW pre-heating tank submerged in the main tank (Fig. 1.4.6, a and b). This design does not require a control device.

Auxiliary energy can be provided by any common method (electricity or fossil-fuel boiler), either directly in the upper part of the pre-heating tank or by an additional back-up device connected in series.

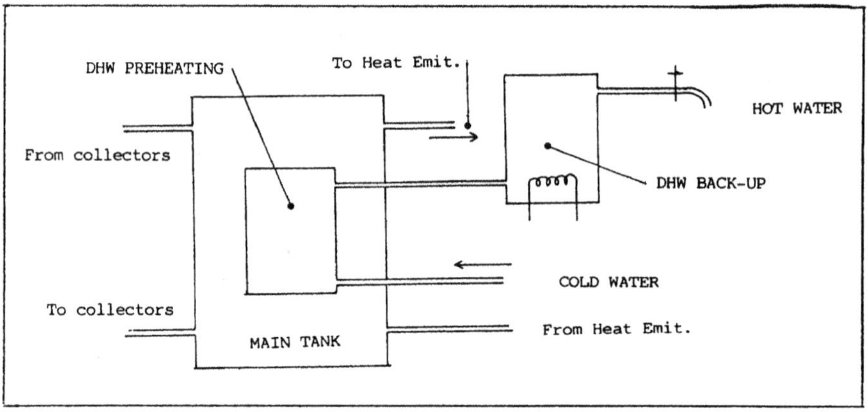

Fig.1.4.6a: Active solar heating system incorporating preheating of domestic hot water (submerged pre-heating tank).

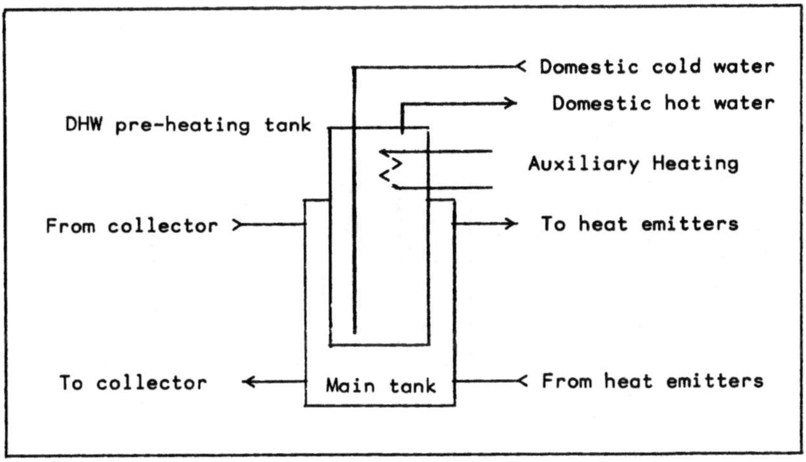

Fig.1.4.6b: DHW submerged pre-heating tank with incorporated auxiliary.

1.4.3 - Examples of working systems

This section presents examples of working space heating (SH) + domestic hot water (DHW) systems. Most of them are described in more detail in the reports of the E.C. Performance Monitoring Group /1.9-10/. Thermal performance has been measured for at least one year and reported by the PMG.

The Bourgoin building (Fig. 1.4.7) comprises 12 flats. Two identical systems have been installed with 152 m2 of flat-plate collectors and 30 m3 of storage. The solar loop incorporates an external heat - exchanger. Distribution is by underfloor heating; auxiliary energy is provided by electric convectors, so the system is of the "parallel-auxiliary" type. The DHW pre-heating circuit is connected to the main tank through a heat-exchanger submerged in the DHW tank.

In Zoetermeer, four slightly different systems have been installed. In House 4 (Fig. 1.4.8) the solar loop (including 35 m2 of flat-plate collectors) is connected to the storage tank (2 m3) via an internal heat-exchanger. Heat distribution is by a conventional air-heating installation which includes the auxiliary gas boiler. This system may be described as a "parallel-auxiliary" type, because there is almost no interaction between the solar part and the conventional boiler. The air temperature at the inlet of the fan-coil is always equal to the internal temperature of the building. A submerged

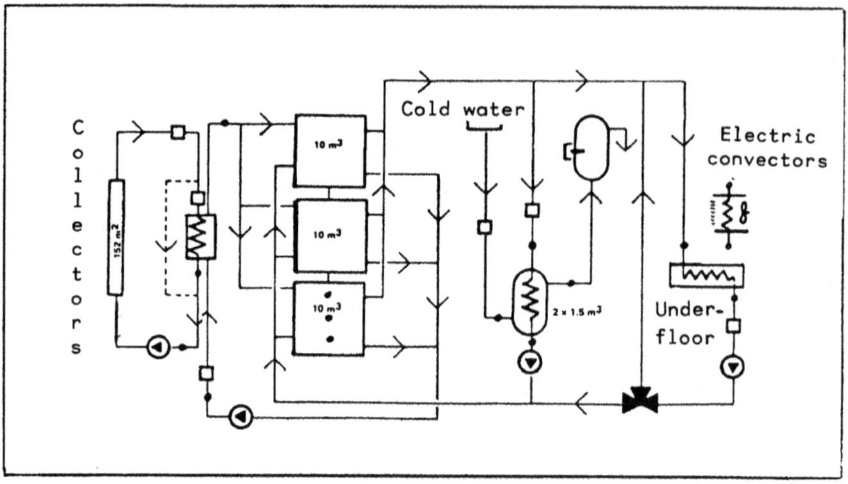

Fig. 1.4.7: Bourgoin solar building (from Godoy /1.10/).

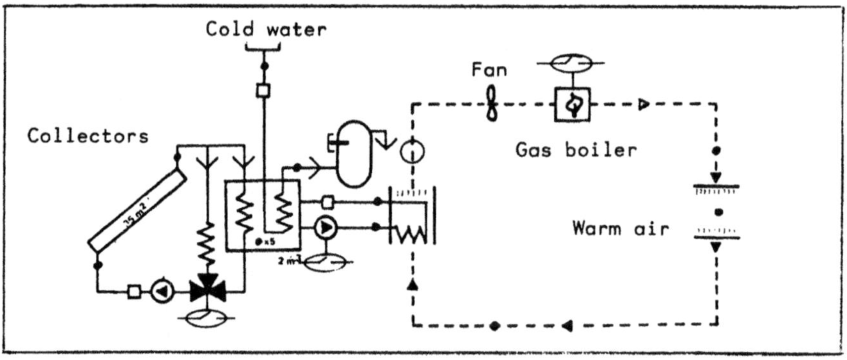

Fig.1.4.8: Zoetermeer solar house. System number 4 (air distribution).

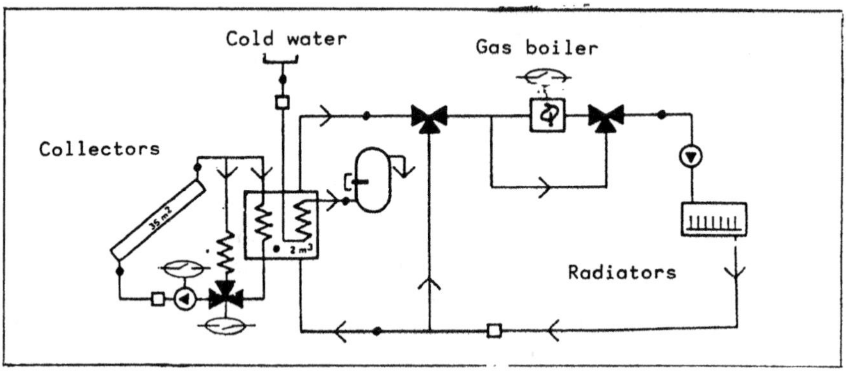

Fig. 1.4.9: Zoetermeer solar house. System number 2 (from /1.10/).

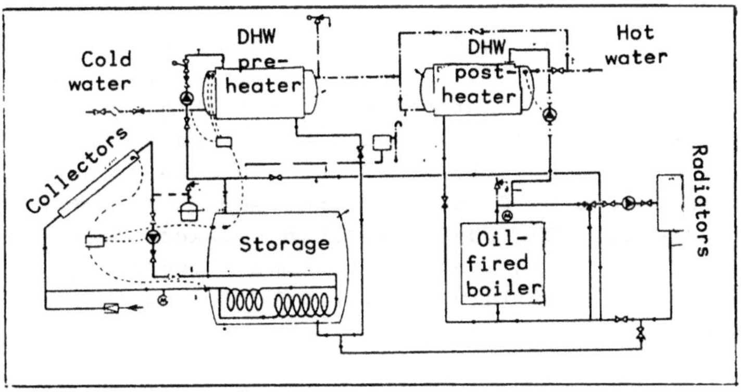

Fig.1.4.10: Gentofte (DK) solar house (from Ferraro /1.9/).

heat-exchanger provides pre-heating of DHW.

Zoetermeer House 2 (Fig. 1.4.9) is a typical "series auxiliary" system with radiators and a gas boiler. Control of the radiators is by a conventional room thermostat with priority to the distribution of solar heat. When the temperature of the water in the tank drops below the level required for a sufficient supply of heat to the radiators, two three-way valves are automatically operated to alter the distribution circuit, so that the circulation bypasses the storage tank and flows through the auxiliary boiler. Whenever solar radiation increases the temperature in the tank sufficiently, the initial "solar" mode of operation is restored.

The Gentofte system is a similar "series-auxiliary" type with radiators and an oil-fired boiler which is also used in winter for DHW back-up (Fig. 1.4.10).

"Direct solar underfloor heating" (coupled with DHW, see Fig. 1.4.11) has been installed in several places in France, for example in Bassens houses A and D /1.9/. Coupling with DHW can be done by means of an automatic three-way valve with priority to the underfloor heating when space heating is required. For simplicity, it is sometimes preferable to replace the three-way valve by two ordinary valves and to distinguish winter (space heating only) and summer (DHW only) operation, the selection being decided by the user. As the mid-seasons, during which the system can meet significant fractions of both the SH and the DHW loads, are very short, the system gains a lot in reliability without a significant decrease in performance.

The last example presented here is a system that has been optimized by the Danish design group of the Performance Monitoring Group (PMG) /1.6/, but not yet built (Fig. 1.4.12). Solar energy for space heating is delivered via a heat-exchanger inserted in the ventilation system. This takes advantage of the low temperature of fresh air and results in better collector efficiencies. The DHW pre-heating takes place in a tank submerged in the main tank; the upper (emerging) part is connected to the auxiliary oil-fired boiler which is switched off in summer.

As air is sometimes used for heat distribution in active solar systems, it should be mentioned that air can be used in the solar loop insted of a liquid. Such systems do not differ significantly from those using a liquid. An air solar loop can incorporate air collectors, ducts and fan with control devices; storage is by a pebble bed; space heating is provided by a conventional air heating system. Mainly owing to the leakage problems encountered for some time in air collectors, these systems have been neglected, but recent experience has demonstrated that inexpensive systems based on air collectors can be built, such as the "roof space collector" /1.20/.

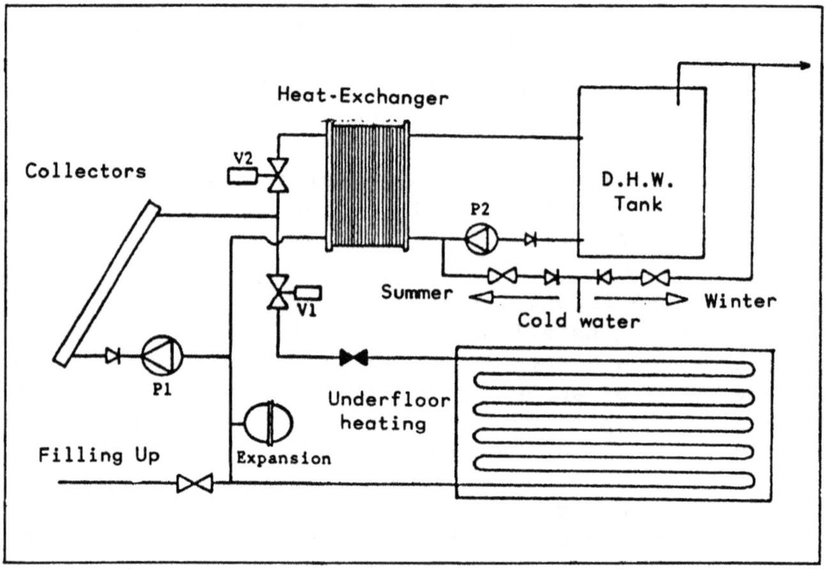

Fig.1.4.11: Direct solar underfloor heating coupled with DHW production (Document Ecole Supérieure d'Ingénieurs de Marseille, ESIM).

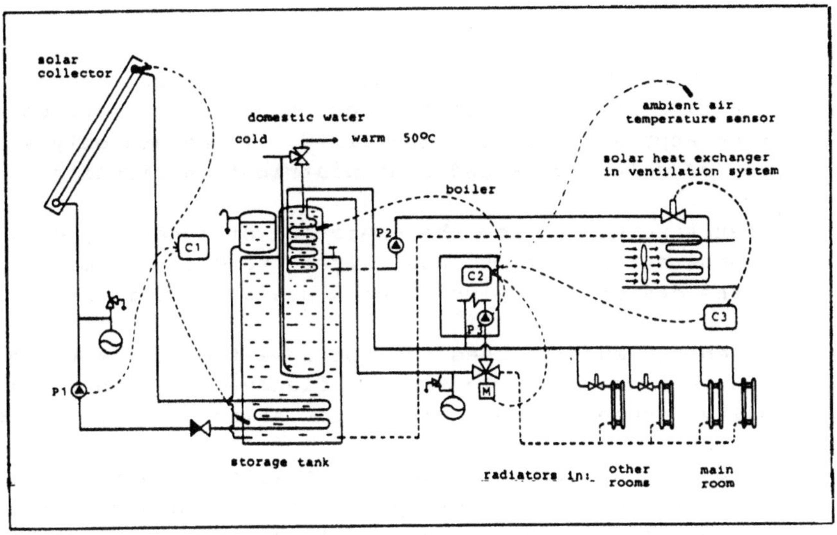

Fig.1.4.12: Active solar SH + DHW system: PMG Danish group variant (from Ferraro /1.6/).

1.5 - CONCLUSION

This chapter has shown that there are a large number of possible active solar system designs and control strategies, but a common approach can be identified. There are only a few basic types of solar loops and heat distribution circuits.

The experience acquired from the operation of carefully monitored first-generation systems has helped to define some simple rules for the design of active solar systems. To summarize:

- Solar energy must be used at the lowest possible temperare, so as to increase collector efficiency (see Chapter 2).
- Priority should be given to solar energy in the distribution.
- It is better not to mix solar and auxiliary energies, thus careful attention should be paid to the design of the distribution loop and its combination with the auxiliary energy supply.
- Stratification should be maintained within the storage tank, both during the solar charge and the discharge.
- The load must be estimated carefully (not over-estimated).

However, the most essential rule is to construct <u>simple</u> systems. Experience has demonstrated that complexity (using sophisticated designs and control) sometimes leads in theory to a small percentage improvement in performance but in practice always introduces much more important losses owing to faulty assembly of components or the incorrect setting of control parameters, breakdowns, etc.

Furthermore, all the usual plumbing, building, and electrical regulations and practical rules have to be applied.

Simplicity has in fact become a basic feature of recent systems. There is also a tendency for systems to be undersized (seldom more than 20 m2 of collector for a single family house) in order to optimise the specific energy production (useful solar output per square metre of collector) and capital investment.

These general aspects of solar system design are discussed in several documents. Those produced by the European Performance Monitoring Group are particularly recommended /1.6/.

REFERENCES OF CHAPTER 1

/1.1/ P.ACHARD and R.GICQUEL. European Passive Solar Handbook. CEC DG XII, Brussels, Report EUR 10683 (1986).

/1.2/ J.ADNOT et al. Comportement en régime transitoire et régulation de boucles de captage. Centre d'Energétique ARMINES. Rapport Plan-Construction (1981).

/1.3/ S.FURBO and S.E.MIKKELSEN. Advances in Solar Energy Technology (ISES Congress, Hamburg, 1987), Vol.1 pp.962-966. Pergamon, Oxford (1988).

/1.4/ M.J.CARVALHO et al. SOLAR ENERGY, Vol.41, n.1, pp.33-40 (1988)

/1.5/ WUESTLING et al. ASME J. of Sol. Energy Engineering, Vol.107, pp.215 (1985).

/1.6/ R.FERRARO and R.GODOY. Solar Heating: Performance and Cost Improvement by Design. CEC-DG XII, Doc. EUR 8948 (1983).

/1.7/ A.T.SOLIMAN. Contribution à l'étude des installations solaires à eau de grande taille. Thèse de Docteur-Ingénieur, Univ. de Provence (1986)

/1.8/ A.LEBRU. Comportement hydraulique des champs de capteurs. CSTB Sophia-Antipolis, rapport MPE/411/GR (1985).

/1.9/ R.FERRARO and R.GODOY. Solar Heating: Performance of recent systems. CEC-DG XII, Doc. EUR 8947 (1983).

/1.10/ R.GODOY et al. Solar Space Heating: an analysis of design and performance data from 33 systems. CEC-DG XII, Doc. EUR 8004 (1982).

/1.11/ E.ARANOVITCH et al. Recommendations for Performance and Durability Tests of Solar Collectors and Water Heating Systems. Commission of the European Communities (European Solar Collector and System Testing Group), JRC Ispra. EUR 11606 EN (1989).

/1.12/ D.J.CLOSE. The performance of solar water heaters with natural circulation. SOLAR ENERGY, Vol.6, pp.33-41 (1962).

/1.13/ P.LAVILLE. BATIMENT-ENERGIE, Vol.29, pp.92-93 (1983).

/1.14/ C.BOUSSEMAERE et J.BOUGARD. Performances thermiques et tests de qualification d'un chauffe-eau solaire à stockage intégré. Fac. Polytechnique de Mons (1986)

/1.15/ ZOLLNER et al. ASME J. of Sol. Energy Engineering, Vol.107, pp.265-272 (1985).

/1.16/ P.LAVILLE. BATIMENT-ENERGIE, Vol.28, pp.47-50 (1983).

/1.17/ A.RABL. Active Solar Collectors and their Application. Oxford Univ. Press, New-York (1985).

/1.18/ D.ROUX et al. Calcul des Planchers Solaires Directs. EDISUD (1983)

/1.19/ W.PALZ and T.C.STEEMERS. Solar Houses in Europe: how they have worked. EUR 7109. Pergamon (1981).

/1.20/ S.Ø.JENSEN and V.KORSGAARD. Advances in Solar Energy Technology (ISES Congress, Hamburg, 1987), Vol.1, pp.1135-1139. Pergamon, Oxford (1988).

CHAPTER 2

THE COMPONENTS OF ACTIVE SOLAR SYSTEMS:

DESCRIPTION, TYPES, AND CHARACTERISTICS

Almost all types of active solar heating systems make use of the same basic components. In this chapter these components are analysed under four headings: 2.1 - the collectors themselves, 2.2 - the solar loop, 2.3 - the storage, and 2.4 - the space heating distribution.

Each of these groups of components includes several types. For example, solar collectors may be of the flat-plate type, the evacuated tubular type, or the compound parabolic type, to name just a few. Heat-exchangers may be of the tubular or plate type, and heat distribution networks may take the form of convectors, radiators, or underfloor heating. The physical principles and operating conditions of these components are described. Characteristic parameters are also defined: these are required as input data by the ESM computation methods reported in Chapters 6 and 7. Likewise, several means of derivation are given, either by computation or by measurements from standard test procedures.

2.1 - ACTIVE SOLAR COLLECTORS

The solar collector is the basic component of any active solar system. Its characteristics dominate the overall system performance. Therefore, this component is analysed in detail here.

2.1.1 - Solar collectors: principles and types

Solar collectors convert radiant energy from the sun into heat and transfer it to a working fluid. This conversion has to be made with maximum possible efficiency, in order to minimize heat losses to the environment. Solar collectors used for low-temperature applications, such as those studied in this book, are generally of the flat-plate type. However, others, such as evacuated tubular collectors or compound parabolic collectors with a small concentration factor, will become more widely used in the near future. A typical flat-plate solar collector is shown in Figure 2.1.1.

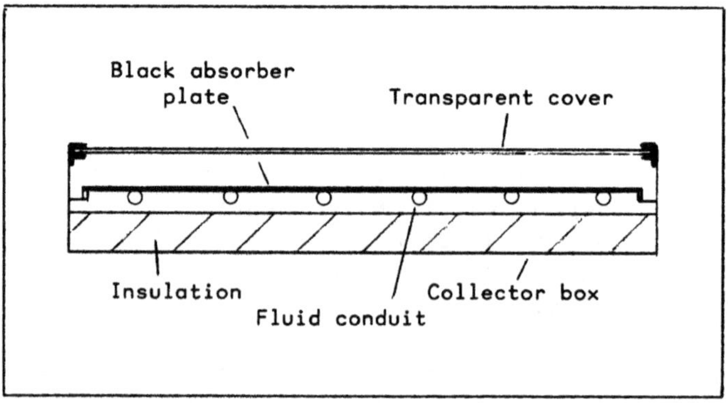

Fig. 2.1.1: Cross-section of a basic flat-plate solar collector.

Thermal conversion takes place in the absorber plate. This plate is typically made of metal (copper, aluminium, or steel), it absorbs the electromagnetic radiation from the sun, and converts it into heat. This plate is in thermal contact with the heat transfer fluid. Solar collector performance depends on several characteristics, two of which are predominant:

- The absorption properties of the collector in the most energetic part of the solar spectrum, which includes not only the visible spectrum but also a significant amount of infra-red radiation. This characteristic accounts for the dark appearance of most solar collectors.
- Good thermal contact between the absorber and the heat transfer fluid.

Conversion cannot be made without thermal losses to the environment. Owing to its greater absorption of solar radiation, the collector is warmer than its surroundings and heat is lost principally by radiation and convection. Thermal insulation is therefore needed. Common insulation materials are used for the back, while a transparent cover is put on the front; this transmits solar radiation and reduces convection losses. Furthermore, common cover materials such as glass are transparent for short wavelength (solar) radiation and opaque for long wavelength (infra-red) radiation emitted by the absorber. Thus, radiation losses are also greatly reduced. This phenomenon is sometimes called the "greenhouse effect". Energy transfers inside the collector are summarized in Fig. 2.1.2. This method for heat loss reduction is used in ordinary "single glazed" collectors.

Losses can be further reduced by adding another transparent cover, giving a "double-glazed" collector. Although heat losses are smaller, less solar radiation is absorbed due to the lower transmittance. Other types of transparent insulation are sometimes used such as honeycomb structure or silica aerogel /2.3/.

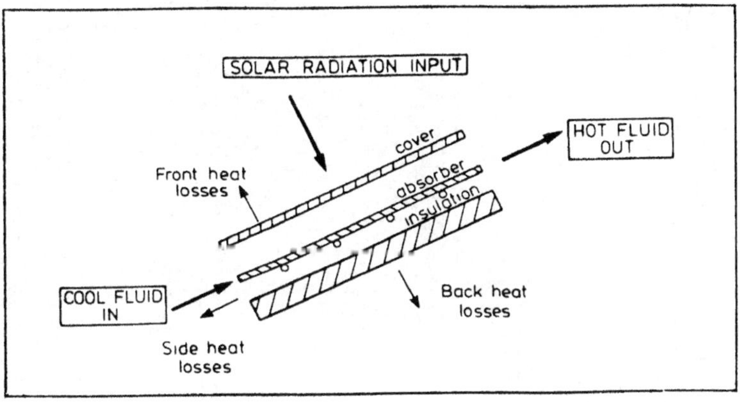

Fig. 2.1.2: Schematic of heat exchanges inside a solar collector /2.4/.

It may be preferable in some cases to reduce the losses by using a selective absorber surface. Such selective surfaces have a high absorption coefficient in the visible and near infra-red part of the spectrum. Unlike the black paints commonly found on the absorber surfaces of ordinary collectors,

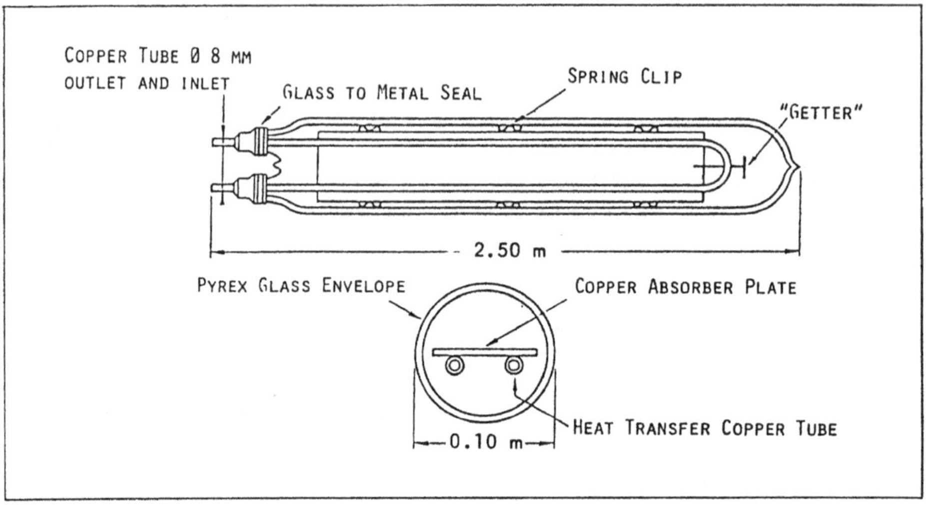

Fig. 2.1.3a: Cross-section of an evacuated tubular collector (Doc. SOLECO).

Fig. 2.1.3b: Evacuated tubular collector (Document from SOLECO, Narbonne, France).

however, selective surfaces have a low emission coefficient beyond a wavelength of 2 micrometres or so. Thus radiative losses are greatly reduced. Nevertheless, the gain in efficiency must be weighed against the higher cost of selective collectors. Typical selective coatings are obtained by deposition of thin films of various metals of metal oxides.

Convective losses can be reduced through vacuum techniques in "evacuated tubular collectors" (ETC), which usually also have a selective surface (Fig. 2.1.3) (it is not yet possible to apply vacuum technique to flat-plate collectors in a reliable way).

Another way to improve collector efficiency is by reducing the area of the absorber surface (and hence the losses) relative to the optical surface; this is done in a concentrating collector such as the one shown in Fig. 2.1.4. This type of collector concentrates the radiation on the absorber located at the focal point. To be collected in the absorber, sunrays must be in planes parallel to the plane formed by the absorber and the line of symmetry of the trough reflector. Therefore concentrating collectors require a sun tracking device, and can utilise only direct radiation and not the diffuse component. For that reason they are restricted to high temperature applications (typically more than 150°C) in sunny climates. Only low concentration collectors, such as compound parabolic collectors (CPC) which can keep a fixed orientation and inclination, find application at low temperatures (Fig. 2.1.5) /2.2/.

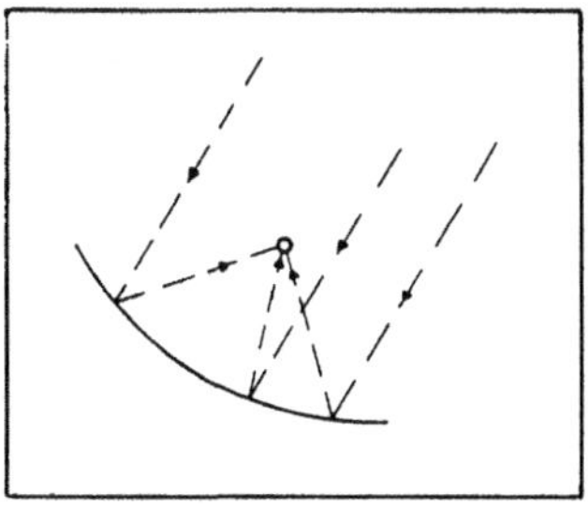

Fig. 2.1.4: Parabolic concentrator (cross-section). Direct radiation is concentrated on the absorber located at the focal point.

It should be noted that, for some particular applications at temperatures close to ambient, heat losses are no longer a problem and the absorber plate can be used directly as a collector. Such collector types are termed "unglazed" collectors (or "atmospheric" collectors), and are widely used nowadays for heating swimming pools.

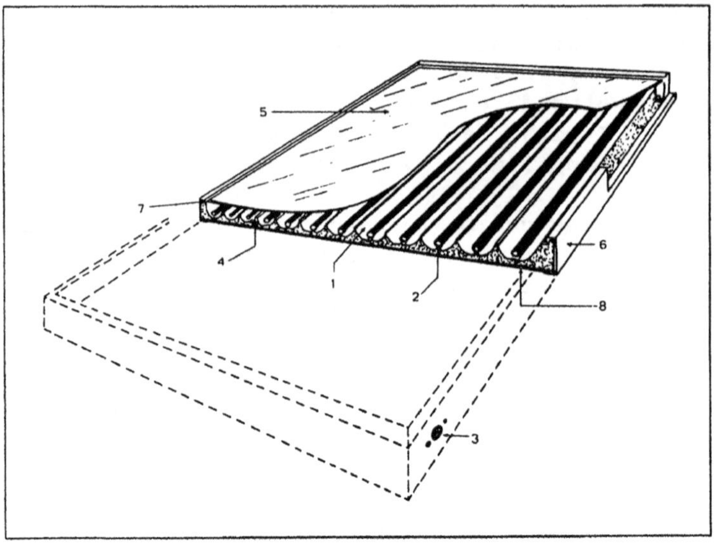

Fig. 2.1.5: Schematic of a compound parabolic concentrator CPC (Document from TECSO, Portugal).

1 - Reflector
2 - Tubing grid
3 - Piping connection
4 - Insulation
5 - Glazing
6 - Collector casing
7 - Gasket
8 - Back pan

2.1.2 - Solar collector theory

The steady-state thermal behaviour of a flat-plate solar collector can be described with a model deduced from simple thermal considerations. Furthermore, this model may be generalized to other types of collectors such as evacuated tubes, CPC's, or even concentrating collectors.

A flat-plate collector of aperture (or net area) A is considered. As defined by the European Collector Testing Group /2.4/, "the aperture area is the opening or projected area of a collector through which the unconcentrated solar energy is admitted". This definition of the surface area is used here. The gross area is sometimes used (e.g., in the USA /2.5/), therefore when values of collector parameters (introduced below) are given, the corresponding definition of the surface area must be stated (net area or gross area), so that calculations are consistent. Collector parameters defined from the net area must not be used with the gross area.

The collector is irradiated by a solar flux I. A mass flow rate $\dot{m}_0$ of fluid circulates through the collector, with a heat capacity c_0; the product of these two terms is referred to as the collector fluid capacitance rate. We have seen that two types of losses take place in a collector; optical losses and thermal losses. A simple analysis can be summarized as follows. There are energy gains, which are proportional to the solar flux, and depend on the transmittance τ of the transparent cover and on the absorptance α of the absorber plate. There are also heat losses from the absorber plate, proportional to the temperature difference between it and the environment (T_{abs} - T_a). A heat loss coefficient U can be defined.

The thermal balance of the collector can be written in order to compute the power received by the fluid through the collector (the fluid temperatures at collector inlet and outlet are denoted by T_{in} and T_{out})

$$P = \dot{m}_0 c_0 \cdot (T_{out} - T_{in}) \qquad (2.1.1)$$

$$= A \cdot \tau\alpha \cdot I - A \cdot U \cdot (T_{abs} - T_a) \ . \qquad (2.1.2)$$

As shown by several authors /2.1-2/ /2.6/, this model may be modified in order to replace the absorber plate temperature T_{abs} by the average fluid temperature T_{av} where

$$T_{av} = (T_{out} + T_{in}) / 2 \ . \qquad (2.1.3)$$

The average fluid temperature is easier to measure than the absorber plate temperature.

The output energy from the collector becomes

$$P = A.F'.\tau\alpha.I - A.F'.U. (T_{av} - T_a) \qquad (2.1.4)$$

where F' is a heat transfer factor (or fin efficiency factor), the calculation of which is indicated in section 2.1.3.

In practice, when measurements are made on a collector using some standard procedure, only the two products $F'.\tau\alpha$ and $F'.U$ are known. They define the "optical efficiency", η_0, and the overall heat loss coefficient, U_L

$$\eta_0 = F' . \tau . \alpha \qquad (2.1.5)$$

$$U_L = F' . U \qquad (2.1.6)$$

Eq. (2.1.4) can then be rewritten as

$$P = A . \eta_0 . I - A . U_L . (T_{av} - T_a) \ . \qquad (2.1.7)$$

Sometimes Eq. (2.1.7) is used as an efficiency relationship:

$$\eta = \dot{m}_0 c_0 \cdot (T_{out} - T_{in}) / (A \cdot I) \qquad (2.1.8)$$

$$= \eta_0 - U_L \cdot (T_{av} - T_a) / I \ . \qquad (2.1.9)$$

This equation is often represented as an efficiency curve (Fig. 2.1.6), which is a straight line if U_L is constant.

Both of these characteristic parameters of a solar collector can be determined either by theoretical analysis or by a suitable test procedure. These two approaches are described in section 2.1.3.

Collector	T range (°C)	η_0 -	U_L ($Wm^{-2}K^{-1}$)
Unglazed	10 - 40	.90	15-25
Single glazed	10 - 60	.80	7
Double glazed	10 - 80	.65	5
Single glazed, selective	10 - 80	.80	5
Evacuated tubular collector	10 -130	.70	2

Table 2.1.1: Typical values of characteristic parameters of common solar collectors and working temperatures. The heat loss coefficient of unglazed collectors is dependent on wind speed.

A more detailed thermal analysis has shown that the overall heat loss coefficient U_L depends on the temperature of the absorber plate and is not constant /2.4/. This dependence is represented by the approximate equation

$$U_L = U_1 + U_2 \cdot (T_{av} - T_a) \qquad (2.1.10)$$

Typical values of η_0 and U_L for common solar collectors are given in Table 2.1.1 (see also Fig. 2.1.9).

Eq. (2.1.7) may be modified again in order to use the fluid inlet temperature instead of the average fluid temperature /2.1-2/:

$$P = A \cdot F_r \cdot \eta_0 \cdot I - A \cdot F_r \cdot U_L \cdot (T_{in} - T_a) \qquad (2.1.11)$$

where F_r is the collector flow-rate correction factor (sometimes called heat removal factor) computed from

$$F_r = \dot{m}_0 c_0 \cdot [1 - EXP (- A.U_L/\dot{m}_0 c_0)] / (A.U_L) \quad . \tag{2.1.12}$$

When the flow rate is large enough ($A.U_L / \dot{m}_0 c_0 < 0.25$), F_r can be approximated by

$$F_r = 1 - 0.5 \cdot A \cdot U_L / \dot{m}_0 c_0 \quad . \tag{2.1.13}$$

The variation of F_r with $A.U_L/\dot{m}_0 c_0$ is represented in Fig. 2.1.7.

Fig. 2.1.6: Efficiency curve of a solar collector.

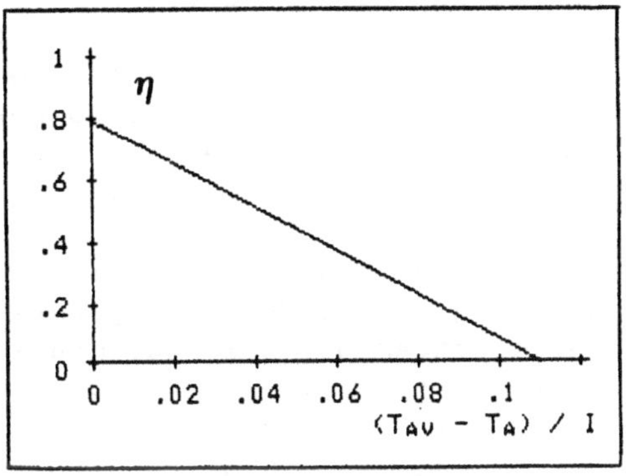

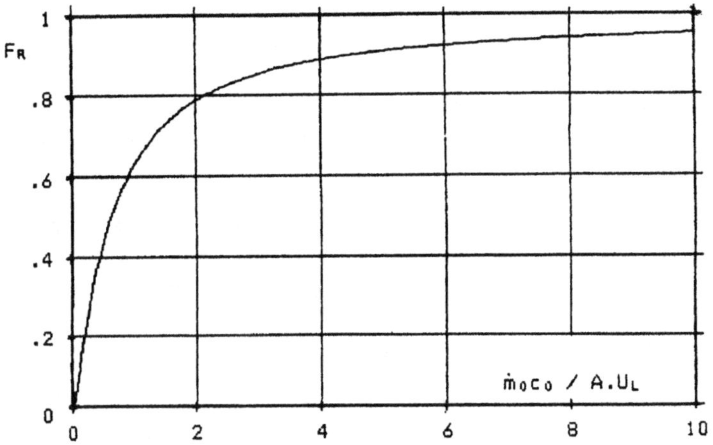

Fig. 2.1.7: Collector flow-rate correction factor F_r as a function of the dimensionless parameter $\dot{m}_0 c_0 / AU_L$.

2.1.3 - Collector characteristic parameters

Both of the collector characteristic parameters, η_0 and U_L (as defined above), may be obtained either by theoretical calculations or by collector testing.

2.1.3.1 - Theoretical approach

Only the basic items needed for the estimation of η_0 and U_L for flat-plate collectors are given here. More detailed information can be obtained from specialized textbooks such as Duffie and Beckman's /2.1/ or Rabl's /2.2/.

As indicated by Eq. (2.1.5), the optical efficiency η_0 is the product of the transparent cover transmittance (τ), the absorption coefficient of the absorber plate (α), and the heat transfer factor (F'); α and τ depend on the materials used and are taken at normal incidence.

Corrections can be made for multiple reflexions and absorption by the glazing. They are generally small (typically 2%) and are really necessary only for double glazed-collectors. By noting

α_{cov}, the absorption coefficient of the transparent cover,

ρ_d, its diffuse reflectance, and

$U_{cov,amb}$, the heat transfer coefficient between transparent cover and ambient,

a better approximation of η_0 will be (from ref. /2.2/)

$$\eta_0 = F' . \tau\alpha / [1 - (1-\alpha) . \rho_d] + \alpha_{cov} . U / U_{cov,amb} . \qquad (2.1.14)$$

The heat loss coefficient U_L is the product of the absorber plate heat loss coefficient U and the heat transfer factor F' (see Eq. 2.1.6).

The absorber plate heat loss coefficient U is obtained as the sum of a top loss coefficient, a back loss coefficient, and a thermal bridge and edge heat loss coefficient:

$$U = U_{top} + U_{back} + U_{edge} . \qquad (2.1.15)$$

The top loss coefficient can be obtained by using a thermal network to represent the collector and requires an iterative computation. Practical formulae have been proposed by Klein /2.1/.

The back loss coefficient is simply computed from the thermal resistance of the back insulation (conductivity, k, and thickness, e)

$$U_{back} = (e / k + R_{surf})^{-1} \qquad (2.1.16)$$

where R_{surf} is the equivalent convective-radiative surface thermal resistance. Additional losses from collector edges and thermal bridges exist; they are difficult to estimate but are not negligible in most cases. Corrections between 0.2 and 0.5 $W.m^{-2}.K^{-1}$ are likely to occur.

The heat transfer coefficient F' characterizes the efficiency of heat transfer between absorber plate and fluid, and includes fin effects in the absorber plate. Fin effects are caused by the fact that the thermal conductivity of the absorber material (even metal) is not infinite. Therefore a temperature gradient is observed in the direction perpendicular to the fluid flow (and not only in the direction of the flow). For simple geometries (without a fin effect), F' can be defined as /2.6/

$$F' = U' . A' / (U' . A' + U . A) \qquad (2.1.17)$$

where U' is the heat transfer coefficient between the fluid and the absorber and A' is the area of contact surface between them. This formula can be used as a first approximation even with fin effect.

For a general plate and tube absorber configuration (as represented in Fig.2.1.1), fin effect occurs and F' can be computed using formulae or charts in the literature /2.1-2/.

Similar calculations can be done for other types of collectors such as evacuated tubular or compound parabolic collectors (see Rabl /2.2/).

2.1.3.2 - Solar collector testing

The second method of obtaining solar collector characteristic parameters is by collector testing. In most countries, national standard test procedures have been adopted /2.7-9/; these standards are close to each other and a synthesis has been proposed by the International Organization for Standardization (ISO) /2.10/. Each national testing standard determines environmental conditions (air temperature, solar radiation, fluid flow-rate and temperature) to be achieved during experimental measurements on the solar collector. The Commission of the European Communities has also produced its own recommendations from the work of the Collector Testing Group /2.4/.

Table 2.1.2: Solar collector testing laboratories in Europe.

			Test Rig	
			Outdoors	Indoors
Joint Research Centre	Ispra	Italy	*	*
DSETL	Copenhagen	Denmark	*	*
CSTB	Sophia-Antipolis	France	*	
Ecole des Mines	Sophia-Antipolis	"	*	
CETIAT	Villeurbanne	"	*	
CEA	Cadarache	"	*	*
Univ. Stuttgart	Stuttgart	Germany	*	
TUV Bayern	Munich	"	*	
CONPHOEBUS	Catania	Italy	*	
TPD-TNO-TH	Delft	Netherlands	*	*
LNETI	Lisbon	Portugal	*	*
INTA	Madrid	Spain	*	
University College	Cardiff	U.K.		*

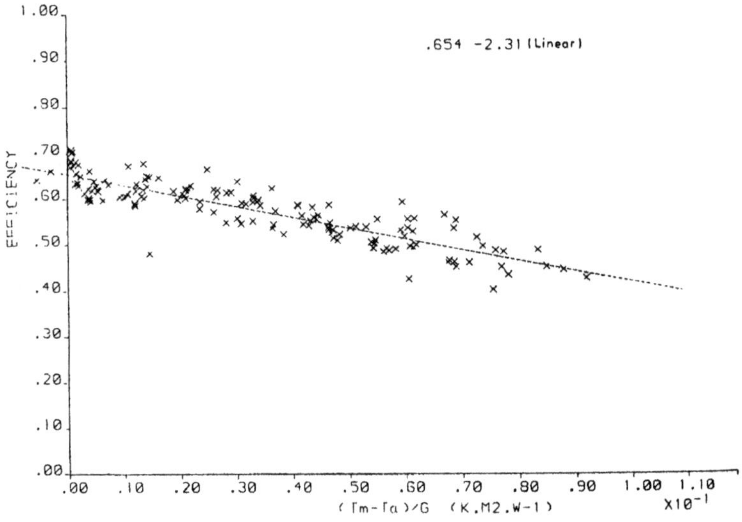

Fig. 2.1.8: Example of solar collector test results (from CSTG).

In each European country, laboratories are approved for solar collector testing. Some of these laboratories are listed in Table 2.1.2. Testing solar collectors is often mandatory for certification purposes.

Collector parameters, η_0 and U_L (or U_1 and U_2 of Eq. 2.1.10), are computed by a simple regression analysis of experimental results.

They are directly available for further use and some intermediate parameters such as the heat transfer factor F' are no longer needed. Values obtained from testing correspond to a fixed flow rate, but they may be used for a large range of flow rate values, because the heat transfer factor F' is not very dependent on the flow rate.

Experimental collector testing results are shown in Fig. 2.1.8. Significant scattering may be observed, but the linear form of Eq. (2.1.9) seems to be a good approximation for practical calculation purposes.

The efficiency curves of various types of collectors are compared in Fig.2.1.9. An important conclusion to be drawn from this comparison is that it does not matter very much whether a collector is "better" than another one; what is important is to select a collector which is suited to a particular application and the operating temperature range within which energy is required. And, of course, the collector price cannot be ignored.

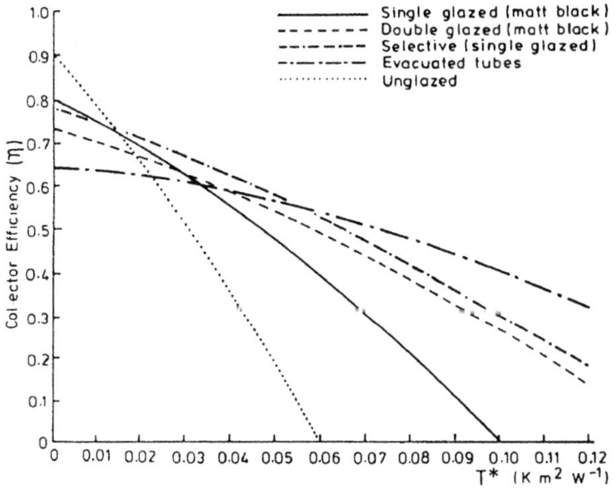

Fig. 2.1.9: Typical instantaneous efficiency curves (from Gillett /2.4/).

2.1.4 - Other characteristics

Certain solar system design methods require the additional collector characteristics mentioned below. Not all of them are needed by ESM correlations, but it is advisable to check that values lie within certain ranges.

Thermal capacity effects tend to decrease the collector efficiency during transient periods i.e., flow-rate changes introduced by the control, rapid variations in solar irradiance, etc. These effects can be characterized by two parameters; the collector heat capacity itself, C, and the collector time constant.
Theoretical determination of heat capacity is not easy. Adnot /2.11/ proposes two extreme values; the minimum value corresponds to the heat capacity of the absorber plate added to the fluid heat capacity. The maximum value further includes half the heat capacity of the rest of the collector (frame, cover, insulation).
Standard testing procedures propose experimental methods for the determination of this heat capacity.

The collector time constant t_{col} can be defined as the ratio of the heat capacity C to the heat loss coefficient:

$$t_{col} = C / (A . U_L) . \qquad (2.1.19)$$

It can also be determined experimentally.

ESM design methods account for the effects of collector heat capacity for time constant values within the range 600 to 3600 seconds (0.15 to 1 hour). A smaller time constant will result in an improved performance (by a few percent).

The incidence angle modifier is an optical corrective term which is introduced to allow for the variations of optical properties of both the transparent cover and the absorber plate with the incidence angle; both τ and α vary with the incidence angle, introducing a decrease of η_0, whenever incidence is non -normal. This variation can be computed theoretically /2.1-2/ or it can be analysed by measurement (Fig. 2.1.10). This variation is also represented by the incidence angle modifier $K(\theta)$ which is applied to the collector optical efficiency η_0.

For a practical range of incidence angles (from 0° to 60°), $K(\theta)$ is close to one and can generally be computed with the following approximation

$$K(\theta) = 1 - b_0 \cdot (1/\cos\theta - 1) \qquad (2.1.18)$$

according to ASHRAE Standard /2.5/ (Fig. 2.1.11). If an experimental $K(\theta)$-curve is known for a given collector, the value of b_0 can be estimated by a simple regression in the 0°-60° incidence angle range. For common types of solar collectors, the following default values may be adopted:

Single-glazed (selective or non-selective)	$b_0 = 0.10$
Double-glazed	$b_0 = 0.17$

Values of b_0 for double walled, ribbed covers are given in Tab. 2.1.3. For evacuated tubular collectors, no standard value exists and sometimes negative b-values may be obtained when back-reflectors are included (at some non-normal incidence angles, the optical efficiency is enhanced) /2.2/.

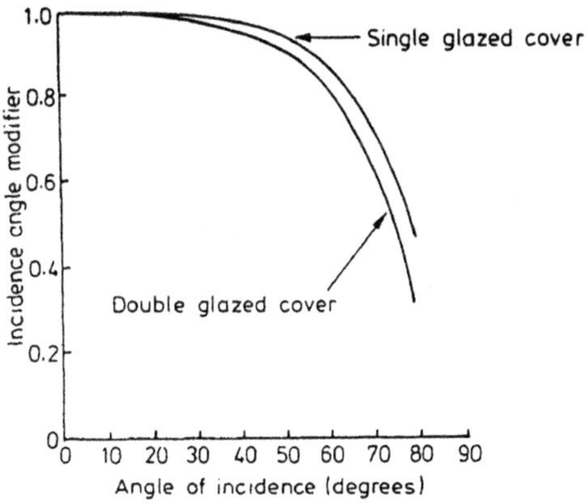

Fig. 2.1.10: Typical incidence angle modifier (from Gillett /2.4/).

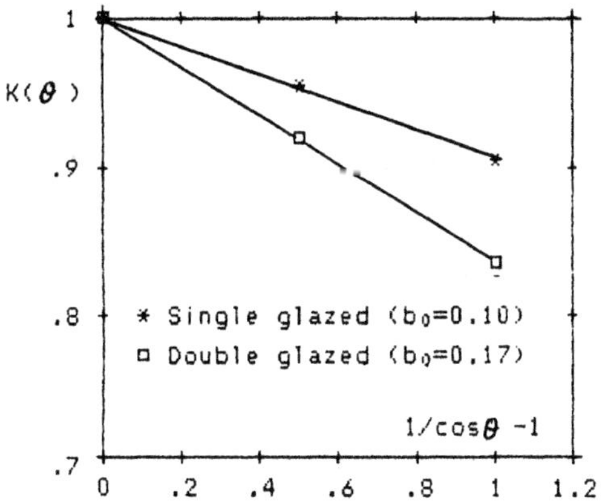

Fig. 2.1.11: Incidence angle modifier according to ASHRAE Standard 93-77.

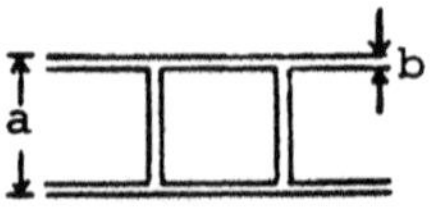

a mm	3	5	10
b mm	0.3	0.45	0.55
Incidence angle modifier	0.24	0.24	0.28
Transmittance for solar radiation (0.3 - 2.5 μm)	0.84	0.84	0.84

Table 2.1.3: Incidence angle modifier b_0 for double walled ribbed covers (values measured at Danish Solar Energy Testing Laboratory). The cover material is polycarbonate.

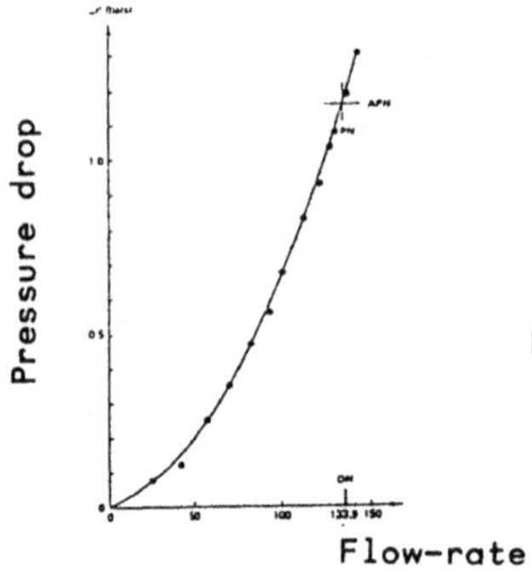

Fig. 2.1.12: Pressure drop in solar collector versus flow-rate (Doc. SOLECO, France).

The <u>pressure drop</u> within the collector must be known for pump sizing. It is given by manufacturers as flow-rate dependent curves (Fig. 2.1.12).

2.1.5 - Practical Considerations

Collector design is beyond the scope of this handbook. Interested readers will find useful aspects in two textbooks, by Gillett /2.4/ and Rabl /2.2/.

Nevertheless, some practical information concerning the installation of solar collectors is given here.

Shading on the collector array from buildings, trees, mountains, or from other collectors should be avoided as far as possible, at least from 0900 to 1500 (solar time) in any season. The best orientation is of course facing due South (North in the Southern hemisphere), but small deviations from this orientation (± 20°) have little influence at mid-latitudes (and even less at low latitudes).

The optimum tilt angle depends on the solar energy use; a tilt angle close to the latitude is the best for year-round energy loads such as water heating with a constant demand. Larger tilt angles (latitude + 15°) are preferred for space heating. As optima are generally flat, this factor is not a drastic constraint and the designer has a certain freedom (±10° or sometimes more).

Collectors must be installed in accordance with the usual plumbing regulations as well as the manufacturer's requirements /2.4/. Protection against freezing and overheating is essential.

Interconnection of collector modules should ensure a homogeneous flow-rate distribution between the collectors. Thus, the reverse return connection (Tickelman loop) is generally the best (Fig. 2.1.13).

Regular system maintenance is necessary. Solar collectors should be cleaned periodically, although natural cleaning by rain is often sufficient.

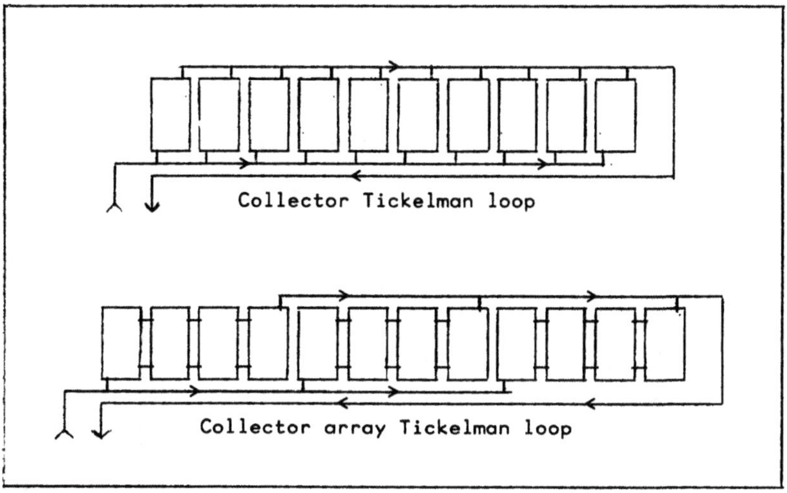

Fig.2.1.13: Connection of collectors with Tickelman loop (reverse return).

2.2 -SOLAR LOOP: PIPES, HEAT-EXCHANGER AND CONTROL SYSTEMS

Besides the solar collectors, which have been described in the preceding section, the "solar loop" of an active solar thermal system includes a number of other components, such as connecting pipes or ducts, the heat-exchanger, pumps, valves, and the control system (see Fig. 2.2.1).

Pipes and ducts conduct the heat transfer fluid from the collector to the storage. A heat-exchanger is necessary when different fluids are used in the collectors and the storage (e.g., water + anti-freeze in collectors, water in storage). Pumps are required to circulate the fluid.

Only those components which affect the system performance are analysed in this section.

2.2.1 - Pipes and ducts.

Pipes are needed to conduct the working fluid from collector to storage, but heat loss occurs during this transfer. This loss often causes a significant reduction in system performance and must be estimated as accurately as possible. In all ESM calculations (see Chapters 6 and 7), a distinction is made between pipes located outdoors (whose losses are included with the collector) and pipes located indoors (whose losses are included with the storage). However, the calculation of their respective heat loss coefficients is essentially the same.

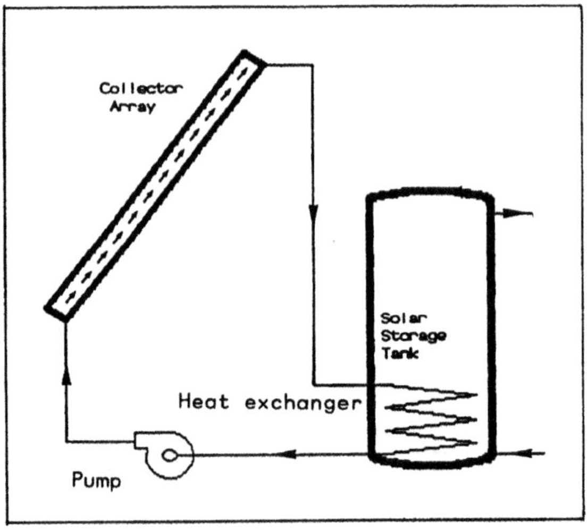

Fig. 2.2.1: Solar loop schematic.

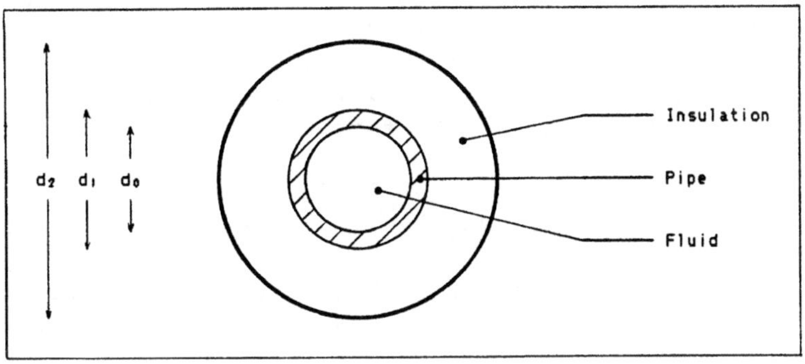

Fig.2.2.2: Pipe and insulation schematic.

2.2.1.1 - Piping loss

If the value of the piping heat loss coefficient is not provided by the manufacturer, it can be calculated with the following equations.

Let us consider a pipe with internal diameter d_0 and external diameter d_1; the pipe material has a thermal conductivity, k_{pip}. Thermal insulation of thermal conductivity k_{ins} is put around the pipe, with an external diameter d_2 (see Fig. 2.2.2). The piping heat loss coefficient u (per metre of pipe length) is given by

$$\frac{1}{u} = \frac{1}{\pi} \left(\frac{1}{h_i d_0} + \frac{1}{k_{pip}} \cdot \mathrm{Ln} \frac{d_1}{d_0} + \frac{1}{k_{ins}} \cdot \mathrm{L} \frac{d_2}{d_1} + \frac{1}{h_e d_2} \right) \qquad (2.2.1)$$

h_i and h_e are the internal (between fluid and pipe) and the external (between insulation and environment) heat transfer coefficients.

Assuming a turbulent flow inside the tube, MacAdams' formula gives the internal heat transfer coefficient /2.12/

$$Nu = 0.023 \; Re^{0.8} \; Pr^{0.4} \qquad (2.2.2)$$

depending on the dimensionless numbers:

Nusselt	$Nu = h_i \cdot d_0 \; / \; k_{fluid}$	(2.2.3)
Reynolds	$Re = \rho \; V \; d_0 \; / \; \mu$	(2.2.4)
Prandtl	$Pr = \mu \; c_p \; / \; k_{fluid}$	(2.2.5)

where ρ, μ, c_p, and k_{fluid} are respectively density, absolute (or dynamic) viscosity (unit: Pa.s), thermal capacity, and thermal conductivity of the fluid inside the pipe (typical values are given in Appendix C). V is its velocity.

If this fluid is water at temperature T_w (°C), then h_i (W/m2/K) is given by

$$h_i = 2040 \; (1 + 0.015 \; T_w) \; V^{0.87} \; / \; d_0^{0.13} \quad . \tag{2.2.6}$$

In most cases, however, this heat transfer coefficient is between 2000 and 8000 W/m2/K and the corresponding thermal resistance may be neglected.

The external heat transfer coefficient into the environment, h_e, is separated into two terms, a convective one and a radiative one.

$$h_e = h_e^{conv} + h_e^{rad} \quad . \tag{2.2.7}$$

For pipes located inside a buiding (in calm air) at about 20°C, the convective term depends on the temperature difference between the pipe (exterior) and the air (d_2 is expressed in metres)

$$h_e^{conv} = 1.31 \; (\Delta t \; / \; d_2)^{0.25} \quad . \tag{2.2.8}$$

The radiative term depends on the average temperature, T_m (K), (of the environment and the external pipe surface) and on the external surface emissivity ϵ

$$h_e^{rad} = 4 \; \sigma \; \epsilon \; T_m^3 \quad . \tag{2.2.9}$$

σ is the Stephan-Boltzman constant ($\sigma = 5.67 \; 10^{-8} \; Wm^{-2}K^{-4}$).

	Pipe diameter d_2 Small (<50 mm)	Large (>50 mm)
Uninsulated pipes	15	12
Insulated pipes with metal shield	4	3
without metal shield	8	7

Table 2.2.1: Typical values (W/m2/K) of the external heat transfer coefficient, h_e.

Typical values of h_e are given in Table 2.2.1. Insulated pipes with an external metal shield have lower h_e values due to the low emissivity of polished metal (and thus, reduced radiative heat transfer).

The associated thermal resistance may be neglected in the case of insulated pipes, but not for uninsulated pipes.

In most cases, due to the high conductivity of metals and the small thickness of the pipes, one additional term of Eq. 2.2.1 may be neglected and the heat loss coefficient per unit length of pipe may be expressed as:

$$u = \pi . d_2 \ . \ h_e \qquad \text{(Uninsulated pipes)} \qquad (2.2.10)$$

$$u = \frac{\pi . k_{ins}}{\text{LOG} \ (\ d_2 \ / \ d_1 \ )} \qquad \text{(Insulated pipes)} \ . \qquad (2.2.11)$$

Outdoors, only insulated pipes are used, and the former value of u (for insulated pipes) has to be increased by 10 to 20% to account for wind velocity effects.

In certain countries, regulations stipulate a minimum insulation for pipes in thermal plants (not specifically solar ones). In the F.R.Germany, for instance, an insulation thickness equivalent to about the pipe diameter has been required since 1982.

The solar loop (like any circulation loop) includes several elements such as clips or valves, generally poorly insulated. For the calculation of heat losses, it is therefore necessary to add to the actual pipe length some "equivalent" pipe lengths

Component	Pipe diameter	Equivalent pipe length
Clip	25 mm	0.2 m
	100 mm	0.5 m
	300 mm	1.5 m
Valve	25 mm	0.5 m
	100 mm	1.2 m
	300 mm	3.0 m

Table 2.2.2: Additional losses of a circulation loop due to thermal bridges: Equivalent pipe length. (from Recknagel /2.13/).

to account for the thermal bridges from these components. Typical values are given in Table 2.2.2 for a fluid temperature of 50°C (these values must be doubled if the pipe is located outside or if the fluid is at 100°C).

2.2.1.2 - Modifying collector characteristics to account for pipe losses

Pipes located outdoors are considered in this section. Collector inlet pipes and collector outlet pipes must be separated; their respective global heat loss coefficients, $A_{in}U_{in}$ and $A_{ou}U_{ou}$ are computed using the above heat loss coefficient per unit length, u, and the total equivalent pipe length.

It has been demonstrated by Beckman /2.1/ that these losses may be integrated into the collector characteristics. In this section a more universal, yet equally simple, formulation of Beckman's correction is proposed.

The loop schematic is given in Fig. 2.2.3. The loop comprises three parts: inlet pipes, collector, and outlet pipes. Temperatures at the inlet and outlet of the loop are denoted as T_0 and T_1. Using the usual "heat-exchanger" equation for the three parts (see section 2.2.2), it becomes:

$$T_{in} - T_0 = (T_a - T_0) \cdot [\, 1 - EXP\,(-A_{in}U_{in}/\dot{m}_0 c_0)\,] \qquad (2.2.12)$$

$$T_{out}-T_{in} = AF_rU_L \cdot (T_a + \eta_0 \cdot I/U_L - T_{in})/\dot{m}_0 c_0 \qquad (2.2.13)$$

$$T_1 - T_{out} = (T_a - T_{out}) \cdot [\, 1 - EXP\,(-A_{ou}U_{ou}/\dot{m}_0 c_0\,)\,] \,. \qquad (2.2.14)$$

Eliminating T_{in} and T_{out} leads to an equation for the loop outlet-inlet temperature difference, equivalent to that of the collector (Eq. 2.1.11) with the modified characteristics, $\eta_0{}^*$

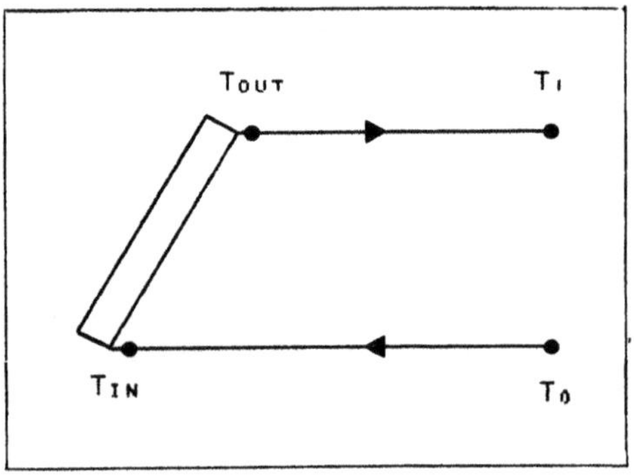

Fig. 2.2.3: Collector + Pipe loop schematic and nomenclature.

and U_L^*

$$T_1 - T_0 = (T_a + \eta_0^*.I/U_L^* - T_0) . AF_r^*U_L^*/\dot{m}_0c_0 \qquad (2.2.14')$$

$$U_L^* = (A.U_L + A_{in}U_{in} + A_{ou}U_{ou}) / A \qquad (2.2.15)$$

$$F_r^* = \frac{\dot{m}_0c_0}{AU_L^*} . [1 - EXP (- A.U_L^*/\dot{m}_0c_0)] \qquad (2.2.16)$$

$$\eta_0^* = \eta_0 . \frac{F_r}{F_r^*} . EXP (- \frac{A_{ou}U_{ou}}{\dot{m}_0c_0}) \qquad . \qquad (2.2.17)$$

For ordinary flat-plate solar collectors, the difference from Beckman's formulation is small. However, this difference cannot be neglected when high performance collectors are used.

2.2.2 - Heat-exchangers

Anti-freeze protection of the collectors generally requires the use of two different fluids; water + anti-freeze in the collectors and water alone in the rest of the system. A heat-exchanger is therefore needed between collector and storage (see Fig. 2.2.1). Consequently, system performance is adversely affected and this component must be sized carefully. Heat-exchangers are sometimes found in other parts of the system (e.g., water/air coil for air distribution systems).

2.2.2.1 - Heat-exchanger types and characteristics

Three types of heat-exchangers may be encountered in an active solar system:

- Tubular heat-exchangers (shell and tubes) (Fig. 2.2.4).
- Plate heat-exchangers (more compact) (Fig.2.2.5).
- Immersed coils (internal heat-exchanger, immersed inside the storage tank) (Fig. 2.2.6).

Tubular heat-exchangers are generally used in large plants and immersed coils in smaller systems such as single-family solar water heaters.

The specialized theoretical and technological aspects of heat-exchangers are not discussed in this handbook. Additional information can be found in several sources /2.14-18/.

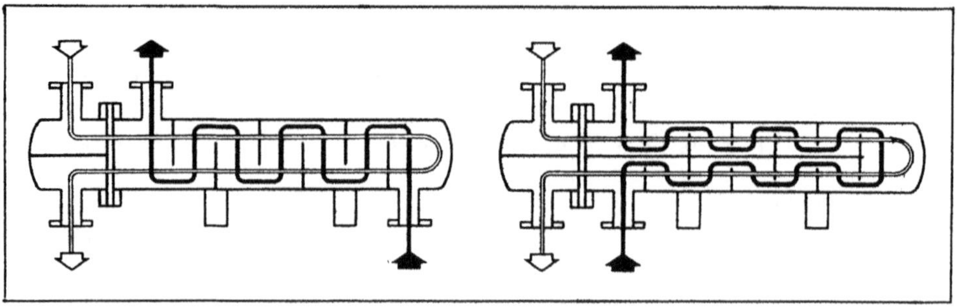

Fig. 2.2.4: Tubular heat-exchangers (Document from CIAT, Culoz, France).

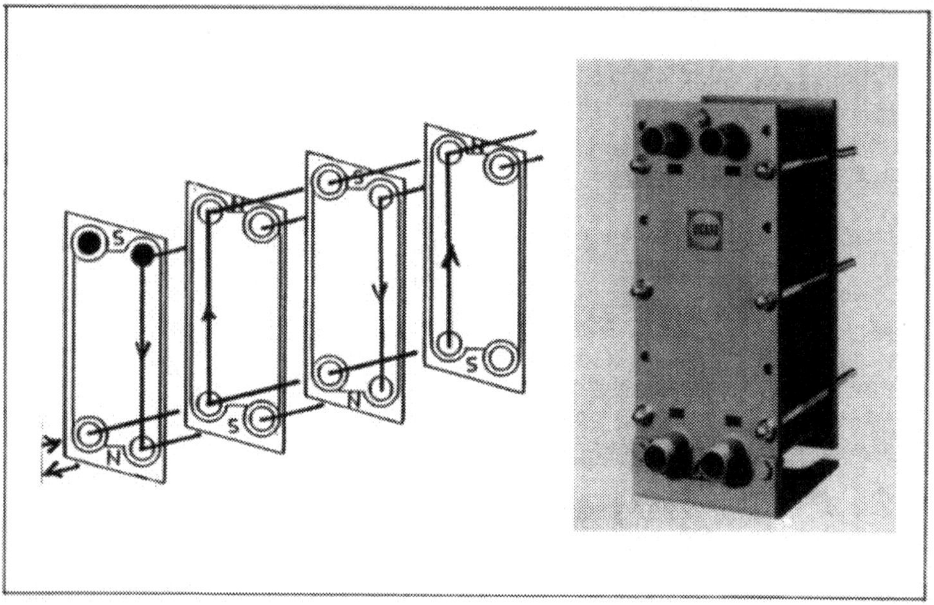

Fig. 2.2.5: Plate heat-exchangers (Doc. VICARB, St Martin d'Hères, France).

From a thermal performance point of view, a heat-exchanger can be characterized by three parameters. The first two parameters are the fluid capacitance rates on the two sides of the heat exchanger, i.e. the mass flow rate multiplied by the thermal capacity.

As we are considering here principally the heat-exchanger of the solar loop (between collector and storage) we shall denote

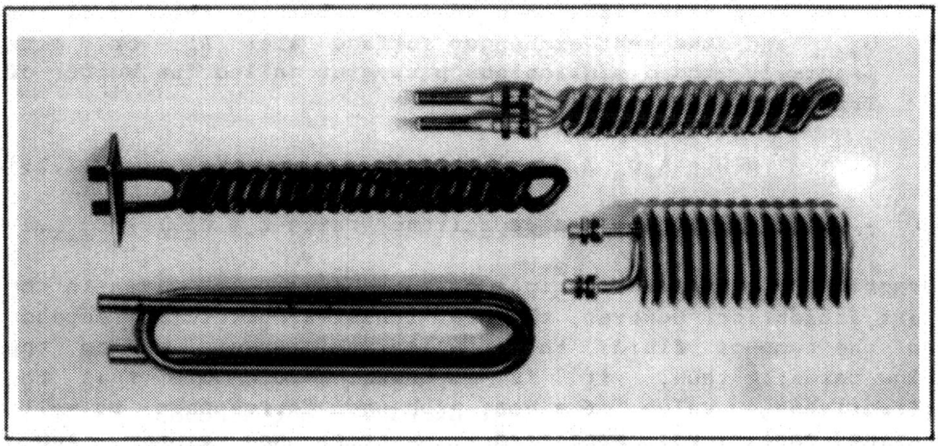

Fig. 2.2.6: Immersed coils (Document from WIELAND, Germany).

$\dot{m}_0$, the mass flow rate of fluid in the collector side of the heat-exchanger (which is also the collector flow rate), with the corresponding thermal capacity c_0 (primary flow-rate)

$\dot{m}_1$, the mass flow rate in the storage side of the heat-exchanger (flow-rate between heat-exchanger and storage; or secondary flow-rate). The fluid on this side is the storage fluid (generally water), the thermal capacity of which is denoted c.

Hence, the thermal capacitance rates on the two sides of the heat-exchanger are $\dot{m}_0 c_0$ and $\dot{m}_1 c$ respectively; the smaller of these two values is denoted C_{min} (and the larger C_{max}).

The third parameter is the heat-exchanger effectiveness, e_1, defined as the ratio of the thermal power exchanged between the two sides to the maximum possible exchanged power. By denoting the fluid inlet temperatures on the two sides of the heat-exchanger as T_0 and T_1 (as in Fig. 2.2.7), the effectiveness can be expressed as

$$e_1 = \frac{P}{C_{min} \cdot (T_1 - T_0)} \qquad (2.2.18)$$

For a given type of exchanger, the effectiveness is a function of:

- The mean heat transfer coefficient between the two fluids U_e, and the heat-exchanger surface area A_e, or, more precisely, the dimensionless parameter called the Number of Transfer Units (NTU)

$$NTU = A_e U_e \, / \, C_{min} \quad . \tag{2.2.19}$$

- The ratio of the two capacitance rates $r = C_{min} \, / \, C_{max}$.

Practical formulae for typical heat-exchangers are given in the next subsection. However, the heat transfer coefficient depends on the type of fluid, the temperature ranges, and on the flow-rates; thus, it is important to note that the effectiveness value of a heat-exchanger corresponds to well-defined conditions: specified flow-rates and fluids, and a limited temperature range.

Given these limitations, Eq. (2.2.18) can be used to determine the effectiveness of a heat-exchanger from the manufacturer's data. This equation can also be transformed in order to estimate the transferred thermal power under different inlet temperatures (as defined in Fig. 2.2.7)

$$P_e = C_{min} \cdot e_1 \cdot (T_1 - T_0) \tag{2.2.20}$$

together with the thermal balance equations

$$P_e = \dot{m}_0 c_0 \cdot (T_0' - T_0) \tag{2.2.21}$$

$$P_e = \dot{m}_1 c \cdot (T_1' - T_1) \tag{2.2.22}$$

However, if the operating conditions are different from the manufacturer's, Eq. (2.2.20) must not be used unless the effectiveness value has been modified, either from new manufacturer's data or by specific calculation. This calculation can be made (for common types of heat-exchangers only) from the equations and charts in the next section, if only flow-rates are changed.

Another formulation of the exchanged power is sometimes used

$$P_e = A_e U_e \cdot F' \cdot \Delta T_m \tag{2.2.23}$$

where ΔT_m is the mean logarithmic temperature difference:

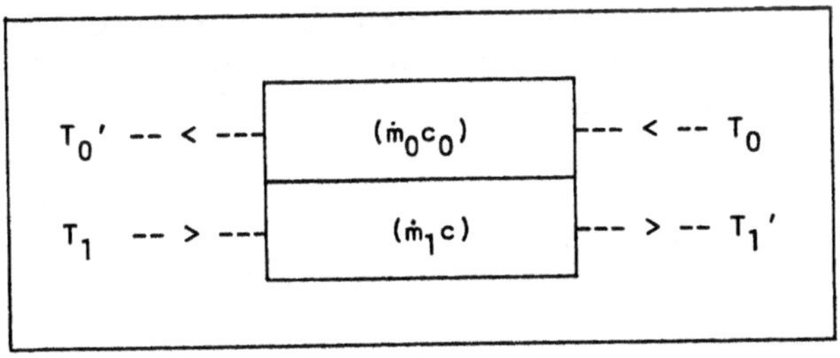

Fig. 2.2.7:
Heat-exchanger schematic and nomenclature.

$$\Delta T_m = \frac{(T_0 - T_1') - (T_0' - T_1)}{\mathrm{Ln}\left(\dfrac{T_0 - T_1'}{T_0' - T_1}\right)} \quad . \qquad (2.2.24)$$

For simple cases (parallel flow or counterflow) $F' = 1$, and for other cases F' remains close to 1, so Eq. (2.2.23) is generally used to size the heat-exchanger for given operating conditions. In most common thermal or chemical processes, the operating temperatures are known and constant, but this is not the case in solar processes (see Chapter 3).

2.2.2.2 - Effectiveness calculation

Effectiveness equations are given in Table 2.2.3 for the most common heat-exchangers (and the easiest to model) /2.19/. They are represented by the curves of Figs. 2.2.8 and 2.2.9.

Charts for estimating F' (used in Eq. 2.2.23) are also available from other sources /2.16/.

The effectiveness relationship for the common counterflow heat-exchanger should be noted:

$$e_1 = \frac{1 - \mathrm{EXP}\,[\,-\,\mathrm{NTU}\,(\,1 - r\,)\,]}{1 - r\,.\,\mathrm{EXP}\,[\,-\,\mathrm{NTU}\,.\,(\,1 - r\,)\,]} \quad . \qquad (2.2.25)$$

Limiting cases are important; if $r = 1$ (same capacitance rate on both sides)

$$e_1 = \mathrm{NTU} \,/\, (\,1 + \mathrm{NTU}\,) \quad . \qquad (2.2.26)$$

If $r = 0$, one flow-rate tends to infinity; this occurs if one side of the heat-exchanger is at a uniform temperature e.g., immersed coils, heating radiators, pipes (considered as heat-exchangers with the environment for loss calculations). Let us assume that $\dot{m}_1 c$ tends to infinity (as is the case with immersed coils)

$$e_1 = 1 - EXP\ (\ - A_e U_e\ /\ \dot{m}_0 c_0\ )\quad . \tag{2.2.27}$$

	Heat-Exchanger Type	Heat-exchanger efficiency e
1	Parallel flow	$\frac{1 - e^{-\beta\,(1+\alpha)}}{1+\alpha}$
2	Counter flow	$\frac{1 - e^{-\beta\,(1-\alpha)}}{1 - \alpha\,.\,e^{-\beta\,(1-\alpha)}}$
3	Shell and tube (1 shell pass, an even number of tube passes)	$\frac{2}{1+\alpha+\sqrt{1+\alpha^2}\,\frac{(1+e^{-\beta\sqrt{1+\alpha^2}})}{(1-e^{-\beta\sqrt{1+\alpha^2}})}}$
4	Cross-flow with both fluids mixed	$\frac{\beta}{\frac{\beta}{1-e^{-\beta}} + \frac{\alpha.\beta}{1-e^{-\alpha\beta}} - 1}$
5	Cross-flow with both fluids unmixed (approximate expression)	$1 - e^{-\left(\frac{1-e^{-\alpha\,.\,n\,.\,\beta}}{\alpha\,.\,n}\right)}$ with $n = \beta^{-0.22}$
6	Cross-flow with 1 fluid unmixed, 1 fluid mixed, $C_{unmixed} = C_{min}$	$\frac{1 - e^{-\alpha\,(1-e^{-\beta})}}{\alpha}$
7	Cross-flow with 1 fluid unmixed, 1 fluid mixed, $C_{mixed} = C_{min}$	$1 - e^{-\frac{1-e^{-\alpha.\beta}}{\alpha}}$

Table 2.2.3: Effectiveness of different heat-exchangers (from /2.19/).

$C_{min} = MIN\ (\ \dot{m}_0 c_0,\ \dot{m}_1 c)$ $\qquad$ $C_{max} = MAX\ (\ \dot{m}_0 c_0,\ \dot{m}_1 c)$

$\alpha = C_{min}\ /\ C_{max} = r$ $\qquad$ $\beta = A_e U_e\ /\ C_{min} = NTU$

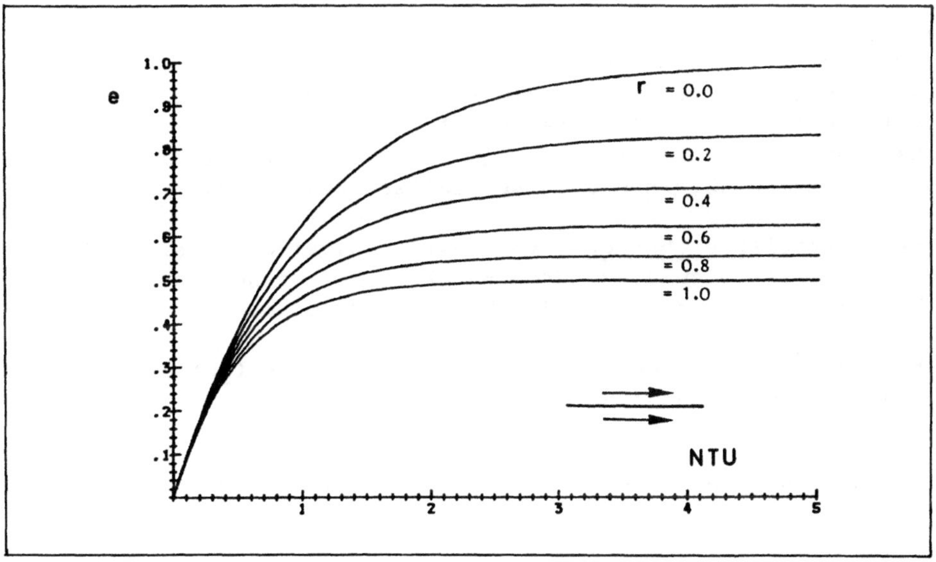

Fig. 2.2.8: Effectiveness of parallel flow heat-exchangers (from Dutré /2.19/).

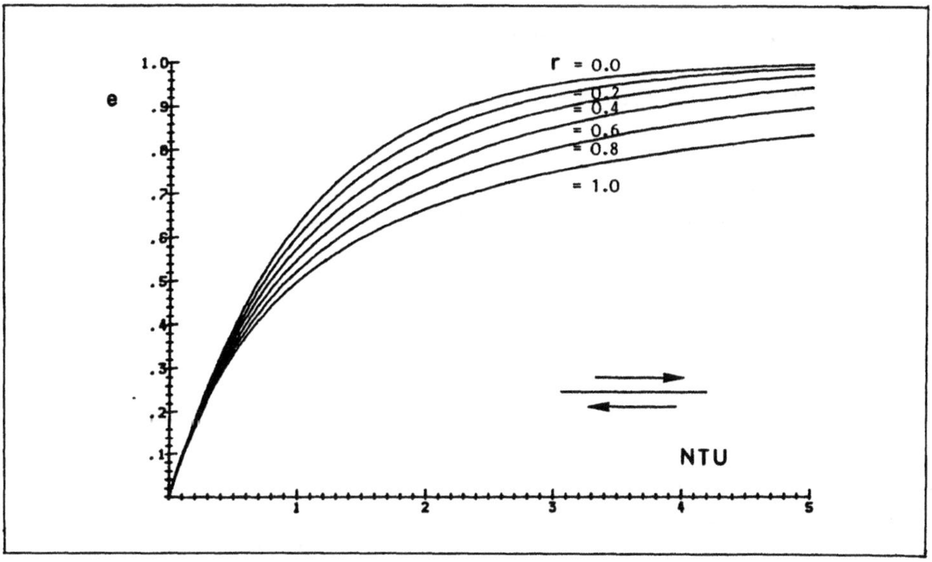

Fig. 2.2.9: Effectiveness of counterflow heat-exchangers (from /2.19/).

2.2.2.3 - Heat transfer coefficient

The global heat transfer coefficient of the heat-exchanger can be obtained from the heat transfer coefficient on each side of the partition wall. A few relationships for common cases, to check orders of magnitude are given here. More detailed information may be obtained from specialized textbooks.

For a tubular exchanger, the heat transfer coefficient inside the tube is calculated from Eq. 2.2.2 (see section 2.2.1.1). This equation is also valid for immersed coils.

The heat transfer coefficient around the tubes (outside perpendicular flow) depends on their arrangement

$$Nu = 0.33\ Re^{0.6}\ Pr^{1/3} \qquad \text{(Staggered)} \qquad (2.2.28)$$

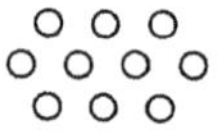

$$Nu = 0.26\ Re^{0.6}\ Pr^{1/3} \qquad \text{(Aligned)} \qquad (2.2.29)$$

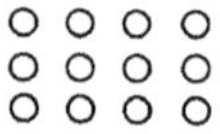

The dimensionless Nusselt, Reynolds, and Prandtl numbers are also defined according to Eq. 2.2.3 to 2.2.5, but using the external diameter of the tubes as the characteristic length. The fluid velocity in the Reynolds number is estimated at the narrowest space between tubes.

If transverse baffles are used, a correction (of about 0.6) must be made to account for dead spots.

For immersed coils, free convection is the dominant heat transfer mode. The heat transfer coefficient can be approximated with the following correlation (synthesized from various experiments /2.20/)

$$Nu = 0.4\ (Gr.Pr)^{0.25} \quad . \qquad (2.2.30)$$

The Grashof number is defined as

$$Gr = g\ \beta\ \Delta T\ d^3\ \rho^2\ /\ \mu^2 \qquad (2.2.31)$$

where g is the acceleration due to gravity,
d is the external diameter of the tube,
ΔT is the average temperature difference between the storage tank and the exterior coil surface,
β is the volume coefficient of expansion of the fluid.

Fluid properties are evaluated at film temperature, and defined as the average of the storage tank temperature and the exterior coil surface temperature.

An iterative procedure is required for the estimation of heat transfer coefficient. From an approximate value of the heat transfer coefficient, exterior coil surface temperature and storage tank temperature are estimated, enabling a better computation of the heat transfer coefficient.

As solar systems do not work under constant conditions, average and extreme cases should be considered. Daily average temperatures can be estimated with the help of the closed-form models discussed in Chapter 3.

2.2.2.4 - Solar loop (collector+heat-exchanger) correction factor

The effect of the heat-exchanger on the collector performance can be summarized, very simply, by a solar loop correction factor, F_r' (introduced by de Winter /2.21/); the solar loop (collector + heat-exchanger) energy output at inlet temperature T_0 is given by

$$P = A\, F_r' \,[\, \eta_0 . I - U_L .(T_0 - T_a) \,] \qquad (2.2.32)$$

similar to Eq. 2.1.11 for the collector, where F_r has been replaced by F_r' (we assume that η_0, U_L, and F_r have been corrected for pipe losses)

$$F_r'/F_r = \left[1 + \frac{A F_r U_L}{\dot{m}_0 c_0} .\left(\frac{\dot{m}_0 c_0}{e_1 . C_{min}} - 1 \right) \right]^{-1} . \qquad (2.2.33)$$

The variations of F_r' are represented by the curves of Fig. 2.2.10.

Some consequences can be pointed out, in the case of an immersed coil (Fig 2.2.11). If the collector flow-rate is quite large ($\dot{m}_0 c_0$ tends to infinity), F_r' has a limiting value

$$F_x = \frac{1}{1 + A.U_L/A_e U_e} \qquad , \qquad (2.2.34)$$

and for any collector flow-rate, F_r' can be approximated by

$$F_r' = MIN\,(F_r \,;\, F_x) \quad . \qquad (2.2.35)$$

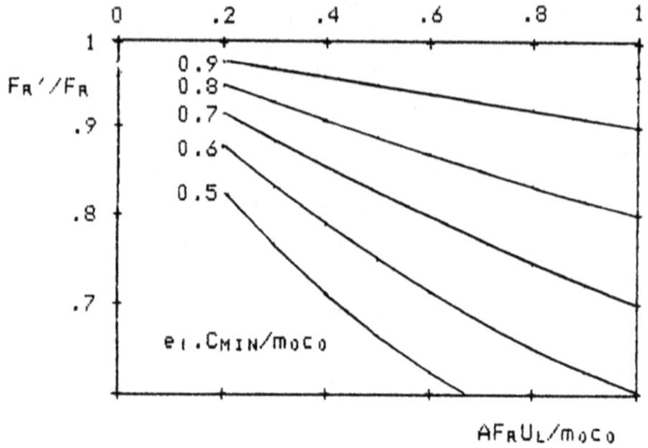

Fig. 2.2.10: Solar loop (collector/heat exchanger) correction factor F_r'/F_r

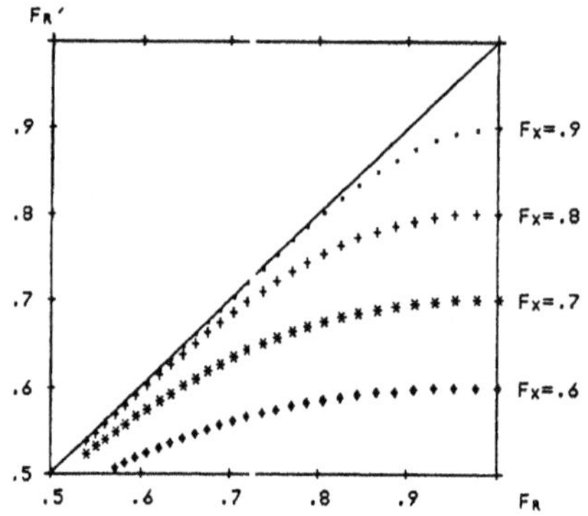

Fig. 2.2.11: Collector-heat-exchanger correction factor F_r' for immersed coils.

Both the factors F_r (for the collectors) and F_x (for the heat-exchanger) represent a decrease in system efficiency, but the term which leads to the lower decrease may be neglected.

This can help the designer in choosing system parameters /2.22/. As the heat transfer coefficient is not known precisely, it will be specified, for a given type of collector and a given flow-rate, in order not to introduce additional loss of performance.

For a reasonably high flow-rate, the heat transfer coefficient must satisfy $A_eU_e > 2\ \dot{m}_0c_0$ (which ensures $F_x > F_r$)

EXAMPLE: $U_L = 7$ W/m2/K
$\dot{m}_0c_0 / A = 70$ W/m2/K (60 l/hour/m2)

From Eq. 2.1.12, $F_r = 0.952$

The immersed coil is sized in such a way that the performance will not be decreased. This requires $F_x > 0.952$

and can be obtained with a total heat transfer coefficient of at least 150 W/K per square metre of collector (according to Eq. 2.2.34).

For A_eU_e / A = 150 W/K/m2, the effectiveness is e_1 = 0.883 (from Eq. 2.2.27) and the exact value of the solar loop correction factor (from Eq. 2.2.33) is F_r' = 0.940; the additional decrease in performance due to the immersed coil is negligible (1.3%).

For an external heat-exchanger, Eq. 2.2.35 can also be used to estimate F_r', but with a smaller level of confidence, modifying F_x

$$F_x = \frac{1}{1 + AU_L / F_eA_eU_e} \qquad (2.2.36)$$

with

$$F_e = [\ 1 - \text{EXP}\ (\ - A_eU_e/\dot{m}_1c)\ ]\ .\ \dot{m}_1c\ /\ A_eU_e \quad . \qquad (2.2.37)$$

2.2.3 - Solar loop control

The solar collector must not be allowed to operate as a "radiator" to the environment. When the solar irradiance is smaller than some threshold value (depending on the working temperature), an appropriate control is needed for the solar loop.

The most common type is a differential control which compares temperatures at the collector outlet and at the bottom of the storage tank. When the difference is greater than some starting differential, d_{on}, the control switches the pump on. When it becomes smaller than a stopping differential, d_{off}, the control stops the pump. In order to avoid unstable pump working, with frequent starting and stopping, d_{on} is taken to be between 5 and 10 K, and d_{off} is much smaller (1 or 2 K). The exact criterion to be satisfied by the differentials is as follows:

$$d_{off} < d_{on} \cdot AF_rU_L / \dot{m}_0c_0 \quad . \qquad (2.2.38)$$

For systems which include an external heat-exchanger, a secondary pump is required. It may be operated by the same control as for the primary or, more often, run continuously during the day (operated by a day-night command, for example).

Other types of controls also exist, for example with a three-way valve, but they generally lead to a similar performance (see section 1.2).

2.3 - STORAGE TANK

In neither space heating nor water heating do loads exactly coincide with solar gains, so energy storage is needed in most active solar systems. It is generally a matter of short-term storage, allowing for a delay of a few hours between supply and demand. Experience shows that most systems rarely store excess solar energy for more than two days.

Long-term or interseasonal storage is still being considered as a promising future option for solar systems, but a number of problems have yet to be solved, principally from the economic point of view. One should bear in mind that we are dealing here with systems which require auxiliary energy, and that energy self-sufficiency is not the objective.

2.3.1 - Storage types

The simplest and the most common type of storage uses the sensible heat of water. Water has a high specific heat and is cheap. Thus, most active thermal solar systems use an insulated water tank for storage. Short-term storage requires about 40 to 100 litres of water per square metre of collector, giving a total of 1 to 4 m3 for a single-family house. Interseasonal storage -which would enable winter space heating from summer solar energy- would require 200 m3 or even more.

Latent heat storage has also been considered as a promising alternative, as some materials have a high heat of fusion at temperatures suitable for space heating, e.g., calcium chloride hexahydrate (melting point: 28°C). An equivalent storage capacity would require a reduced volume compared with water. But for the time being, a number of problems have not been completely solved (heat transfer, stability of the salt...) and the economic advantage has still to be demonstrated.

2.3.2 - Physical aspects

Hot water is lighter than cold water and tends to rise to the top of the storage tank, while cold water remains at the bottom. This is known as thermal stratification. In order not to mix water at different temperatures, this phenomenon is encouraged: the cold water inlet (from heat-emitters or mains water) is generally located at the bottom, as well as the cold

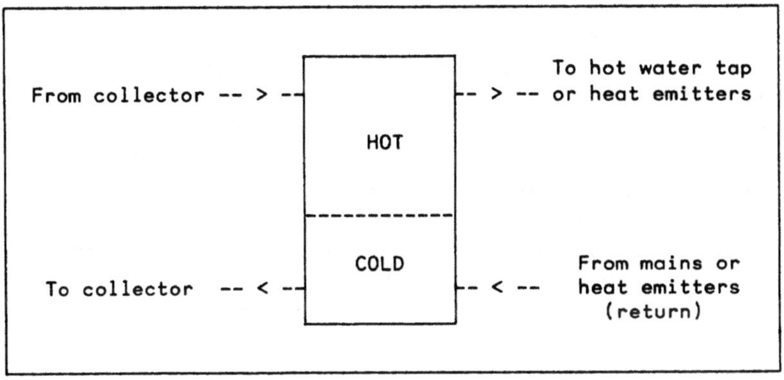

Fig. 2.3.1: Storage and stratification

water outlet to the solar collectors. The hot water inlet (from collectors) and outlet (to heat emitters or DHW taps) are located at the top (Fig. 2.3.1). Baffles are needed when high flow-rates occur e.g., for DHW draw-off.

Thermal stratification is a complex problem which has been investigated in many studies. Detailed simulation programmes enable a comparison between two ideal cases: fully-mixed tanks (with no stratification; the water in the tank is at a uniform temperature at all times) and perfectly stratified tanks (with a large number of water layers at different temperatures).

Stratification affects space heating and domestic hot water systems in different ways. Common heat emitters require a high circulation flow-rate and a small temperature difference (typically 5 to 10 K). Space heating is needed throughout the day and the volume of water circulating through the heat emitters during one day is much greater than the tank capacity. Thus, the water is rather mixed in the storage tank: temperature differences between the top and bottom parts are within a few degrees Celsius in SH applications.

On the other hand, in DHW applications the amount of hot water drawn off each day is equivalent to the tank capacity; hot water is required at a greater temperature difference (30 to 40 K with respect to cold water). DHW is drawn off at particular times during the day, over short periods. Thus thermal stratification, enhanced by low collector flow-rates, has a beneficial effect on DHW systems.

For many years, high flow-rates (of the order of 40-80 l/h per square metre of collector) have been specified to ensure good collector efficiency. However, it has recently become obvious (both from theory and experiment) that much lower collector flow-rates are suitable if a good thermal stratification can be maintained in the storage tank. The corresponding single pass design (in which the exact contents of the storage or the daily demand pass once through the collector during a daily cycle) performs even better if the load occurs during daylight hours. With a flow-rate of 5-10 l/h/m2 an improvement in performance of 10-20% can be observed in DHW systems /1.3-1.5/. Furthermore, such a low flow-rate greatly reduces the parasitic energy consumption (by pumps).

The second important physical aspect of storage is thermal loss, which takes place through the tank wall and decreases the amount of energy stored in the tank.

2.3.3 - Storage characteristics

Two characteristic input parameters of the storage are required by the ESM methods. The first is the effective mass of water in the tank, M (kg). Parts of the tank containing auxiliary heaters should be excluded, as well as dead zones (lower than the heat-exchanger for example). The thermal capacity of water will be denoted c (kJ/kg/K).

The second parameter is the overall heat loss coefficient, A_sU_s (W/K). This can be determined by a theoretical computation, as product of the tank external area and a heat transfer coefficient (the reciprocal of the thermal resistance of the tank wall). It must be corrected to account for cold bridges, particularly with the ground. This correction may sometimes be as much as a factor 2.

A better estimate of the overall heat loss coefficient is obtained by specific standardized tests, whose results are available for some industrial solar tanks /2.23-24/.

The test procedure proposed by the European Storage Testing Group is simple: the tank is filled with hot water at a known uniform temperature; thermal losses are allowed to proceed during a sufficient period of time, typically 24 hours; then the tank is emptied and the average temperature of the water is measured. The overall heat loss coefficient may be determined by a simple calculation from these results /2.25/.

2.3.4 -**Coupling of domestic hot water production and storage**

The essential characteristics of solar hot water production are concerned with the load (daily demand, mains cold water temperature, and required temperature). These are described in Chapter 5. For space heating and DHW coupled systems, another parameter is needed: this concerns the heat transfer between the storage tank and the hot water production. It is called the DHW production effectiveness, e_3.

If an additional solar storage tank is involved (i.e., a special storage tank or part of a tank for DHW without back-up), inside or outside the main store and with a capacity larger than 30% of the daily demand, this effectiveness is taken equal to 1.

For instantaneous DHW production using an internal heat-exchanger (submerged in the main tank; or a simple pipe), e_3 is the effectiveness of the heat-exchanger and is given by

$$e_3 = 1 - \mathrm{EXP}\,(\, - A_{dh}U_{dh} \,/\, \dot{m}_3 c \,) \quad . \qquad (2.3.1)$$

$A_{dh}U_{dh}$ is the average global heat transfer coefficient and $\dot{m}_3$ the average draw-off flow-rate (daily demand divided by the average draw-off time); $\dot{m}_3$ is generally between 200 and 600 litres/hour for a single family and is highly user-dependent. The heat transfer coefficient, U_{dh}, can be computed using the immersed coil equation (2.2.29); it is between 200 and 400 W/K.m2. This results in values of e_3 of 0.4 to 0.2, or even lower. Note that this effectiveness value is equal to the ratio of the difference between the DHW exchanger outlet and inlet (= mains) temperatures, divided by the temperature difference between the storage and the mains water. Thus, with an effectiveness of 0.25, a storage tank at 90°C will preheat mains water at 10°C to a temperature of only 30°C. Clearly, although a system without a pre-heating tank may be cheap, it is of little value in energy terms, except when a large heat-exchange area is involved.

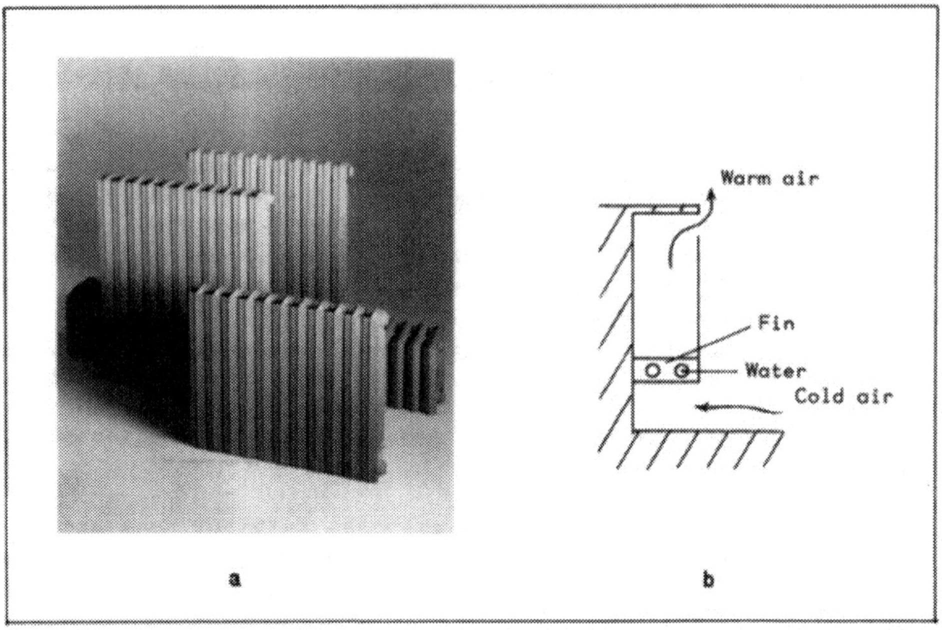

Fig. 2.4.1: Heat emitters: a) radiators (Document from DE DIETRICH, France).
b) convector.

Fig. 2.4.2: Heat emitters: fan coil unit (without casing).
(Document from CIAT, France).

2.4 - SPACE HEATING SYSTEM: HEAT EMITTERS

The characteristics of the solar loop, which influence the long-term system performance, were reviewed in sections 2.1 and 2.2. For systems which meet a space heating load, certain other characteristics are also important, i.e., the load itself (see Chapter 5), the heat distribution connection and control, and the characteristics of the heat emitters.

The heat distribution control, together with the auxiliary supply, has been analysed in Chapter 1. As far as auxiliary energy is concerned, most types of boilers can be included in active solar systems, using any type of energy source i.e., gas, oil, coal, wood, electricity. These energy sources can be connected in "series" with the solar part of the system, or installed in "parallel", with their own distribution system (see section 1.3). Characteristics of the heat emitters are discussed below.

2.4.1 - Heat emitter types

Standard heat emitters may be used in an active solar system. However, it should be borne in mind that a low distribution temperature improves the solar collector efficiency and, thus, the performance of the whole system. Therefore, better active solar space heating systems should be designed employing low temperature heat emitters. With this reservation, common types of emitters may be used, such as water radiators or convectors (Fig. 2.4.1), fan-coil units (Fig. 2.4.2), and water underfloor heating.

2.4.2 - Heat emitter characteristics

Heat emitters are generally designed for given conditions, corresponding to the maximum required power; for an internal temperature T_c and a minimum external temperature, T_{min}, a maximum power P_{max} has to be supplied by the system

$$P_{max} = GV \cdot (T_c - T_{min}) \quad . \tag{2.4.1}$$

Under these conditions, heat emitters are characterized by a set of two temperatures: the water flow temperature, T_f (inlet of emitters), and the return temperature, T_r (outlet of emitters). These values depend on the type of heat emitter e.g., 90°C / 70°C for most radiators, 35°C / 30°C for underfloor heating. Manufacturer's data enable the computation of the required emitter area.

Heat emitters are simply heat exchangers. Nevertheless, as a large part of the exchange takes place as radiation and free convection, the heat transfer coefficient can no longer be considered as constant but varies with the temperature. For a given mean temperature difference between the emitter and the building, ΔT, the heating power is proportional to $(\Delta T)^m$

$$P / P_{max} = (\Delta T / \Delta T_{max})^m \quad . \qquad (2.4.2)$$

For radiators and convectors, m = 1.3, while it is taken equal to 1 for underfloor heating and fan-coil units.

Generally speaking , ΔT is the average logarithmic temperature difference

$$\Delta T = (T_f - T_r) / \mathrm{Ln}\ [(T_f - T_c)/(T_r - T_c)] \qquad (2.4.3)$$

but if $(T_r - T_c) / (T_f - T_c) > 0.7$ the arithmetic mean may be adopted

$$\Delta T = (T_f + T_r) / 2 - T_c \quad . \qquad (2.4.4)$$

The ESM1 method requires heat emitters to be characterized by the water heat capacitance rate, $\dot{m}_2 c$, and by the average effectiveness, e_2. $\dot{m}_2 c_2$ is simply

$$\dot{m}_2 c = P_{max} / (T_f - T_r) \quad . \qquad (2.4.5)$$

For underfloor heating and fan-coil units (m = 1), the effectiveness does not depend on the temperature and is equal to its nominal value

$$e_2 = (T_f - T_r) / (T_f - T_c) \quad . \qquad (2.4.6)$$

For radiators and convectors, the average effectiveness is computed at an average power taken as

$$P = P_{max} / 2 \quad . \qquad (2.4.7)$$

Average flow and return temperatures, T_f' and T_r', can be computed in this case using Eqs. (2.4.2) and (2.4.3), noting that

P is of course proportional to $(T_f' - T_r')$.

This can be summarized by the following equations:

$$a = (P / P_{max})^{(1 - 1/m)} \qquad (2.4.8)$$

$$e_2 = 1 - [(T_r - T_c)/(T_f - T_c)]^a \quad . \qquad (2.4.9)$$

Here, T_r and T_f correspond to P_{max}. In the most common case where $P/P_{max} = 0.5$ and $m = 1.3$, $a = 0.85$.

REFERENCES OF CHAPTER 2

/2.1/ DUFFIE and BECKMAN. Solar Engineering of Thermal Processes. Wiley, New-York (1981).

/2.2/ A.RABL. Active Solar Collectors and their Application. Oxford Univ. Press, New-York (1985).

/2.3/ S.SVENDSEN and K.I.JENSEN. Advances in Solar Energy Technology (ISES Congress, Hamburg, 1987), Vol.1, pp.620-624. Pergamon (1988).

/2.4/ W.GILLETT. Solar Collectors: Test methods and design guidelines. Doc. EUR 9778. REIDEL, 316 p. (1985).

/2.5/ ASHRAE 93-77. Methods of testing to determine the thermal performance of solar collectors (1978).

/2.6/ R.GICQUEL. REVUE GENERALE DE THERMIQUE, Vol.184 (1977).

/2.7/ AFNOR. Norme P.50-501. Capteurs solaires à circulation de liquide: Détermination de la performance thermique. (1980).

/2.8/ DIN 4757/4. Determination of efficiency of solar collectors. Beuth Verlag, Berlin (1982).

/2.9/ BSI. British Standard 6757 (1982).

/2.10/ ISO. Solar collector testing standards.

/2.11/ J.ADNOT. Contribution à l'étude théorique et expérimentale des boucles de captage thermique de l'énergie solaire. Thèse Docteur-Ingénieur, Ecole des mines de Paris (1981).

/2.12/ W.H.McADAMS. Heat Transmission. 3d ed. McGRAW-HILL, New-York (1954).

/2.13/ RECKNAGEL et al. Manuel Pratique du génie Climatique, 2ème ed. PYC Editions, Paris (1986).

/2.14/ KAYS and LONDON. Compact Heat-Exchangers. McGraw Hill (1964).

/2.15/ Heat Exchanger Design Handbook. Hemisphere Pub. Corp. (1983).

/2.16/ TEMA. Standards of Tubular Exchanger Manufacturer Association. USA (1978).

/2.17/ R.VIDIL. Les Echangeurs à plaques et joints. Greth-CETIAT (1984).

/2.18/ AFGAN and SCHLUNDER. Heat-exchangers: Design and Theory Sourcebook. McGRAW HILL (1974).

/2.19/ W.DUTRE. EMGP2: a European transient simulation model for thermal solar systems. EUR 9777. REIDEL (1985).

/2.20/ T.J.FEIEREISEN et al. ASME J. of Solar Energy Engineering, (1983).

/2.21/ DEWINTER. SOLAR ENERGY, Vol.17, p.335 (1975).

/2.22/ BOURGES. Advances in Solar Energy Technology (ISES Congress, Hamburg, 1987), Vol.2, pp.1172-1176. Pergamon (1988).

/2.23/ AFNOR. Ballons solaires: mesure des performances thermiques. P.50-301

/2.24/ S.Ø.JENSEN and S.A.SVENDSEN. Test and Calculation methods used in Denmark for solar collectors, heat storages and solar heating systems. TUD Thermal Insulation Lab, Report 85-7, Lyngby (1985)

/2.25/ H.VanDIJK. The European Solar Storage Testing Group. Solar Energy applications to Building and Solar Radiation Data (Contractors'meeting, Brussels, 1987), Vol.4, pp.42-57 (1988).

CHAPTER 3

PRINCIPLES OF THE SIMPLIFIED METHODS

Active solar systems and their basic components have been discussed in the first two chapters. In this chapter, the calculation of system performance is dealt with. We should bear in mind that the objective of this handbook is to present a set of simplified methods. These methods will be described in detail in chapter 6 (ESM1 for space heating) and chapter 7 (ESM2 for hot water). Both of these methods are based on simple closed-form models giving first estimates, which are then improved by corrective terms.

Closed-form approaches involve a number of assumptions and they do not always result in accurate calculation which is generally the advantage of detailed simulation programmes. However, closed-form approaches are a valuable tool for understanding the basic physical phenomena in a system, and thus may be used to design systems and, for some components, to derive practical sizing rules. These models, which serve as basic algorithms for simplified methods, are described in this chapter.

3.1 - BASIC MODELS AND ASSUMPTIONS

Analytical approaches developed in the following paragraphs are based on simple component models. Most of them have been introduced in Chapter 2 and are summarised here.

3.1.1 - Collector loop

The whole collector loop, including solar collectors, pipes, and heat-exchanger, is represented by a single equation

$$P_{col} = A.F_r'. [\eta_0.I - U_L.(T_{in} - T_a)] \tag{3.1.1}$$

which gives the instantaneous collected power. This is the collector equation (2.1.11) where the collector characteristics, U_L and η_0, are corrected in order to account for pipe losses according to Eqs. (2.2.15-17). F_r' is the solar loop correction factor (Eq. 2.2.33) which accounts for the effect of the heat-exchanger. T_{in} is the fluid temperature at the solar loop inlet.

3.1.2 - Storage losses

Storage tank losses are represented by

$$P_{los} = A_s.U_s . (T_s - T_e) \quad . \tag{3.1.2}$$

T_s is the average storage temperature, T_e the tank environmental temperature. The storage tank has a thermal capacity Mc and a heat loss coefficient, A_sU_s.

3.1.3 - Distribution loop

For space heating systems, the heat emitters are modelled as a heat-exchanger, characterized by their effectiveness e_2 and fluid capacitance rate $\dot{m}_2c$. The power distributed to the building to meet the space heating load is:

$$P_{dis} = \dot{m}_2c . e_2 . (T_f - T_c) \tag{3.1.3}$$

$$P_{dis} = \dot{m}_2 c \, . \, (T_f - T_r) \qquad . \qquad (3.1.4)$$

T_f and T_r are, respectively, the flow (heat emitter inlet) and return (outlet) temperatures. Set-point values of these two temperatures are obtained with a distributed power equal to the required power for space heating of the building, P_{sh} (or rate of heat loss, as defined by Eq. 5.1.8 in chapter 5):

$$T_f^* = T_c + P_{sh} / (\dot{m}_2 c \, . \, e_2) \qquad (3.1.5)$$

$$T_r^* = T_c + P_{sh} \, . \, (1 / e_2 - 1) \qquad . \qquad (3.1.6)$$

A distinction is made between Series connected auxiliary and Parallel auxiliary. In both cases a perfect control, in which the flow temperature is modulated, is assumed to ensure that P_{dis} is exactly equal to the required power. In the first case (series), T_f and T_r are fixed by the auxiliary and equal to the set-points defined above (Eqs. 3.1.5 and 3.1.6).

In the second case (Parallel), the flow temperature is the storage tank temperature T_s (in its top part), and the return temperature is

$$T_r = (1 - e_2) \, . \, T_s + e_2 \, . \, T_c \qquad . \qquad (3.1.7)$$

In both cases, several situations can occur, according to the storage temperature T_s (see Fig. 1.4.1 or 1.4.2). If this temperature is high enough to make complete solar operation possible (without auxiliary; i.e., if $T_s > T_f^*$), the storage is partly by-passed and the flow rate through the tank is controlled so as to obtain $T_f = T_f^*$. If T_s is too low (smaller than the return temperature T_r), no solar contribution is possible and the circulation through the tank is stopped.

Thus, the solar power distributed from the storage tank to the heat emitters can be written as

$$\text{Series} \quad P_{sol} = \begin{cases} P_{sh} & \text{if } T_s > T_f^* \\ \dot{m}_2 c . (T_s - T_r^*) & \text{if } T_f^* > T_s > T_r^* \\ 0 & \text{if } T_r^* > T_s \end{cases} \qquad (3.1.8)$$

$$\text{Parallel} \quad P_{sol} = \begin{cases} P_{sh} & \text{if } T_s > T_f^* \\ \dot{m}_2 c . e_2 . (T_s - T_c) & \text{if } T_f^* > T_s > T_c \\ 0 & \text{if } T_c > T_s \end{cases} \qquad (3.1.9)$$

In most space heating systems, the difference between flow temperature and return temperature is small, generally less than 15 to 20K. The space heating load occurs during the whole day but varies greatly during the heating seasons (as does the return temperature).

A similar equation represents the power distributed for hot water

$$P_{dis} = \dot{m}_3 c \cdot (T_s - T_m) \qquad (3.1.10)$$

where $\dot{m}_3$ is the instantaneous hot water mass demand and T_m the mains cold water temperature. This relationship is true if the solar storage temperature T_s is lower than the required hot water temperature T_h; otherwise,

$$P_{dis} = \dot{m}_3 c \cdot (T_h - T_m) \qquad (3.1.11)$$

Hot water loads require high temperature differences ($T_h - T_m$ = 30 to 50K) and low average draw-off flow rates (but concentrated draw-off at particular times).

3.1.4 - Summary of assumptions

All the component models developed here are linear with respect to temperature differences. Their coefficients (heat transfer coefficient for collector, heat-exchanger, and heat emitters) are assumed to be constant (independent of temperatures and flow rates).

Thermal masses (except storage mass) are neglected in all the components (collector, heat emitters, pipes, etc.). Additional heat losses are either neglected or assumed to be included in those already accounted for. System control is assumed to be perfect (without hysteresis or differential).

These assumptions are needed to derive the simple analytical models presented in this chapter. They are not required by the simplified methods themselves, in which corrective terms can be included to account for additional aspects of the system.

3.2 - THRESHOLD IRRADIANCE AND BASE TEMPERATURE

The phenomenon of threshold irradiance and its consequences, through the concepts of collectable energy, utilizability and base temperature, are analysed in this section. Similarly, an example of the application of threshold irradiance to simple systems, such as industrial process heating (IPH), is given.

3.2.1 - Threshold irradiance

Consider Eq. 3.1.1, giving the useful power coming from a collector, and including a term for solar gains ($AF_r'\eta_0 I$) and a term for thermal losses $AF_r'U_L(T_{in}-T_a)$. In normal operation, gains are greater than losses, but the opposite situation may occur, in which losses are greater than gains, and the collector loses energy to the environment.

Under given temperature conditions, there is a critical value of the solar irradiance for which gains are equal to losses and the collector does not provide energy (P=0); this value is called the threshold irradiance or critical intensity, I_c, and it is defined by

$$I_c = U_L \cdot (T_{in} - T_a) / \eta_0 \quad . \qquad (3.2.1)$$

If the incident solar irradiance is higher than this threshold value, the collector operates normally. If the irradiance is lower than the threshold value, it is better to stop the collector operating, in order not to lose energy. This may be achieved by a control device, which stops the circulation pump if the collector outlet temperature drops below the inlet (or the storage) temperature.

It may be useful on an existing facility to compute I_c and to check if operation (even discontinuous, this is not a problem) starts more or less for the radiation values around I_c.

The power from the collector can thus be written as

$$P_{col} = A \cdot F_r' \cdot \eta_0 \cdot (I - I_c)^+ \quad . \qquad (3.2.2)$$

The superscript + indicates that the term between brackets is taken only if positive; otherwise, a value of zero is adopted.

3.2.2 - Available energy and utilizability

If temperatures T_{in} and T_a are assumed constant over a certain period of time (e.g., hour, day, or month), the useful energy output from the collector during this period can be computed by a simple integration of Eq. 3.2.2

$$E_c = A \cdot F_r' \cdot \eta_0 \cdot \int_0^{t_d} (I - I_c)^+ \cdot dt \qquad (3.2.3)$$

The integral in Eq. 3.2.3 is called the available or collectable energy above the threshold I_c during the period t_d:

$$H_{av} (I_c) = \int_0^{t_d} (I - I_c)^+ \cdot dt \quad . \qquad (3.2.4)$$

This integral depends only on the value of I_c and on the statistical distribution (frequency distribution) of solar irradiance on the collector surface during the period.

The collectable energy above a threshold of zero is merely the incident solar irradiation on the collector surface during the period:

$$H_{av} (0) = H_t \quad . \qquad (3.2.5)$$

Generally, threshold irradiances are positive and the collectable energy is only a fraction of the incident irradiation; this fraction is called utilizability and is defined as

$$\phi (I_c) = H_{av} (I_c) / H_t \quad . \qquad (3.2.6)$$

This concept was introduced by Whillier /3.1/ in 1953 and it has been the basis of a number of applications, a complete review of which can be found in a paper by Klein and Beckman /3.2/.

Both the available energy and the utilizability can be computed from the frequency distribution of solar irradiance during the period in question. Cumulative frequency curves of solar irradiance, as defined in Chapter 4, are a suitable presentation of radiation data for these computations.

In the following sections, the available energy H_{av} will be noted either as a function $H_{av}(I_c)$ of the threshold I_c, or as a function $H_{av}(T_{in})$ of a fluid temperature T_{in}; in the latter case, the threshold irradiance I_c is a necessary intermediate parameter estimated from Eq. 3.2.1.

3.2.3 - Base Temperature

The assumption "T_{in} = constant" made in the previous section is seldom fulfilled in practice: the collector inlet temperature may vary widely, except in special systems, like industrial process heating. The fluid entering the collector comes from the storage, the temperature of which increases during the day due to solar heat collection. Therefore the collector inlet temperature also increases.

Furthermore, the return temperature from the distribution loop can vary according to specific constraints (higher space heating loads imply higher return temperatures; the mains cold water temperature is not the same throughout the year, etc.).

Thus, even an average collector inlet temperature is not known, but usually a lower limit can be given. This minimum value is called the base temperature of the system over the considered period. It is then possible to use Eq. 3.2.4 to calculate an upper limit for the useful energy output.

For hot water systems, the mains cold water temperature is an estimate of the lowest temperature value which will be observed at the collector inlet; for space heating, the average return temperature from heat emitters over the period is the corresponding minimum value. For a series-connected auxiliary,

$$T_r = T_c + L_0 \cdot (1/e_2 - 1) / (\dot{m}_2 c . t_e) \qquad (3.2.7)$$

where t_e is the heat emitter operating time during the period (usually 24 hours per day) and L_0 is the corresponding space heating load.

3.2.4 - Application to industrial process heating

Consider a simple solar industrial process heating system as defined in section 1.3.3 and shown in Fig.3.2.1. The constant load is characterized by two temperatures, "cold" (mains) temperature T_m and "Hot" (demand) temperature T_h. ($T_h > T_m$) and by the process flow-rate $\dot{m}_1 c$ (assumed to be constant over the operating period). The system does not include any storage but a heat exchanger is possible.

The power from the collector loop is given by Eq.3.1.1 (with $T_m = T_{in}$), and the collector loop outlet temperature can be calculated from

$$T_{out} - T_m = A.F_r'.\eta_0.(I - I_c) / \dot{m}_1 c \qquad . \quad (3.2.8)$$

I_c is the initial threshold irradiance

$$I_c = U_L . (T_m - T_a) / \eta_0 \qquad . \quad (3.2.9)$$

The maximum outlet temperature is obtained for the maximum incident irradiance I_{max}. Provided that T_{out} is lower than T_h, all the collected energy makes a useful contribution to the load, but as soon as it becomes higher than T_h, the excess energy is discarded.

As found by Gordon and Rabl /3.3-4/, it is easy to determine the collector area A_{dump} above which energy dumping occurs. For systems without a heat-exchanger, in which the process flow rate is also the collector flow rate, collector area is obtained from Eq. 3.2.8 with $T_{out} = T_h$

$$A_{dump} = - \frac{\dot{m}_1 c}{U_L} . Ln \left[1 - \frac{U_L.(T_h - T_m)}{\eta_0.(I_{max} - I_c)} \right] \qquad . \quad (3.2.10)$$

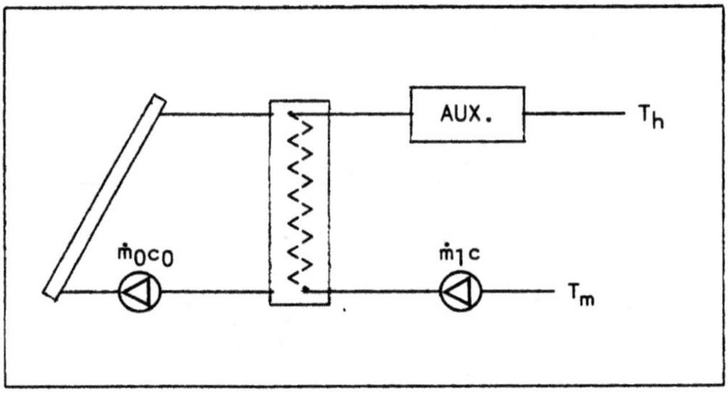

Fig. 3.2.1: Simple solar industrial process heating (IPH) with heat-exchanger. The process fluid circulates directly through the heat-exchanger.

For a system with a heat - exchanger, the collector area is obtained by (3.2.11)

$$A_{dump} = - \frac{\dot{m}_0 c_0}{U_L} . Ln \left[1 - \frac{1}{\dfrac{\eta_0 . \dot{m}_0 c_0 . (I_{max} - I_c)}{U_L . \dot{m}_1 c . (T_h - T_m)} + \dfrac{\dot{m}_0 c_0}{e_1 . C_{min}} - 1} \right]$$

If $A < A_{dump}$, the useful solar energy output over a given period of time may be written:

$$E_c = A . F_r' . \eta_0 . H_{av} (I_c) \quad . \qquad (3.2.12)$$

If energy dumping takes place ($A > A_{dump}$), then

$$E_c = A.F_r'.\eta_0 . [H_{av}(I_c) - H_{av}(I_c')] \qquad (3.2.13)$$

The dumping threshold irradiance, I_c', is given by

$$I_c' = I_c + \dot{m}_1 c . (T_h - T_m) / AF_r'\eta_0 \quad . \qquad (3.2.14)$$

These results have been generalized to systems with storage by Collares-Pereira et. al. /3.4-5/.

3.3 - CLOSED-FORM MODEL FOR A SIMPLE ACTIVE SYSTEM: COLLECTOR + STORAGE

In this section a simple active solar system including a collector loop (with an optional heat exchanger) and a storage tank (Fig. 3.3.1) is considered. Similarly, the system operation during a period in which no load occurs is analysed.

The models described in this section have been developed by Phillips /3.6-7/ and Gordon-Zarmi /3.8/, and recently summarized by Bourges and Adnot /3.9/.

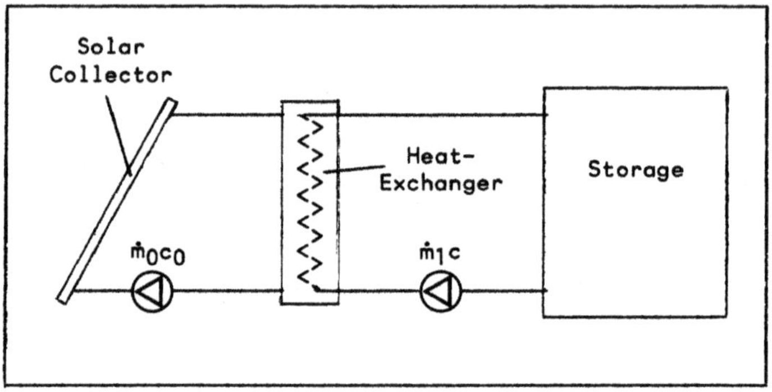

Fig. 3.3.1: Standard schematic of a typical solar collector + storage loop.

3.3.1 - Basic equation

At any time t, the average temperature of the storage T is governed by the simple linear differential equation

$$Mc\ dT/dt = A.F_r'.[\eta_0.I - U_L.(T_{in}-T_a)]^+ - A_sU_s.(T-T_e) \qquad (3.3.1)$$

where T_{in} is the collector inlet temperature (corresponding here to the bottom storage temperature). This equation shows that the variation of energy content in the store corresponds to the difference between collected power P_{col} and heat losses from the store P_{Los}. A period of continuous operation means that the solar irradiance I must always be higher than the threshold irradiance I_c defined at inlet temperature T_{in} (Eq. 3.2.1). The climatic parameters I and T_a may vary during the period.

3.3.2 - Fully mixed storage

3.3.2.1 - General equation and solution

If we assume that the storage tank is fully mixed, i.e., at a uniform temperature T(t) at a time t, Eq. 3.3.1 may be completed by setting

$$T_{in} = T \tag{3.3.2}$$

and it becomes (3.3.3)

$$t_{mix}\, dT/dt + T = (A.F_r'.U_L.T_L + A_sU_s.T_e) / (A.F_r'.U_L + A_sU_s)$$

where t_{mix}, the time constant of the collector - mixed storage loop, is defined by

$$t_{mix} = M.c / (A.F_r'.U_L + A_sU_s) \quad . \tag{3.3.4}$$

Typical values of t_{mix} range from 1 hour (unglazed collector and small storage capacity) to 40 hours (evacuated tubular collector); usual values range from 5 to 20 hours.

The limiting temperature T_L is a function of the climatic parameters

$$T_L = T_a + \eta_0 \cdot I / U_L \quad . \tag{3.3.5}$$

The general solution of Eq. 3.3.3 is (3.3.6)

$$T(t) = T_0.EXP(-t/t_{mix}) + \int_0^t T_L(t').EXP[(t'-t)/t_{mix}] \cdot dt'/t_{mix}$$

where T_0 is the initial temperature (t' is the dummy integration variable from time 0 to t).

This formula may be used directly for any constant or increasing "limiting temperature" function T_L versus time. However, if T_L decreases, it is necessary to check the condition for system operation ($T<T_L$) continuously.

3.3.2.2 - Constant irradiance

The case with constant irradiance or more precisely with constant limiting temperature T_L is the simplest one. Its solution is

$$T(t) - T_0 = (T_L' - T_0) \cdot [1 - EXP (-t/t_{mix})] \tag{3.3.7}$$

with T_L' defined as

$$T_L' = (A.F_r'.U_L.T_L + A_sU_s.T_e) / (A.F_r'.U_L + A_sU_s) \quad . \qquad (3.3.8)$$

The stored energy E_s during a period of time t is written as

$$E_s = Mc . [T(t) - T_0] \qquad (3.3.9)$$
$$= G_{cst} . [A.F_r'.U_L.t.(T_L-T_0) + A_sU_s.t.(T_e-T_0)]$$

or, if we introduce the available solar energy H_{av} above temperature level T_0 on the collector surface:

$$E_s = G_{cst} . [AF_r'\eta_0 H_{av}(T_0) + A_sU_st (T_e - T_0)] \quad . \qquad (3.3.10)$$

This means that the stored energy equals the available energy at the initial temperature T_0 (minus the losses at this temperature) multiplied by a correction factor G_{cst}:

$$G_{cst} = [1 - EXP (- t / t_{mix})] . t_{mix} / t \quad . \qquad (3.3.11)$$

This term lies between zero and one (Fig. 3.3.2). This term has also been introduced by Phillips /3.6/ as a "system heat storage factor" (with the symbol F_s); it is a kind of global system efficiency for hourly or daily periods.

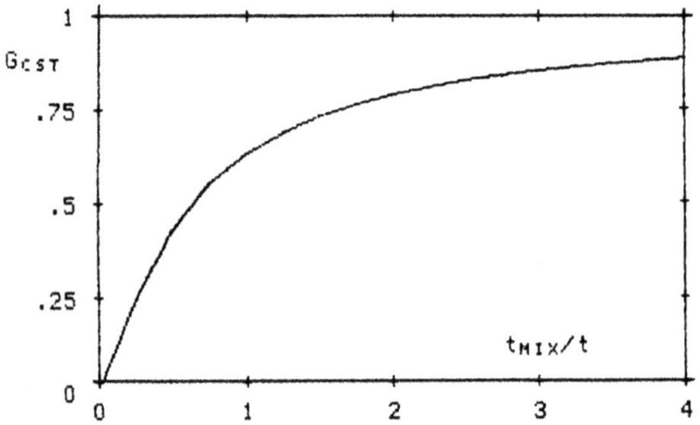

Fig. 3.3.2: System heat storage corrective factor G_{cst} for a fully mixed tank (constant irradiance; Eq. 3.3.11).

3.3.2.3 - Time-dependent irradiance: Sinusoidal irradiance

A typical daily irradiance profile is sinusoidal (both on clear and completely overcast days) and can be represented during the collecting period by

$$T_L - T_0 = u . \sin (w.t) \qquad (3.3.12)$$

with $w = \pi / t_d$ and u derived from the maximal solar irradiance I_{max}:

$$u = T_a - T_0 + \eta_0 \cdot I_{max} / U_L \quad . \tag{3.3.12'}$$

The initial threshold irradiance of Eq. 3.2.1 (with $T_{in} = T_0$) is assumed to be exceeded for a period t_d during the day. Furthermore, it is assumed that there are no losses. Eq. 3.3.3 becomes

$$t_{mix} \cdot dT/dt + T = T_0 + u \cdot \sin(wt) \tag{3.3.13}$$

with the solution (3.3.14)

$$T - T_0 = \frac{u.[w.t_{mix}.EXP(-t/t_{mix}) + \sin(wt) - wt_{mix}.\cos(wt)]}{[1 + (wt_{mix})^2]}$$

Collector operation stops at time t_{stop}, such that

$$T(t_{stop}) - T_0 = u \cdot \sin(wt_{stop}) \tag{3.3.15}$$

t_{stop} is the non-zero solution of the implicit equation

$$w.t_{mix}.\sin(wt_{stop}) = EXP(-t_{stop}/t_{mix}) - \cos(wt_{stop}) \tag{3.3.16}$$

As with constant irradiance, we can try to express the results as a system efficiency with respect to $H_{av}(T_0)$, the solar energy available at the initial temperature T_0:

$$H_{av}(T_0) = 2\, u\, t_d / \pi \quad . \; (U_L / \eta_0) \tag{3.3.17}$$

$$E_s = G_{sin}.A.F_r'.\eta_0.H_{av}(T_0) \quad . \tag{3.3.18}$$

The system efficiency G_{sin} depends only on t_{mix}/t_d and is given by

$$G_{sin} = \pi/2 \cdot t_{mix} / t_d \cdot \sin(wt_{stop}) \tag{3.3.19}$$

G_{sin} is compared with G_{cst} in Table 3.3.1. It can be seen that G_{sin} is always greater than G_{cst}, although the collector operating time is smaller. Both have the same asymptotic straight line when t_{mix}/t_d becomes high enough (>2)

t_{mix}/t_d	t_{stop}/t_d	G_{sin}	G_{cst}
.10	.596	.150	.100
.20	.669	.271	.199
.30	.719	.364	.289
.40	.756	.436	.367
.50	.784	.493	.432
.75	.829	.596	.552
1.00	.855	.664	.632
1.25	.881	.713	.688
1.50	.892	.749	.730
1.75	.901	.777	.762
2.00	.910	.800	.787
2.50	.913	.833	.824
3.00	.915	.857	.850
4.00	.920	.889	.885
5.00	.930	.910	.906
7.50	.950	.938	.936
10.00		.955	.952

Table 3.3.1: Operating time t_{stop} and system heat storage factor G_{sin} for a mixed storage without losses, receiving sinusoidal irradiance, with respect to ratio t_{mix}/t_d (Eqs. 3.3.16 and 3.3.19). Comparison with G_{cst}, system heat storage factor under constant irradiance (Eq. 3.3.11).

3.3.2.4 - Time-dependent irradiance: Time-averaged storage temperature

Any daily irradiance profile can be considered with the help of Eq. 3.3.6. The continuous checking of the condition $T > T_L$ generally requires a numerical solution, but the solution may be approximated by the average storage temperature procedure.

Let us reconsider the case with constant irradiance: the time-averaged storage temperature during the operating period can be written

$$T_s - T_0 = (T_L' - T_0) . (1 - G_{cst}) \qquad (3.3.20)$$

and the stored energy (Eq. 3.3.10) becomes

$$E_s = A.F_r'.\eta_0.H_{av}(T_s) + A_sU_s.t.(T_e - T_s) \quad . \qquad (3.3.21)$$

This means that the stored energy is equal to the collectable energy at T_s minus the losses at T_s.

This result can be used as the basis of an iterative calculation for any irradiance profile:

- Each iteration corresponds to a base temperature T_b, with which is associated a threshold irradiance I_c (Eq. 3.2.1 with $T_{in} = T_b$).

- The available energy H_{av} and the collector operating time t_{op} are computed from I_c and the true irradiance profile.

- The heat storage efficiency factor, G_{cst}, is computed from Eq. 3.3.11 with $t = t_{op}$.

- During the collector operation, the irradiance is assumed to be constant, leading to

$$T_L = T_b + H_{av}(T_b) / t_{op} \cdot (\eta_0 / U_L) \quad . \qquad (3.3.22)$$

- The time-averaged storage temperature T_s is estimated from Eq. 3.3.20.

The first iteration is made with the initial storage temperature ($T_b = T_0$), the second iteration is made with the average storage temperature obtained at the end of the first iteration. The stored energy can then be computed from Eq. 3.3.21 with the values of H_{av}, t_{op}, and T_s obtained during the second iteration.

Results are compared with the exact solution for a sinusoidal irradiance profile (Table 3.3.2). The error is smaller than 2% in this case. A similar accuracy is obtained for any symmetrical irradiance profile (increasing during the morning and decreasing in the afternoon), but errors can become rather

t_{mix}/t_d	b = 1	b = .8	b = .6
.50	1.0	0.5	-1.1
.75	-0.5	-0.8	-1.6
1.0	-0.7	-0.9	-1.4
1.25	-0.7	-0.8	-1.1
1.5	-0.6	-0.7	-0.9
1.75	-0.6	-0.6	-0.8
2.0	-0.5	-0.5	-0.7

Table 3.3.2: Error (in %), in the calculation (on the basis of the average temperature) of the energy stored by a mixed storage in sinusoidal irradiance (of half-period t_d) depending on t_{mix}/t_d and on the parameter $b = A.F_r'.U_L / (A.F_r'.U_L + A_sU_s)$.

large with asymmetric profiles, such as linearly increasing or decreasing ones (never observed in practice) /3.9/.

Nevertheless, this "symmetrical" assumption seems reasonable for daily calculations (and has been kept for ESM2).

3.3.3 - Perfectly stratified storage

Another ideal storage model type is the perfectly stratified storage which is, generally speaking, close to reality. This model is also called a "plug flow" model; heated fluid from the collector flows in at the top of the tank and displaces the storage volume downwards, thus circulating the fluid at the bottom through the collector. Neither mixing nor heat transfer occurs between the different layers. Under constant solar irradiance this model corresponds to two zones (with variable volume), each at a uniform temperature.

A storage cycle is defined as the period during which the whole content of the tank passes through the collector; its duration is t_c:

$$t_c = M / \dot{m}_1 \quad . \tag{3.3.23}$$

With a uniform initial temperature T_0, the bottom storage temperature (to be introduced in Eq. 3.3.1) is approximately constant and equal to T_0 over the first cycle; if more accuracy is required, it can be corrected for heat losses. The space-averaged storage temperature at the end of the first cycle may be used as an initial temperature for a second cycle, and so on. The complete solution has a closed form, but it is rather complex /3.9/.

In order to simplify this solution, consider a storage tank without heat loss and a constant solar irradiance. At the end of the first cycle, the storage is at a uniform temperature T_1

$$T_1 = a \cdot T_0 + (1 - a) \cdot T_L \tag{3.3.24}$$

with

$$a = 1 - A F_r' U_L / \dot{m}_1 c \tag{3.3.25}$$

and at the end of the n-th cycle, it becomes

$$T_n = a^n \cdot T_0 + (1 - a^n) \cdot T_L \quad . \tag{3.3.26}$$

This is the general solution of Eq. 3.3.1 in this simple case, for a whole number of cycles. It can be demonstrated that this solution is equivalent to the solution of the differential equation describing the fully mixed system (Eq. 3.3.3), where F_r' is replaced by the product $K_s.F_r'$.

$$t_{str}\ dT/dt + T = T_L \tag{3.3.27}$$

$$t_{str} = M.c\ /\ (A.F_r'.K_s.U_L) \tag{3.3.28}$$

where t_{str} is the stratified system time-constant.
K_s is a stratification coefficient /3.7/ given by

$$K_s = Ln\ (1\ /\ a)\ /\ (1 - a)\quad . \tag{3.3.29}$$

K_s is greater than 1 but approaches unity when the flow rate $\dot{m}_1c$ becomes very large (and the tank becomes mixed).

Fig. 3.3.3 enables us to understand this procedure. The exact time variation of the stratified storage temperature is represented by a set of segments. The "equivalent mixed storage temperature" is represented by an exponential curve that passes through all the segment ends.

An "equivalent mixed system" can be defined for any stratified storage. The expression is exact for a whole number of cycles, and is quite reliable for any operating period exceeding two cycles. Formulas 3.3.4 to 3.3.11 of the fully mixed storage can be modified in order to account for the stratification coefficient.

Eq. 3.3.20 (giving the time-averaged storage temperature) can also be modified accordingly. It provides only an approximation

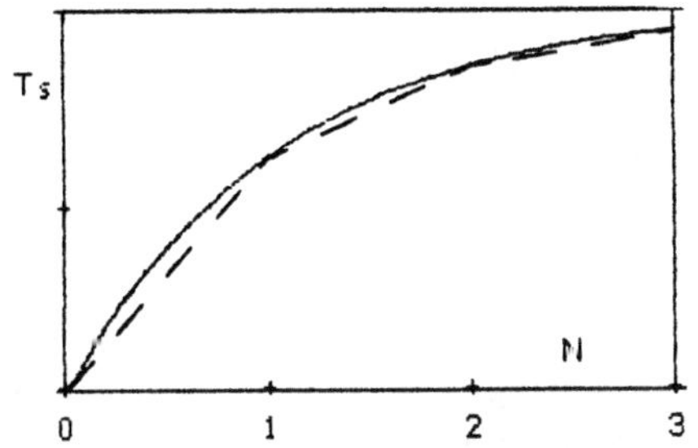

Fig. 3.3.3: Time evolution of the average storage temperature of a "perfectly" stratified" storage (---) and of the "equivalent mixed" storage (——) (versus number of storage cycles, N).

of the real average storage temperature of the stratified tank during the period (the smaller the stratification coefficient the better the approximation); however, it is an interesting intermediate parameter.

The time-averaged collector inlet temperature (or average bottom storage temperature) for a whole number of cycles n is given by:

$$T_{in} = (T_0 + T_1 + \ldots + T_n) / n \quad . \qquad (3.3.30)$$

(This is an exact expression, although subject to the assumptions made).

From Eq. 3.3.26 we obtain:

$$T_{in} - T_0 = (T_L - T_0) \cdot \left(1 - \frac{1 - a^n}{n (1 - a)} \right) \quad . \qquad (3.3.31)$$

The elimination of T_0 between Eq. 3.3.31 and Eq. 3.3.20 (modified by K_s) yields the following expression for the time-averaged collector inlet temperature

$$T_{in} = K_s \cdot T_s + (1 - K_s) \cdot T_L \quad . \qquad (3.3.32)$$

For a fully mixed storage ($K_s = 1$), we find $T_{in} = T_s$ again.

This result provides a way of generalizing the average storage temperature procedure, allowing its computation for any irradiance profile. At the end of the first iteration, T_{in} is calculated from T_s by Eq. 3.3.32; this value is used as a base temperature during the second iteration.

3.3.4 - Consequences of the stratification theory

The analysis of the stratification coefficient shows that for a direct loop (without a heat exchanger),

$$K_s = 1 / F_r \qquad (3.3.33)$$

The physical meaning of this relation is that the stratification exactly compensates the decrease in performance due to the low flow rate. Provided that at least one storage cycle takes place during the collector operating period, the collector flow rate has no influence on the system performance. This results in recommendations of lower collector flow rates than formerly used. For example, for a DHW solar system with 3 m2 of collector and a 150 litre tank, having a daily operating period of about 10 hours without draw-off, a flow rate of 15 litres/hour can be adopted. This is one-tenth of the usually recommended values and implies that the thermal performance of a conventional system may be obtained with a much lower parasitic energy consumption. Theoretical evidence of this fact was published by Gordon and Zarmi /3.10/. We shall even see (section 3.4) that small flow rates sometimes give rise to optimum performance.

This result can be generalized to a collector loop incorporating an external heat-exchanger /3.11/. An optimum primary/secondary flow rate ratio can be found leading to a maximum value of the $K_s.F_r'$ product (Fig. 3.3.4):

$$(K_s.F_r')_{max} = 1 \;/\; (\; 1 + A\; U_L \;/\; A_e U_e \;) \quad . \qquad (3.3.34)$$

This maximum value is obtained if $\dot{m}_1 c/\dot{m}_0 c_0 = (K_s.F_r')_{max}$.

Under these conditions, the flow rates can be ignored; they no longer have an influence on the thermal performance of the system. However, the heat transfer coefficient $A_e U_e$ of the heat exchanger still has an influence on the thermal performance of the system.

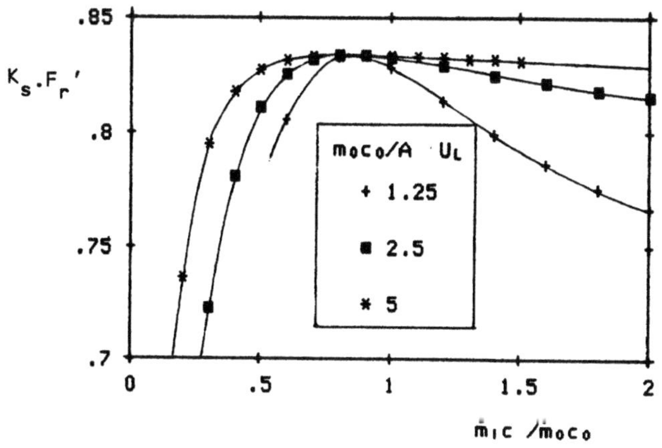

Fig. 3.3.4: Variation of the $K_s.F_r'$ product versus heat-exchanger flow-rate ratio $\dot{m}_1 c/\dot{m}_0 c_0$ $(A.U_L/A_e U_e = 0.2)$.

This limiting value corresponds to the collector/heat exchanger factor F_r' obtained with a submerged heat exchanger and an infinite collector flow rate, and a fully mixed tank (see section 2.2.2.4). This factor is a guide for the sizing of a heat exchanger: a 5% reduction would require the ratio A_eU_e / AU_L equal to 20.

Another example of a direct application of this theory is in the sizing of the storage for a hot water system. Eq.3.3.10 (or better, Eq. 3.3.18 if storage heat loss can be neglected) makes it possible to simply estimate the maximum temperature reached in the storage tank at the end of a day.

3.4 - GENERALIZATION TO ACTIVE SOLAR SYSTEMS WITH DISTRIBUTION LOOP

The models developed in section 3.3 have been derived for collector + storage system (sometimes restricted to storage without heat loss). However, these models can be generalized (with more complicated formulas) to systems with a distribution loop.

3.4.1 - General equation

In order to make an analytical solution possible, an additional assumption is required, that of a constant draw-off flow rate $\dot{m}_3$ during the collector operating period. This assumption avoids swings between different states, each of which would be represented by a particular differential equation. This makes it difficult to apply the results to space heating, for which oscillations of state of the distribution subsystem are frequent.

The corresponding system is represented in Fig. 3.4.1. The cold water from the mains passes directly through the collector. For a mixed tank, this configuration is more efficient than a simpler arrangement with mains water inlet to the tank and draw-off taken from the tank. This simpler arrangement may be considered as a particular case of Fig. 3.4.1 in which the secondary flow rate $\dot{m}_1$ tends to infinity. For a mixed tank, we call the schematic of Fig. 3.4.1 "half-mixed".

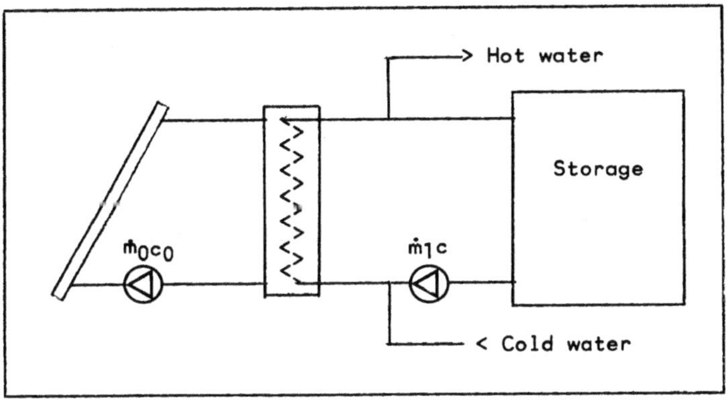

Fig. 3.4.1: Schematic of the modelled system ("half-mixed" storage).

For a perfectly stratified tank, this representation is identical to the usual arrangement with the mains water inlet to the bottom of the tank.

The general behaviour of this system can be described by a set of relationships, starting from the bottom storage temperature T_{bot}, used for the computation of the collector inlet temperature T_{in}

$$T_{in} = (1 - z) \cdot T_m + z \cdot T_{bot} \tag{3.4.1}$$

with

$$z = 1 - \dot{m}_3 / \dot{m}_1 \quad . \tag{3.4.2}$$

The collector outlet temperature is

$$T_{out} = a \cdot T_{in} + (1 - a) \cdot T_L \tag{3.4.3}$$

with the factor a, computed from Eq. 3.3.25, and T_L from Eq. 3.3.5. Generally speaking, T_L is time-dependent.

The power distributed to the load is

$$P_{dis} = \dot{m}_3 c \cdot (T_{out} - T_m) \tag{3.4.4}$$

and the differential equation for the storage temperature is written as (3.4.5)

$$Mc.dT/dt = \dot{m}_1 c \cdot z \cdot (T_{out} - T_{bot}) - A_s U_s \cdot (T - T_e) \quad .$$

3.4.2 - <u>Mixed tank</u>

The system with a fully mixed tank is modelled by the preceding equations with $T_{bot} = T$ and it leads to the storage differential equation similar to 3.3.3

$$t_{mix} \cdot dT/dt + T = T_k \tag{3.4.6}$$

with the new definition of the time constant of the system (consistent with the former definition given in Eq. 3.3.4)

$$t_{mix} = M.c \; / \; [\; A_s U_s + z \cdot (\; A.F_r'.U_L.z + \dot{m}_3 c \;)\;] \tag{3.4.7}$$

and with T_k (replacing T_L)

$$T_k = \frac{A_sU_s.T_e + z.A.F_r'.U_L.T_L + a.z.\dot{m}_3c.T_m}{A_sU_s + z.(A.F_r'.U_L.z + \dot{m}_3c)} \quad . \qquad (3.4.8)$$

With constant solar irradiance and ambient temperature, the former solution (Eq. 3.3.7) can be generalized to

$$T(t) - T_0 = (T_k - T_0) . [1 - EXP(-t / t_{mix})] \qquad (3.4.9)$$

and the energy accumulated in the storage tank after time t can be written as (3.4.10)

$$E_s = G_{cst}.t.[z.A.F_r'.U_L.(T_L-T_b) + z.\dot{m}_3c.(T_m-T_0) + A_sU_s.(T_e-T_0)]$$

with a base temperature

$$T_b = z . T_0 + (1 - z) . T_m \quad . \qquad (3.4.11)$$

G_{cst} is the system heat storage factor (obtained from Eq. 3.3.11, but using the new definition of the time constant).

The time - averaged storage temperature is similar to Eq. 3.3.20 and written as

$$T_s - T_0 = (T_k - T_0) . (1 - G_{cst}) \quad . \qquad (3.4.12)$$

The stored energy becomes (3.4.13)

$$E_s = [z.A.F_r'.U_L.(T_L-T_b') + z.\dot{m}_3c.(T_m-T_s) + A_sU_s.(T_e-T_s)].t$$

with the average collector inlet temperature (modified base temperature)

$$T_b' = z . T_s + (1 - z) . T_m \quad . \qquad (3.4.14)$$

The energy distributed to the load during the same period is (3.4.15)

$$E_d = (1 - z).A.F_r'.U_L.t.(T_L - T_b') + z.\dot{m}_3c.(T_s - T_m).t \quad .$$

All the formulas developed in section 3.3.2 can be easily generalized to the case with load. The iterative procedure for the "time-averaged" storage temperature (used in calculations with any irradiance profile) can also be generalized in the same way: The base temperature T_b for each iteration is replaced by T_b' from Eq. 3.4.14.

3.4.3 - Stratified tank

The same reasoning used in section 3.3.3 for a perfectly stratified storage tank without draw-off is used here. However, the storage cycle duration becomes:

$$t_c = M / (\dot{m}_1 - \dot{m}_3) = M / (\dot{m}_1 . z) \qquad (3.4.16)$$

and the mean storage temperature at the end of the first cycle can be written as

$$T_1 = a_1.T_0 + a_2.T_L + a_3.T_m + (1-a_1-a_2-a_3).T_e \qquad (3.4.17)$$

with the following parameters:

$$a_1 = a . z . EXP (- t_c / t_{loss}) \quad . \qquad (3.4.18)$$

$$a_2 = (1 - a) . G_{loss} \quad . \qquad (3.4.19)$$

$$a_3 = a . (1 - z) . G_{loss} \quad . \qquad (3.4.20)$$

$$t_{loss} = M.c / A_sU_s \quad . \qquad (3.4.21)$$

$$G_{loss} = [1 - EXP(- t_c/t_{loss})] . t_{loss}/t_c \quad . \qquad (3.4.22)$$

For a whole number of cycles and a constant irradiance, it can be demonstrated (after some rather tedious calculations) that again an equivalent mixed tank system can be defined with the following differential equation (derived from Eqs. 3.4.6-8)

$$t_{str}.dT/dt + T = T_k' \qquad (3.4.23)$$

including the generalized time constant

$$t_{str} = M.c / [z.K_1.(\dot{m}_3c + A.F_r'.U_L.z) + A_sU_s.K_2] \qquad (3.4.24)$$

and the limiting temperature (3.4.25)

$$T_k' = \frac{(K_1.z. (A.F_r'.U_L.T_L + \dot{m}_3c.a.T_m) + K_2.A_sU_s.T_e)}{[K_1.z.(A.F_r'.U_L.z + \dot{m}_3c) + K_2.A_sU_s]} \quad .$$

K_1 and K_2 are two stratification coefficients:

$$K_1 = G_{loss} . Ln (1 / a_1) / (1 - a_1) \quad . \qquad (3.4.26)$$

$$K_2.A_sU_s = K_1.\dot{m}_1c.z.[\ (1-a_1)/F_p + a.z - 1\ ] \quad . \qquad (3.4.27)$$

In the case without heat loss and draw-off, we find $K_1 = K_s$ as before.

As in section 3.3.3, this approximation is valid whenever one storage cycle takes place during the collector operating time.

An approximation of the time-averaged storage temperature during time t is found by

$$T_s^* - T_0 = (\ T_k' - T_0\ ) \cdot (\ 1 - G_{str}\ ) \qquad (3.4.28)$$

with

$$G_{str} = [\ 1 - EXP\ (\ - t\ /\ t_{str}\ )\ ] \cdot t_{str}\ /\ t \quad . \qquad (3.4.29)$$

Eq. 3.4.17 (and its generalization to n cycles) can be used in order to determine the exact time-averaged bottom storage temperature over n cycles (the enthusiastic reader will find the complete relationship in ref. /3.9/). Elimination of T_0 between this relationship and Eq. 3.4.28 leads to the following equation (which is exact for a whole number of cycles and a good approximation otherwise):

$$\begin{aligned} T_{bot} = \ & K_1.T_s^* + a_2 \cdot (G_{loss}-K_1) \cdot T_L\ /\ (1-a_1) \qquad (3.4.30) \\ & + a_3 \cdot (G_{loss}-K_1) \cdot T_m\ /\ (1-a_1) \\ & + T_e \cdot [\ 1 - G_{loss} + (1-a_1-a_2-a_3) \cdot (G_{loss}-K_1)\ /\ (1-a_1)\ ] \end{aligned}$$

This last equation is not directly dependent on time.

3.4.4 - Application and daily simulation

The model of Eq. 3.4.17 can be applied to study the sensitivity of the system performance to the collector loop flow rate, through the number of storage cycles. It demonstrates that the daily system efficiency has a maximum value for one cycle, if draw-off occurs during collector operation (Fig. 3.4.2).

The iterative procedure defined in section 3.3.2 for the time-averaged storage temperature can be extended to the general system described by Eqs. 3.4.16-30. This generalisation includes all possible cases: Stratified or mixed tank, with or without heat losses, and with or without draw-off. This generalisation also uses the daily irradiance profile to determine the collector operating time and the daily utilizability.

For longer periods, such as a month, this approach can lead

to another calculation procedure based on a typical meteorological day (or using the cumulative frequency curve of solar irradiance defined in Chapter 4). First the daily simulation is extended to include night draw-off and heat losses. This 24-hour simulation can be repeated until a periodic state is obtained, when the final storage temperature is equal to its initial temperature. This procedure has been used as the basis of the ESM-2 method for solar domestic hot water systems and will be described in detail in Chapter 7.

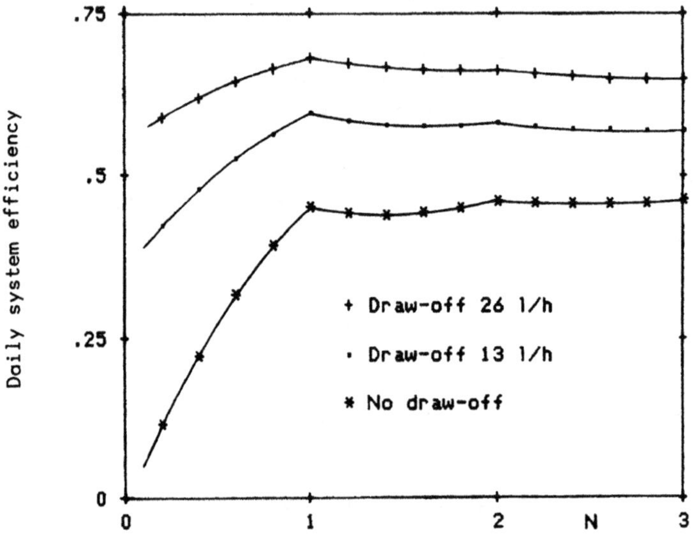

Fig. 3.4.2: Daily system efficiency of a stratified system as a function of the number of cycles, N (direct system).

REFERENCES OF CHAPTER 3

/3.1/ A.WHILLIER. Solar Energy Collection and its Utilization for House Heating. Ph.D. Thesis, Mech.Eng., MIT, Cambridge, Mass. (1953).

/3.2/ S.A. KLEIN and W.A. BECKMAN. Review of Solar Radiation Utilizability. ASME J. of Solar Energy Engineering, Vol. 106, pp.393-402 (1984).

/3.3/ J.M.GORDON and A.RABL. Design, Analysis and Optimization of Solar Industrial Process Heating plants without storage. SOLAR ENERGY, Vol.28, pp.519 (1982).

/3.4/ A.RABL. Active Solar Collectors and their Application. Oxford Univ. Press, New-York (1985).

/3.5/ M.COLLARES-PEREIRA et al. Design and Optimization of Solar Industrial hot water systems with storage. SOLAR ENERGY Vol.32, n.1, p.121-134 (1984).

/3.6/ W.F. PHILLIPS. Integrated performance of liquid-based solar heating systems. SOLAR ENERGY, Vol. 26, pp. 287-295 (1981).

/3.7/ W.F. PHILLIPS and R.N. DAVE. Effects of stratification on the performance of liquid-based solar heating systems. SOLAR ENERGY, Vol. 29, No. 2, pp. 111-120 (1982).

/3.8/ J.M. GORDON and Y. ZARMI. An analytic model for the long-term performance of solar thermal systems with well-mixed storage. SOLAR ENERGY, Vol. 35, No. 1, pp. 55-61 (1985)

/3.9/ B.BOURGES and J.ADNOT. A generalized closed-form model for solar hot water systems. SOLAR ENERGY (To be published)

/3.10/ J.M. GORDON and Y. ZARMI. Thermosyphon systems: single vs multi-pass. SOLAR ENERGY, Vol. 27, No. 5, pp. 441-442 (1981)

/3.11/ BOURGES. Advances in Solar Energy Technology (ISES Congress, Hamburg, 1987), Vol.2, pp.1172-1176. Pergamon (1988).

CHAPTER 4

SOLAR RADIATION DATA

To a large extent, the local solar resource determines the long-term performance of any type of solar system. This implies that solar radiation must be carefully estimated and the data presented in a suitable form. ESM methods require specific presentation of solar radiation data as cumulative frequency curves of solar irradiance (CFC's). All the elements needed for CFC calculations are given in this chapter. First, basic properties of solar radiation processes are reviewed. Then CFC's are defined, as well as a method of derivation. Additional items, such as average incidence angle modifier, are also described. When using directly ESM programmes for personal computers (ESM 1.4 or 2.1), these data are already available on disk files and no solar radiation computation is required.

4.1 - SOLAR RADIATION PROCESSES

The solar radiation received by a solar collector on earth (on any time scale) depends on five factors:

- The place where the collector is located, namely the geographic latitude.
- The time of the year, which determines the solar declination and hence the angle of incidence of the sun's rays.
- Local climates introducing atmospheric effects on solar radiation, the most important being that of clouds.
- The orientation and tilt angle of the collector.
- The coefficient of reflection (albedo) of the ground in the local environment.

In order to estimate the local solar resource, it is necessary to know the basic physical aspects of solar radiation at the earth's surface. Most of the items reviewed here will be used for practical CFC's computation, in section 4.3.

4.1.1 - The sun and the earth

The sun produces a large amount of energy, due to the continuous nuclear fusion in its centre, at temperatures of many millions of degrees. This energy is transferred out to its surface and is radiated into space at an average rate of 66

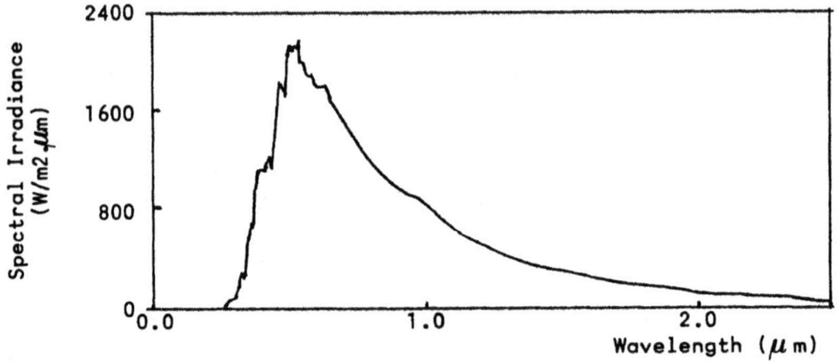

Fig. 4.1.1: Standard extraterrestrial solar spectrum.

MW/m2. Wavelengths of solar radiation range from 250 to 5000 nm (0.25 to 5 micrometres); the peak-value is about 500 nm. The solar spectrum (Fig. 4.1.1) is about that of a black body radiating at 5800 K. The visible range (400 to 800 nm) represents only 45% of the radiated energy.

The earth is situated at a mean distance of 149.6 million km from the sun. From this distance, the sun is seen as a disk of 32' of arc. Above the earth's atmosphere, the mean solar flux is only 1367 W/m2 on a surface perpendicular to the sun's rays /4.3-4/. This value is called the Solar Constant, B_0. Its value is presently known from satellite measurements with an accuracy within 0.5 %

The earth moves about the sun in an elliptical orbit (Fig. 4.1.2) and the distance between sun and earth changes during one year by ± 1.7%; thus, the solar constant must be corrected according to the day of the year (J, counted from 1 January) by the following factor /4.2/

$$r = 1 + 0.034 \cdot \cos(0.986 \cdot J - 2°) \quad . \qquad (4.1.1)$$

The plane of the earth's orbit is called the <u>Ecliptic</u> .

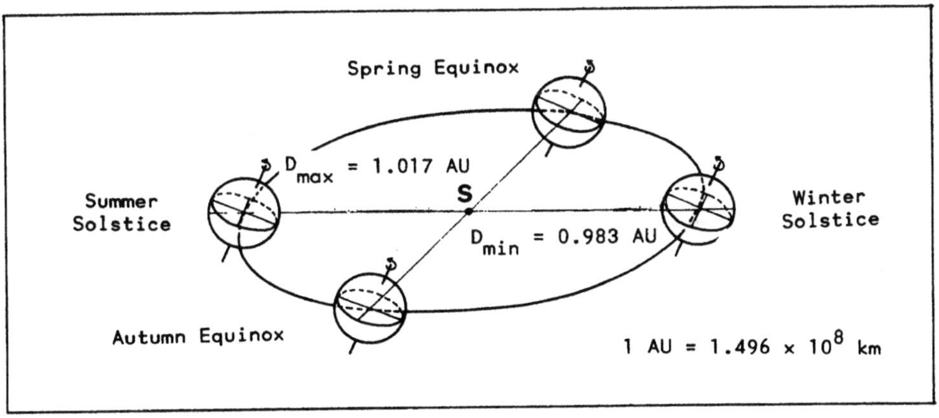

Fig. 4.1.2: Motion of the earth about the sun (from Perrin de B. /4.2/).

4.1.2 - Geometrical aspects

In addition to the earth's motion about the sun, the earth rotates once per day about its polar axis. This rotation causes the day/night sequence. The equatorial plane, perpendicular to the polar axis, is inclined at an angle of 23.45° with respect to the ecliptic plane. This angle causes the different seasons. In a given place, the apparent motion of the sun in the sky varies according to the day of the year; the daylength varies and so does the angle of incidence of the sun's rays.

In order to better understand these phenomena, one must consider their geometrical aspects. For this purpose, some definitions must first be given.

The earth's diameter is neglected and the local sphere is defined as a sphere, with its centre in the considered place and with unit radius. Two coordinate systems are used (Fig. 4.1.3):

- Hourly coordinates, whose reference planes are equator and local meridian. Each point in the sky is defined by two angles: its declination δ and its hour angle ω.
- Horizontal coordinates, whose origin planes are the horizontal plane and the meridian. Each point is characterized by its azimuth ψ and altitude γ (or its complementary, zenith angle).

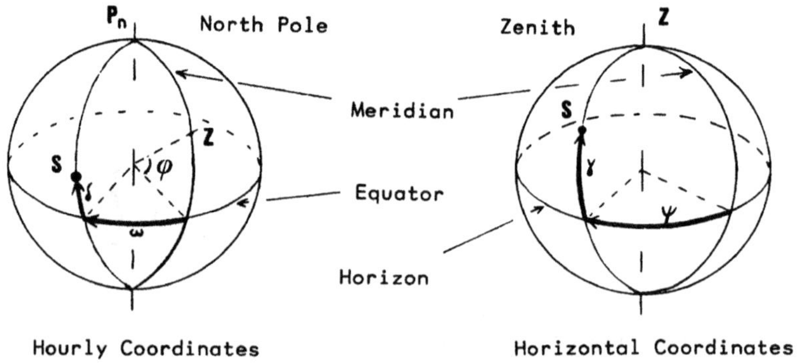

Fig. 4.1.3: Spherical co-ordinate systems.

The angle between zenith and equatorial plane is the latitude φ of the place.

Solar declination δ is the angle between the sun-earth axis and the equatorial plane. It varies from +23°27' at the summer solstice (22 June) to -23°27' at the winter solstice (22 December). It is equal to a zero value at the equinoxes (spring, 21

March, and autumn, 23 September).

At the J-th day of the year (from 1 January), the solar declination may be computed with the approximate equation ($\pm 1°$)

$$\delta = 23.45 \cdot \sin (0.986 \cdot J - 80°) \quad . \qquad (4.1.2)$$

A more accurate formula has been proposed, which is dependent on the year considered /4.5/ (4.1.3)

$$\delta = 0.3723 + 23.2567.\sin wt + 0.1149.\sin 2wt - 0.1712.\sin 3wt$$
$$- 0.7580.\cos wt + 0.3656.\cos 2wt + 0.0201.\cos 3wt$$

where $w = 360 / 365.2422$ (4.1.4)

and (4.1.5)

$$t = J - 79.301 - 0.2422.(YEAR-1969) + INT [0.25.(YEAR-1969)].$$

This formula is accurate within 0.02°.

The sun's hour angle ω is related to local time (in hours)

$$\omega = 15 \cdot (t - 12) \quad . \qquad (4.1.6)$$

Spherical trigonometry enables one to pass from one coordinate system to another and yields equations of solar altitude and azimuth:

$$\sin \gamma = \sin \varphi \cdot \sin \delta + \cos \varphi \cdot \cos \delta \cdot \cos \omega \quad . \qquad (4.1.7)$$

$$\sin \psi \cdot \cos \gamma = \cos \delta \cdot \sin \omega \quad . \qquad (4.1.8)$$

$$\cos \psi \cdot \cos \gamma = \sin \varphi \cdot \cos \delta \cdot \cos \omega - \cos \varphi \cdot \sin \delta . \qquad (4.1.9)$$

Sunrise and sunset times are obtained when $\sin \gamma = 0$ and the corresponding astronomical sunshine duration (number of hours between sunrise and sunset) is given by

$$S_0 = (t_h / 180) \cdot \cos^{-1} (- tg \varphi \cdot tg \delta) \quad . \qquad (4.1.10)$$

In order to determine the solar irradiance received by a solar collector, it is necessary to know the sun's incidence angle on the collector surface. This incidence angle is the angle between the beam radiation and the normal to the surface (Fig. 4.1.4). If the surface is characterized by its azimuth angle α (deviation of the projection on a horizontal plane of the normal to the surface from the local meridian; east negative) and its slope β, the angle of incidence θ is computed by the general equation

$$\cos\theta = \cos\gamma \,.\sin\beta \,.\cos(\psi - \alpha) + \sin\gamma \,.\cos\beta \quad . \qquad (4.1.11)$$

In some cases, a simpler form may be used, e.g., for surfaces facing due south ($\alpha = 0°$) (4.1.12)

$$\cos\theta = \sin\delta \,.\, \sin(\varphi - \beta) + \cos\delta \,.\, \cos(\varphi - \beta) \,.\, \cos\omega$$

or for vertical surfaces ($\beta = 90°$)

$$\cos\theta = \cos\gamma \,.\, \cos(\psi - \alpha) \quad . \qquad (4.1.13)$$

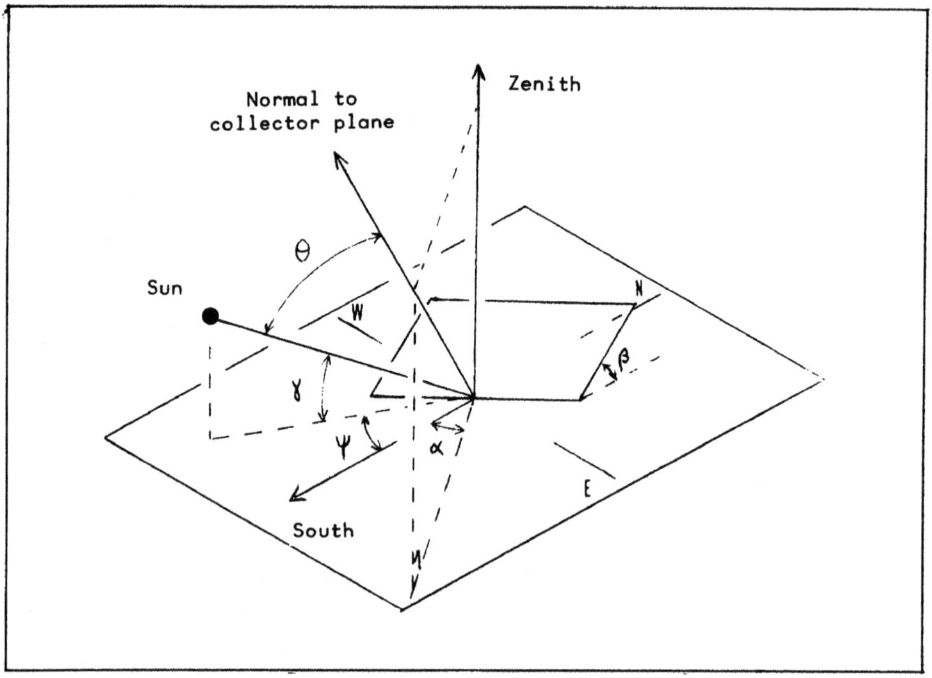

Fig. 4.1.4: Sun's incidence angle on a collector surface with azimuth α and slope β (γ: solar altitude; ψ: solar azimuth).

Extraterrestrial irradiance (with no atmosphere) on any surface is simply the product of the solar constant (modified by the sun-earth distance corrective factor, r) and the cosine of the incidence angle of a sun ray with respect to the surface

$$B_0 (\alpha, \beta) = B_0 \cdot r \cdot \cos\theta \quad . \qquad (4.1.14)$$

For a horizontal plane, $\cos\theta$ becomes $\sin\gamma$.

Daily extraterrestrial irradiation on a horizontal surface is computed by integrating Eq. (4.1.14) with respect to time between sunrise and sunset:

$$B_{0d} = t_h / 2\pi \cdot B_0 \cdot r \cdot \int_{-w_0}^{w_0} \sin\gamma \cdot d\omega \qquad (4.1.15)$$

$$B_{0d} = t_h / \pi \cdot B_0 \cdot r \cdot (\pi / 180 \cdot w_0 \cdot \sin\varphi \cdot \sin\delta + \sin w_0 \cdot \cos\varphi \cdot \cos\delta) \quad . \qquad (4.1.16)$$

The sunset hour angle is given by

$$w_0 = \cos^{-1} (- tg\,\varphi \cdot tg\,\delta) \quad . \qquad (4.1.17)$$

4.1.3 - <u>Effects of atmosphere and clear sky irradiance</u>

Not all the extraterrestrial radiation is available at ground level due to various atmospheric effects such as:

- Scattering of short-wavelength radiation by atmosphere molecules (Rayleigh scattering).
- Absorption by atmospheric gases (ultra-violet radiation is absorbed by ozone, and infra-red by water vapour and CO_2).
- Scattering by aerosols and dust.

Furthermore, reflection, scattering and absorption by clouds need to be considered if the sky is overcast (even partially).

These phenomena cause solar radiation received at ground level to be the sum of two components: beam (or direct) radiation, B , coming directly from the solar disk, and diffuse radiation from the celestial hemisphere (often called "diffuse sky radiation"), D.

A third component has to be accounted for, as far as tilted surfaces are concerned: this is the ground-reflected diffuse radiation, which depends on the tilt angle of the surface and the local ground albedo (reflection coefficient of the ground in the vicinity of the collector), ρ_g.

The sum of these three components is called global radiation.

The diffuse fraction may vary widely from 20 to 100% of the global irradiance, according to cloud conditions (from clear to overcast sky). On average,the diffuse fraction represents about 50% of the global irradiation on a horizontal plane in Europe.

It is recalled that irradiance is the power received by a unit area of the surface at a certain time and is expressed as (W/m2). Irradiation is an energy per unit area, namely irradiance integrated over some interval of time. Irradiation units are (MJ/m2) or (kWh/m2); the integration time interval must be specified as hourly, daily, or monthly irradiation.

The rate at which the intensity of solar radiation decreases as it traverses the atmosphere is a function of the quantity of molecules crossed. It is proportional to air mass AM (equal to one at sea level with the sun at zenith) and given approximately by

$$AM = (1 - 0.1 . z) / \sin \gamma \quad . \qquad (4.1.18)$$

at an altitude z above sea level (Unit: km).

Clear sky absorption and scattering phenomena are generally characterized by a single simplified parameter: the Linke turbidity factor, T_L. This factor is defined as the number of clean and dry atmospheres (with no water vapour or aerosols) which would lead to the same amount of beam radiation as measured. This factor generally lies between 2 for a clean sky, and 6 for a humid and polluted atmosphere. Several methods for estimating T_L can be found in references /4.2/ /4.21/.

The Linke turbidity factor can be used for the calculation of clear sky irradiance on a given plane, through Kasten's formula for direct radiation at normal incidence:

$$B_c = B_0 . r . EXP [- AM . T_L / (0.9 AM + 9.4)] \quad . \qquad (4.1.19)$$

For global radiation on a horizontal plane, Perrin de Brichambaut proposed

$$G_c = r . (1270 - 56 . T_L) . (\sin \gamma)^{(T_L+36)/33} \quad . \qquad (4.1.20)$$

Diffuse sky radiation on a horizontal surface can be obtained from the difference between global and direct radiation.

Variations are represented at the solstices in Figure 4.1.5.

The ratio of global irradiance on a horizontal plane G to the extraterrestrial value on a horizontal plane is called the "clearness index", K_t

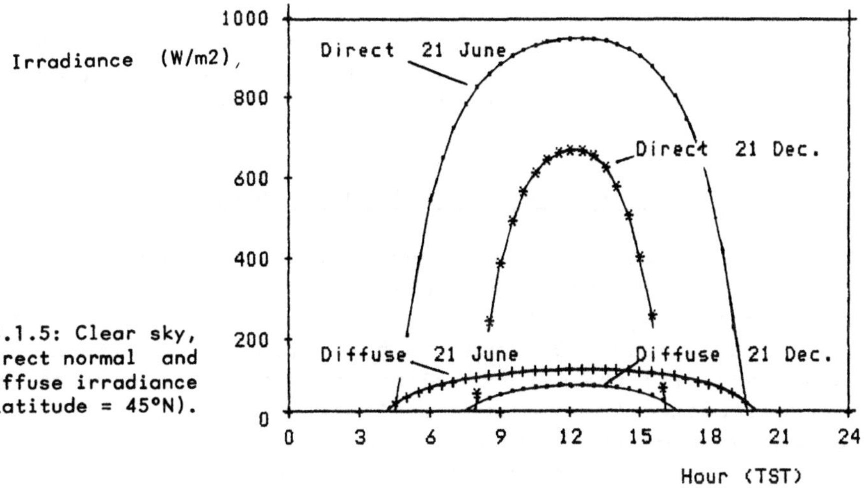

Fig. 4.1.5: Clear sky, direct normal and diffuse irradiance (Latitude = 45°N).

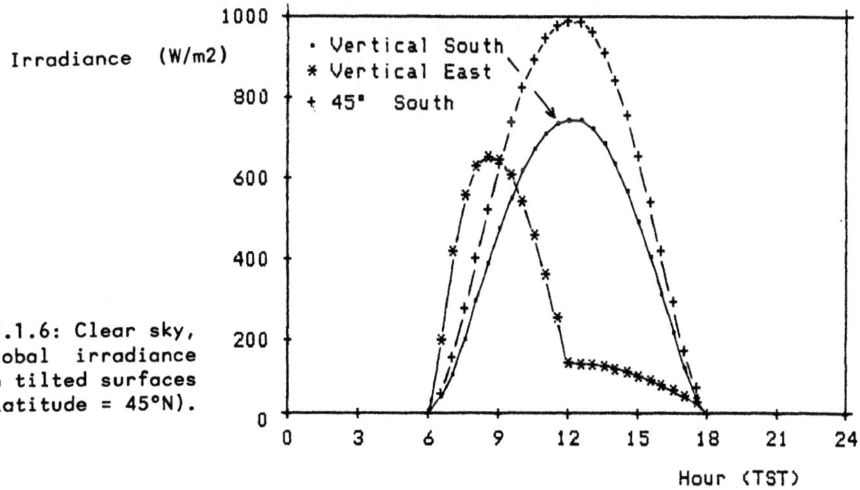

Fig. 4.1.6: Clear sky, global irradiance on tilted surfaces (Latitude = 45°N).

$$K_t = G / (B_0 \cdot r \cdot \sin \gamma) \qquad (4.1.21)$$

This index ranges between 0.05 (for a heavily overcast sky) and 0.75.

A daily (or monthly) clearness index can also be defined as the ratio between the measured global daily (or monthly) irradiation on a horizontal surface and the corresponding extraterrestrial irradiation (see Eq. 4.1.16). The clearness index is a very useful parameter as will be seen in the next sections.

4.1.4 - Global irradiance on tilted surfaces

Solar radiation data measured at meteorological stations comprise: i) the global radiation on a horizontal surface, G, measured by a pyranometer; ii) the diffuse sky radiation on a horizontal surface, D, measured by a pyranometer with a shadow-band; and iii) the direct radiation at normal incidence, B, by a pyrheliometer (less often available). An hourly time-step is generally used for integration and recording; results are presented as hourly irradiations (unit: J/cm2 or kJ/m2) or mean hourly irradiances (unit: W/m2).

Solar collectors are generally installed on tilted surfaces facing due South (in the Northern hemisphere). The global irradiance on tilted surfaces is seldom available from measurements, thus it must be computed from the commonly measured parameters. The position of the sun is estimated in the middle of the hour (e.g., at 13.30 for 13.00-14.00).

The three components of solar radiation are related at any time by the equation

$$G = B \cdot \sin \gamma + D \quad . \qquad (4.1.22)$$

An assumption is needed to compute the angular distribution of diffuse sky and ground-reflected radiation. Generally, both are assumed to be isotropic, which leads to the simple formula for global irradiance on any tilted plane (4.1.23)

$$G(\alpha, \beta) = B.\cos\theta + D.(1 + \cos\beta)/2 + \rho_g.G.(1 - \cos\beta)/2 \ .$$

Ground albedo, ρ_g, depends on the nature of the local environment. A value 0.2 is often adopted. However, particular conditions, such as snow cover, introduce much higher values (0.6 to 0.8). Typical values are given in Table 4.1.1

The assumption of isotropic distribution of diffuse sky radiation has been largely questioned. This assumption seems to introduce systematic underestimation for south-oriented planes in the range of 10%, and an even greater underestimation at low solar altitudes /4.6/.

Other diffuse sky models have been proposed, the simplest being that of Hay - Willmott /4.7/ which considers diffuse sky radiation as the sum of a circumsolar and an isotropic component. The diffuse isotropic fraction is defined as

$$F = \begin{cases} 1 - K_t \cdot (1 - D/G) & \text{if } \cos\theta \,/\, \sin\gamma > (1 + \cos\beta)/2 \\ 1 & \text{otherwise} \end{cases} \qquad (4.1.24)$$

and

$$G(\alpha, \beta) = [B + (1-F).D/\sin\gamma].\cos\theta + F.D.(1 + \cos\beta)/2 + \rho_g.G.(1 - \cos\beta)/2 \quad . \qquad (4.1.25)$$

Systematic underestimation still occurs, but within an acceptable range of 4 to 5 %. Another model, proposed by Perez /4.8/ /4.32/, accounts for a horizon band and leads to long-term errors within 1%, but is more complicated.

Type of ground surface	Albedo
Dry sand	.25
Meadow	.25
Cultivated fields	.20
Fresh snow	.80
Old snow	.60
Sea ($\gamma > 30°$)	.05
$\gamma = 20°$	.15
$\gamma = 10°$	.35
White paint	.80
Black mat paint	.05

Table 4.1.1: Typical ground-albedo values (from Perrin de B. /4.2/).

If only global radiation is available, the diffuse component is estimated for each hour by a correlation model such as the Orgill and Hollands formula /4.10/:

$$D/G = \begin{cases} 1 - 0.247 \cdot K_t & \text{if } K_t < 0.35 \\ 1.557 - 1.84 \cdot K_t & \text{if } 0.35 < K_t < 0.75 \\ 0.177 & \text{if } K_t < 0.75 \end{cases} \qquad (4.1.26)$$

Other similar models are also available e.g., by Erbs /4.9/. The beam component can be estimated from Eq. 4.1.22.

4.1.5 - Monthly global irradiation on tilted surfaces

Monthly global irradiation on the collector surface is the basic meteorological input for solar system design methods. It can be estimated by several means:

- Models developed in the previous section can be used whenever hourly data files are available with at least global radiation on horizontal surfaces. Calculation is done hour by hour and requires a computer. Typical meteorological files are available for a set of representative European locations as test reference years (TRY) /4.37/.

- Simplified methods are also available, using as input the monthly global irradiation on a horizontal plane or the monthly sunshine duration. A number of models and pieces of software have been published for that purpose in recent years e.g., Perrin de Brichambaut /4.2/ /4.34/, Klein/4.21/, or the CASAMO software /4.22/. A recent model, by Page /4.20/, has been used to derive the European Solar Radiation Atlas /4.23/ /4.17/.

In addition to these methods and several data sources (see section 4.3.3), simple charts are proposed in this section. They are intended for estimation of the monthly global irradiation on planes facing due south and tilted at angles of 30°, 45° and 60° to the horizontal. These charts (Fig. 4.1.7 to 4.1.12) give the conversion factor R of global irradiation from horizontal to tilted planes versus the monthly clearness index

$$R = G_m(\alpha, \beta) \; / \; G_m(0,0) \quad . \qquad (4.1.27)$$

These charts have been derived by Bourges /4.35/ from a regression analysis of data from the European Solar Radiation Atlas; corresponding formulas are also available and have been validated with respect to measured values /4.36/.

The monthly global irradiation on a horizontal plane can be estimated with Angstrom - type correlations from the sunshine duration S:

$$K_{tm} = a + b \cdot S / S_0 \quad . \qquad (4.1.28)$$

a and b are regression parameters and depend on climate and season. Typical ranges are 0.20-0.25 for a and 0.50-0.55 for b. Sunshine duration is a parameter more widely recorded by meteorological services than global radiation (and for longer periods). Sunshine duration characterizes the time during which beam radiation is observed.

4.1.6 - Available solar radiation data

Solar radiation is measured at a number of meteorological stations around the world. Measurements are difficult, particularly as far as diffuse or beam radiation are concerned. Thus, measurements are limited to a few tens of locations in each country. Mean hourly irradiances are generally available, but sometimes only daily irradiations.

In France, for example, global radiation on a horizontal plane is measured at 31 stations (14 stations with more than 10 years' data); diffuse radiation is measured at 12 stations (only 2 stations with more than 5 years' data); and direct radiation at only 3 stations. Global and diffuse radiation on various tilted surfaces have been measured only during temporary 3-year campaigns at 2 locations.

The situation is similar in all European and North-American countries.

In many other countries, particularly the developing countries, few systematic meteorological network measurements are conducted. In those countries, solar radiation has to be estimated from sunshine duration, cloud cover information, or, more recently, through satellite image processing. Only long-term statistics, such as mean monthly irradiation, can be derived from these technics.

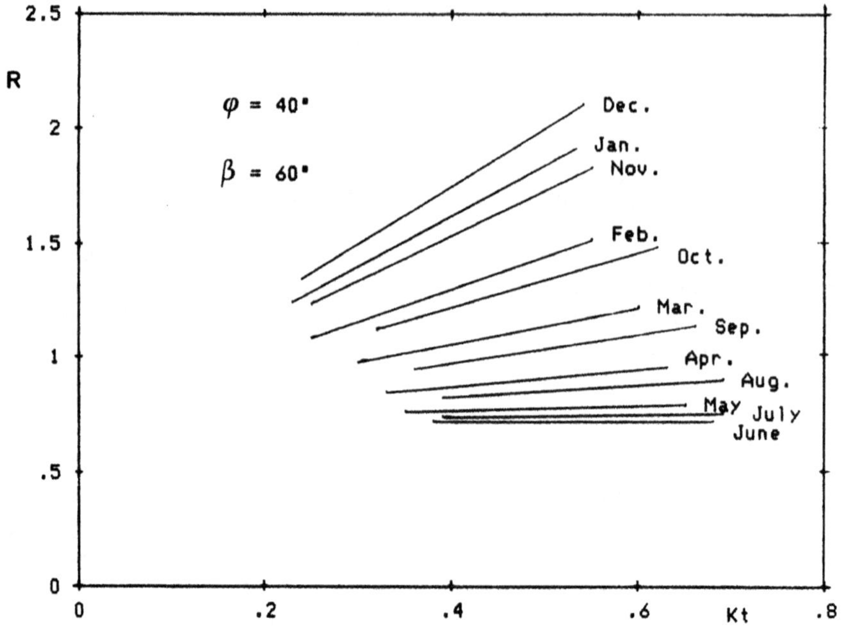

Fig. 4.1.7: Conversion factor R from monthly global irradiation on a horizontal plane to monthly global irradiation on a tilted plane (oriented towards the South) (from /4.35/).

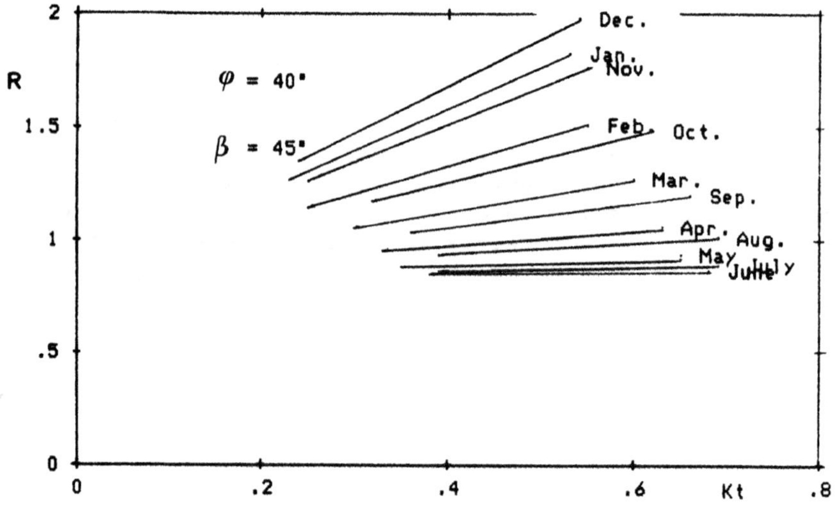

Fig. 4.1.8: Idem Fig. 4.1.7.

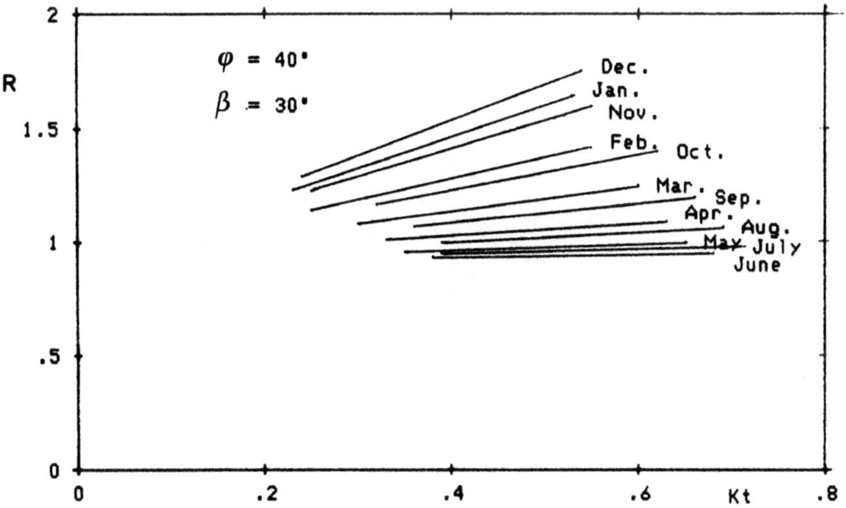

Fig. 4.1.9 and 4.1.10: Idem Fig. 4.1.7.

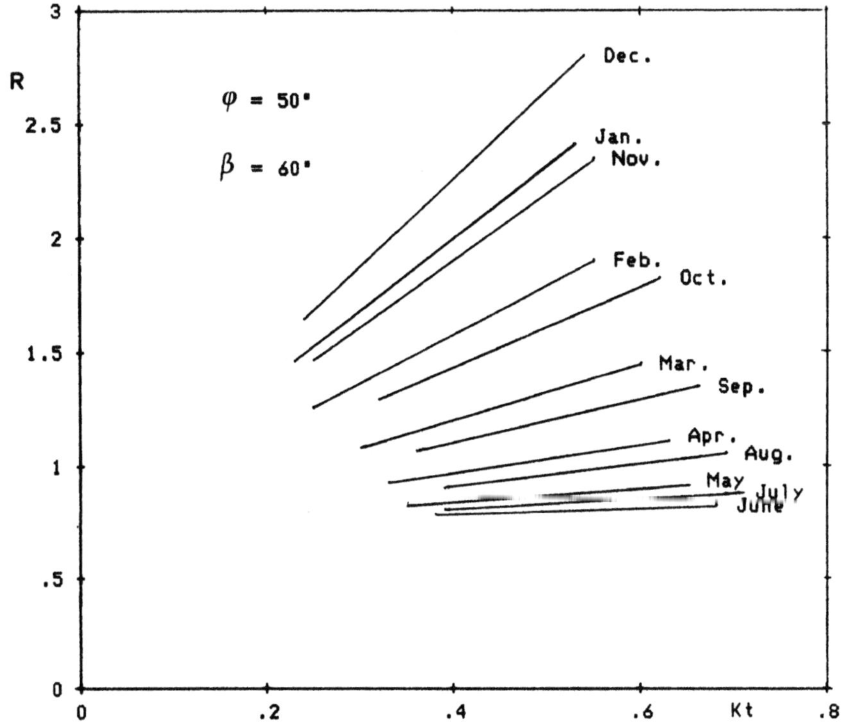

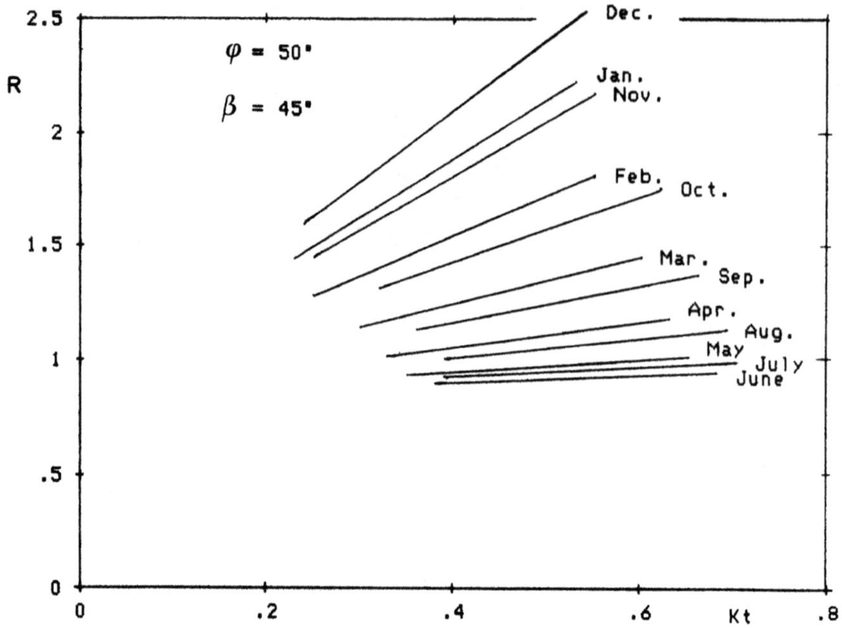

Fig. 4.1.11 and 4.1.12: Idem Fig. 4.1.7.

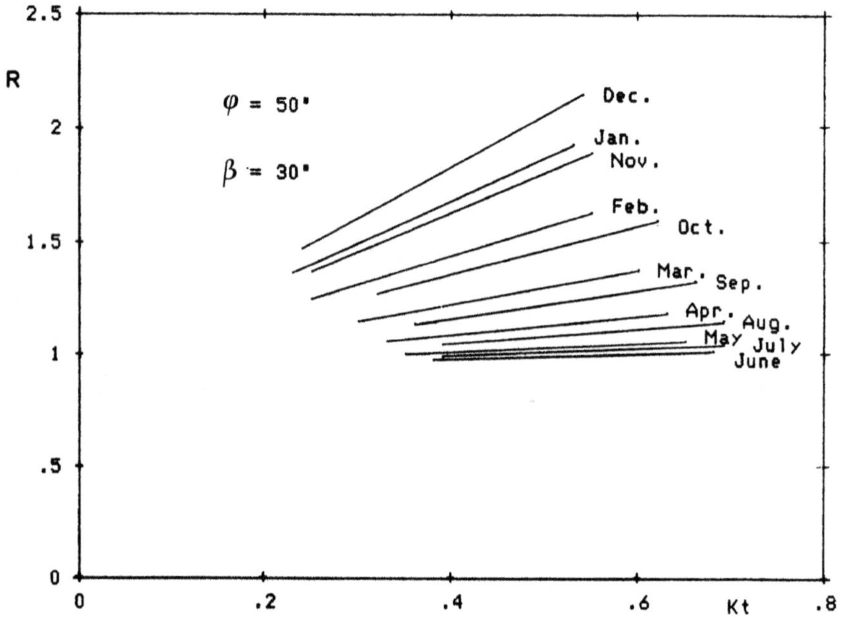

4.2 - CUMULATIVE FREQUENCY CURVES OF SOLAR IRRADIANCE: DEFINITIONS

ESM calculation methods start from the basic premise of the importance of threshold irradiance in solar thermal active systems. For space heating, a minimum temperature of the heat transfer fluid must be reached before energy can be supplied to the dwelling. Furthermore, at a given storage temperature, a minimum solar irradiance on the collector surface is required for collector operation. These two conditions imply that only a fraction of the incident irradiation during a period of time can be used by the solar system; this fraction is called "utilizability" (see section 3.2).

To account for this factor in the calculation, the frequency distribution of solar global irradiance on the collector surface for the period considered must be known. This frequency distribution is presented here as monthly cumulative frequency curves (CFC's) of solar irradiance.

4.2.1 - CFC's: Definitions and properties

Cumulative frequency curves of solar irradiance are defined as curves giving the time, n_h, (mean daily number of hours during the month considered) during which the solar irradiance on a given plane has exceeded a value I_c (threshold irradiance or critical intensity).

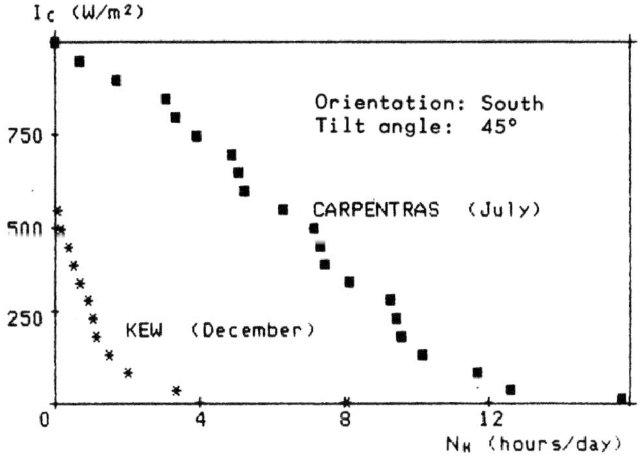

Fig.4.2.1: Examples of cumulative frequency curves (CFC) of solar irradiance (global).

Examples are given in Fig.4.2.1: The time, n_h, is read on the horizontal axis (unit: hours/day); the critical intensity, I_c, is read on the vertical axis (Unit: W/m^2).

From this example, it can be seen that intensities greater than 300 W/m^2 on a tilted plane (45°) facing due South, were observed in Carpentras for a period of 9.2 hours/day in July, and only for 0.9 hour/day at Kew during the month of December.

A CFC is defined for a particular location, month, orientation, and slope, and generally concerns global irradiance (for flat-plate solar collectors). However, CFC's could be drawn for any type of collector, such as concentrating collectors (direct irradiance) or CPC's (direct + a fraction of diffuse). The time interval of one month has been chosen because it is convenient for most of the applications: one month represents a homogeneous period as regards the solar declination and the climate, as well as the space heating load. For some applications, yearly CFC's could be defined.

Some properties of CFC's should be noted. The CFC intercept with the horizontal axis is denoted t_d (see Fig. 4.2.2) and is the mean number of hours during which a non-zero irradiance has been measured. The CFC intercept generally corresponds to the mean number of hours between sunrise and sunset (astronomical sunshine duration), but it can sometimes be shorter (e.g., for concentrating collectors).

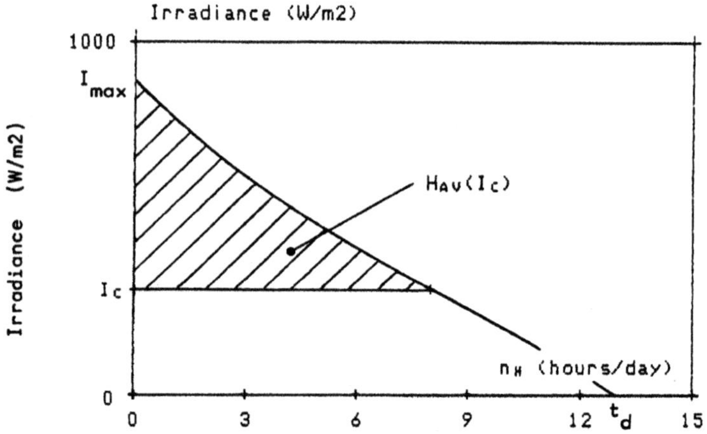

Fig. 4.2.2: Calculation of available energy above a given critical intensity I_c (here 300 W/m^2) from the CFC.

If we normalize abscissas to one (by division by t_d), CFC's reduce to the usual cumulative probability function.

The intercept with the vertical axis (Fig.4.2.2) represents the maximal solar irradiance, I_{max}, which can be achieved during the month on the plane considered. For south-oriented planes, I_{max} is generally observed at solar noon on clear days.

The area delimited by the CFC and the two axes is the mean daily sum of irradiation, H_t, incident on the plane for the period considered.

If the threshold irradiance, I_c, of a solar system (as defined in section 3.2) is known, the CFC makes it possible to obtain the collector operating time, t_{op}, as $n_h(I_c)$ directly.

The area delimited by the CFC, the vertical axis, and the horizontal line $I = I_c$, represents the solar energy available above the critical intensity I_c (dashed area in Figure 4.2.2). The solar energy available is given by the simple integral: (unit: $Wh/m^2.day$)

$$H_{av}(I_c) = \int_{I_c}^{I_{max}} n_h (I) . dI \qquad . \qquad (4.2.1)$$

As noted previously, for the zero threshold

$$H_{av} (0) = H_t \qquad . \qquad (4.2.2)$$

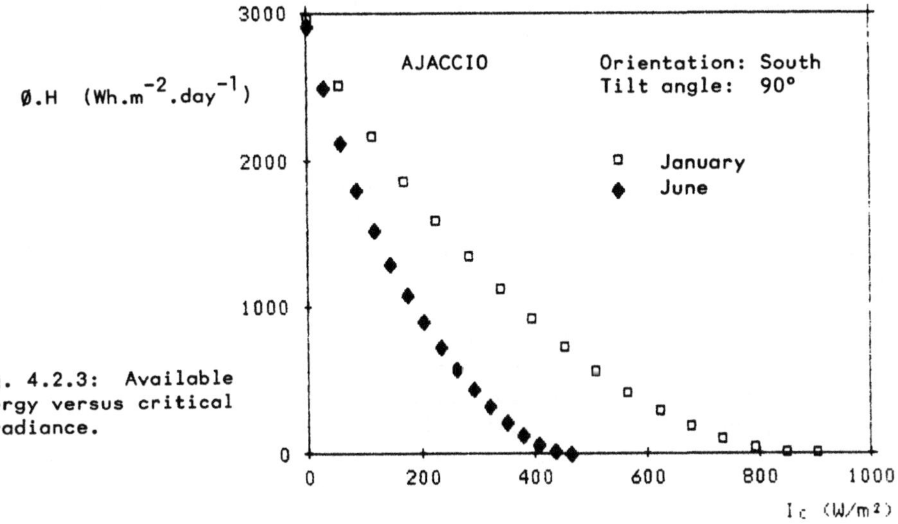

Fig. 4.2.3: Available energy versus critical irradiance.

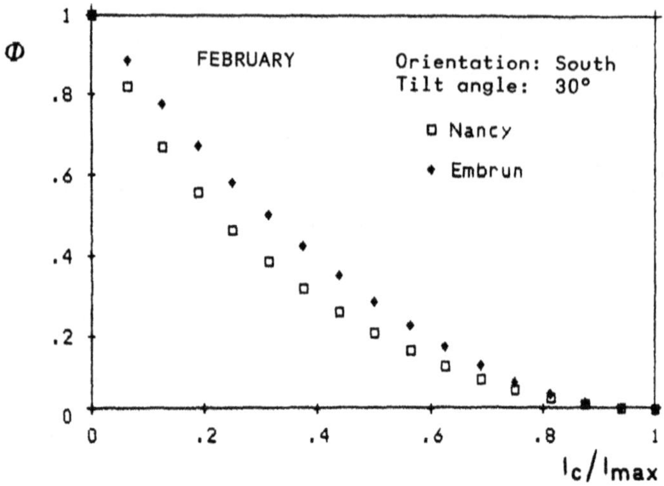

Fig.4.2.4: Utilizability versus reduced critical intensity.

Utilizability is therefore defined as

$$\phi\ (I_c) = H_{av}\ (I_c)\ /\ H_t \quad . \qquad (4.2.3)$$

Available energy or utilizability curves may be drawn versus critical intensity, I_c, or reduced critical intensity, $y_c = I_c/I_{max}$.

Utilizability curves vanish at the maximal irradiance I_{max} (y_c=1) (Fig.4.2.3 and 4.2.4).

4.2.2 - CFC computation from solar radiation data files

Only hourly irradiation data, corresponding to mean hourly irradiances, are available. Thus, CFC's are not exactly cumulative frequency curves of solar irradiance but of mean hourly irradiance.

The first step in a CFC calculation is the computation of the complete file of hourly values of irradiance on the plane of interest. Models and equations of section 4.1.4 are used for that purpose (global irradiance).

The second step enables one to obtain the CFC by a simple sorting of the previously discussed hourly data file. Irradiance bins of 25 W/m^2 are used, in order to get sufficient accuracy of the curve, with a sample of 30 to 50 points.

The corresponding FORTRAN programme, TRIFC, is described in

Appendix G for the computation of CFC's of global irradiance on tilted planes.

4.2.3 - CFC fitting and practical equations

In order to summarize the information from CFC's and use it practically in system performance estimates, it is necessary to fit the CFC's with an analytical formula. CFC's are first normalized with reduced time (similar to a probability)

$$x = n_h \; / \; t_d \qquad (4.2.4)$$

and reduced irradiance

$$y = I \; / \; I_{max} \quad . \qquad (4.2.5)$$

These reduced parameters are only defined in the range (0;1).

An orthogonal polynomial fit has been chosen, because it enables one to use a simple identification procedure (without matrix inversion) and allows one to determine the number of polynomials used according to the desired accuracy /4.11/.

A CFC is defined by the equation

$$x = f_0(y) + a_1.f_1(y) + \ldots + a_6.f_6(y) \qquad (4.2.6)$$

Each CFC is characterized by the scale parameters, t_d and I_{max}, and six parameters a_i; the $f_i(y)$ polynomials are given in Table 4.2.1. For most cases, four polynomials are enough. A curve fitting sub-routine, POLORT, is available /4.45/.

Scale parameters of the fit are:

- I_{max}, the highest irradiance level observed at least three times a month, and
- t_d, the mean monthly astronomical number of hours between sunrise and sunset (for CFC's of global irradiance).

The quantity v, defined as (4.2.7)

$$v = 0{,}5 + a_1.(5/6)^{1/2} + a_3.(1/10)^{1/2} + a_5.(13/420)^{1/2}$$

is the reduced integral of the CFC

$$v = H_t \; / \; (\; t_d \; . \; I_{max} \;) \quad . \qquad (4.2.8)$$

This equation shows that the mean monthly irradiation is not independent of a_i-values.

The curve fitting is not perfect and there are small errors in the computation of H_t from the a_i. However, these errors can be neglected since they are always smaller than 1%. The root mean square error (RMSE) of the curve fit is generally small (within 0.5 to 1%).

The Utilizability Ø can be computed by

$$Ø = [\ g_0(y) + a_1 \cdot g_1(y) + \ldots + a_6 \cdot g_6(y)\ ] \ / \ v \qquad (4.2.9)$$

where $g_i(y)$ polynomials are integrals of the $f_i(y)$ as

$$g_i(y) = \int_y^1 f_i(t)\ .\ dt \qquad . \qquad (4.2.10)$$

The $g_i(y)$ polynomials are also given in Table 4.2.1.

Several tables with values for CFC fitting parameters have been established for a sample of 12 European locations (surfaces oriented towards the South; slope = 0° to 90°) and can be used directly with these formulas /4.11/ /4.14/.

They have been computed from the test reference years (TRY) data files. TRY data files contain a series of 12 actual months, taken from different years but each representative of long-term radiation data. Perez's algorithm has been used for the computation of diffuse sky radiation on tilted planes. A ground albedo of 0.20 has been assumed.

Part of these data are given in Appendix A.

Long-term CFC data (5 to 10 years) are also available for 12 French locations /4.13/.

All the data referred to in this section are available on mini-disks and are suitable for input with ESM programmes.

Similarly, long-term CFC data sets are being established for other European locations within the framework of the CEC project EUFRAT /4.38-39/. Results from that project will concern global radiation on tilted surfaces (including non-southerly orientations) and solar radiation for Compound Parabolic Collectors (CPC's) and concentrators.

Table 4.2.1: Useful formulae for CFC and Utilizability computations from the reduced critical intensity $y = I/I_{max}$ $(0 < y < 1)$ and the a_i-parameters. The orthogonality of the f_i can be summarized by the definition relationship:

$$\int_0^1 f_i(y) \cdot f_j(y) \cdot dy = 1 \quad \text{if} \quad i = j \quad \text{(0 otherwise)}$$

Functions g_i are defined as follows

$$g_i(y) = \int_y^1 f_i(t) \cdot dt$$

$$x = n_h / t_d$$
$$= f_0(y) + a_1.f_1(y) + \ldots + a_6.f_6(y)$$

$$\emptyset = [\, g_0(y) + a_1.g_1(y) + \ldots + a_6.g_6(y)\,] / v$$

$$v = 0.5 + a_1.(5/6)^{1/2} + a_3.(1/10)^{1/2} + a_5.(13/420)^{1/2}$$

$$f_0(y) = 1 - y$$
$$f_1(y) = y.(1-y).30^{1/2}$$
$$f_2(y) = y.(1-y).(-1+2y).210^{1/2}$$
$$f_3(y) = y.(1-y).(9-42y+42y^2).10^{1/2}$$
$$f_4(y) = y.(1-y).(-1+8y-18y^2+12y^3).2310^{1/2}$$
$$f_5(y) = y.(1-y).(2-24y+90y^2-132y^3+66y^4).1365^{1/2}$$
$$f_6(y) = y.(1-y).(-18+300y-1650y^2+3960y^3-4290y^4+1716y^5).35^{1/2}$$

$$g_0(y) = 0.5\ (1-y)^2$$
$$g_1(y) = (5/6)^{1/2}.(1-y)^2.(2y+1)$$
$$g_2(y) = (105/2)^{1/2}.(1-y)^2.y^2$$
$$g_3(y) = (1/10)^{1/2}.(1-y)^2.(1+2y-42y^2+84y^3)$$
$$g_4(y) = (1155/2)^{1/2}.(1-y)^2.y^2.(4y^2-4y+1)$$
$$g_5(y) = (13/420)^{1/2}.(1-y)^2.(1+2y-207y^2+1404y^3-2970y^4+1980y^5)$$
$$g_6(y) = (35/4)^{1/2}.(1-y)^2.y^2.(18-176y+605y^2-858y^3+429y^4)$$

4.3 - CUMULATIVE FREQUENCY CURVES OF SOLAR IRRADIANCE: A CORRELATION METHOD

Although CFC's are already available for a large sample of meteorological stations, CFC's are not available for any one location. However, solar system designers usually study systems to be installed anywhere in the world. Therefore, a reconstitution method for CFC's is necessary from commonly available parameters. A method using the mean monthly irradiation on the collector surface and the maximal clear sky irradiance is proposed in this section. This method was developed for CFC's of global irradiance and only concerns applications of flat-plate collectors. If another type of solar radiation is considered (e.g., beam radiation for concentrators), other methods should be applied (see section 4.4.1).

4.3.1 - CFC correlation

It has been found from earlier research /4.15/ /4.11/, that normalized CFC's (in their reduced form) are well described by their integral value, v:

$$v = H_t / (t_d \cdot I_{max}) \quad . \qquad (4.3.1)$$

CFC's characterized by the same v value are very close to each other.

A statistical analysis has been performed and correlations have been established in order to derive a practical relationship between a_i and v /4.16/.

These formulae are suitable for the interval $0.15 < v < 0.55$ (see Table 4.3.1). Reduced CFC's and utilizability charts have been drawn in Fig.4.3.1 and 4.3.2, using v as a parameter. More accurate formulae have been made available recently from the EUFRAT project /4.40/.

From the formulae given in Table 4.3.1, it can be concluded that an estimate of CFC's is possible from three parameters; t_d (number of hours between sunrise and sunset), I_{max} (maximal irradiance), and H_t (mean monthly irradiation on the collector plane). These three parameters enable an estimate of the CFC's reduced integral, v, and from v, the CFC may be computed.

Computation of these 3 parameters for CFC's of global irradiance (flat-plate collectors) is described in the following three subsections.

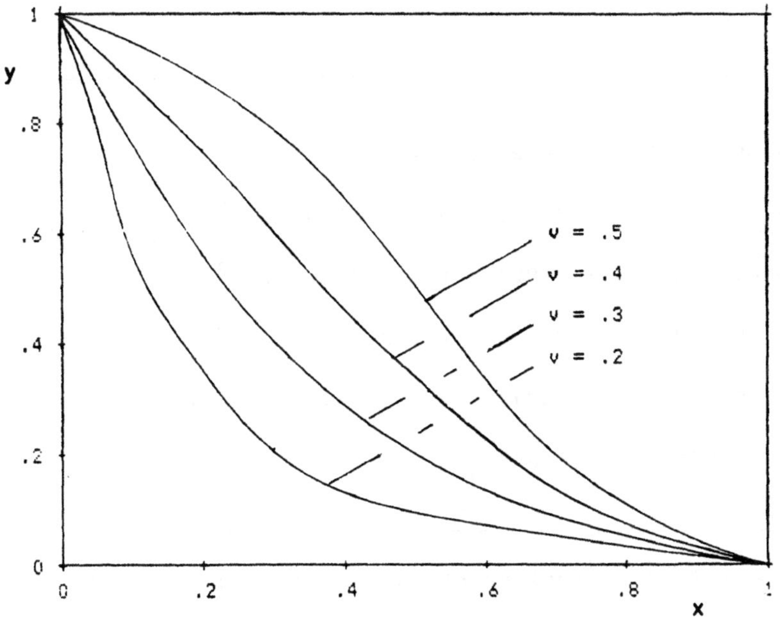

Fig. 4.3.1: Reduced CFC charts.

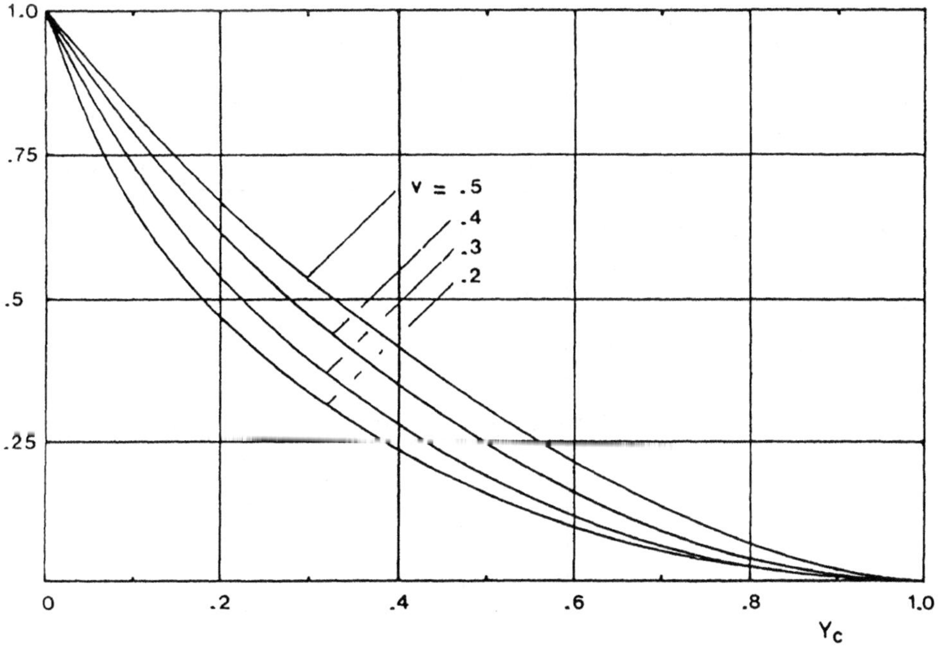

Fig. 4.3.2: Reduced Utilizability charts.

$$a_1 = -0,42297 + 0,16183 \cdot v + 2,48691 \cdot v^2 - 2,23697 \cdot v^3$$

$$a_2 = 0,34623 - 1,47635 \cdot v + 1,83694 \cdot v^2$$

$$a_3 = -0,27890 + 1,91022 \cdot v - 4,77807 \cdot v^2 + 4,11137 \cdot v^3$$

$$a_4 = 0,18865 - 1,36236 \cdot v + 3,37632 \cdot v^2 - 2,75771 \cdot v^3$$

$$a_5 = -0,11845 + 0,98872 \cdot v - 2,77182 \cdot v^2 + 2,53303 \cdot v^3$$

$$a_6 = 0,06968 - 0,60008 \cdot v + 1,73839 \cdot v^2 - 1,65775 \cdot v^3$$

Table 4.3.1: Correlations of the parameters a_i of the CFC's.

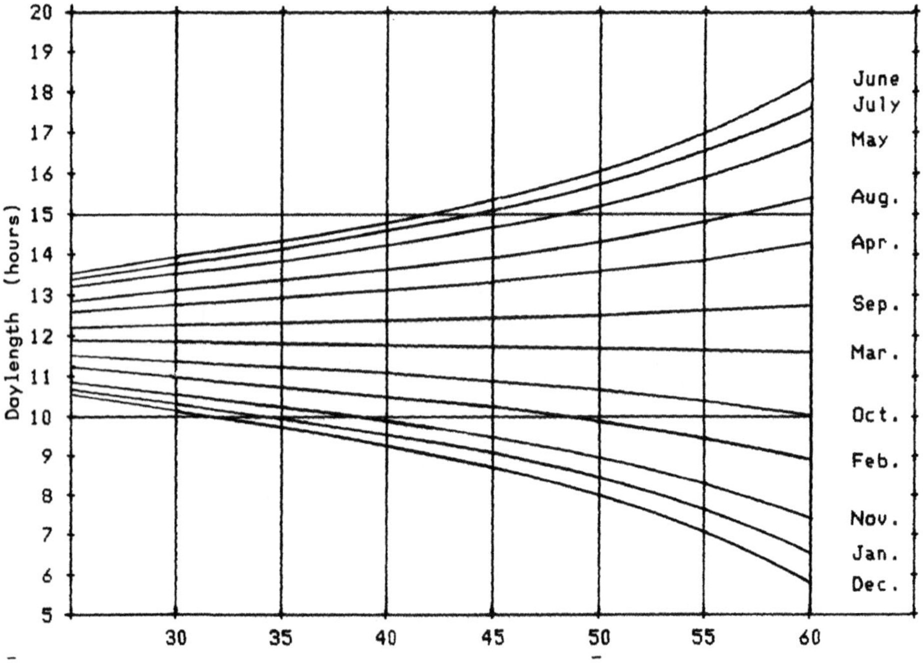

Fig. 4.3.3: Mean monthly number of hours between sunrise and sunset versus latitude.

4.3.2 - Number of hours between sunrise and sunset (t_d)

The astronomical equation given above (Eq.4.1.17) is used (time: hours; angles: degrees)

$$t_d = (t_h / 180) . \cos^{-1} (- tg\, \varphi . tg\, \delta) \quad . \quad (4.3.2)$$

An average value of solar declination is taken for each month (Table 4.3.2). t_d-charts in relation to latitude are given in Fig.4.3.3.

Month	Jan.	Feb.	Mar.	Apr.	May	June	July	Aug.	Sep.	Oct.	Nov.	Dec.
Decl.	-20.8	-13.1	-2.0	9.6	18.7	23.0	21.3	13.9	3.2	-8.5	-18.3	-23.0

Table 4.3.2: Mean monthly solar declination used for computing mean number of hours between sunrise and sunset.

4.3.3 - Monthly global irradiation on tilted planes (H_t)

In this section, monthly CFC's of global irradiance are considered; thus, the mean monthly irradiation on the collector plane H_t (area delimited by the CFC and the axes, see Fig. 4.2.2) is simply the mean monthly global irradiation on the collector plane, denoted $G_m(\alpha, \beta)$, if the collector surface has an azimuth angle α and a slope β (see section 4.1.5).

As mentioned previously, mean monthly irradiation values on a horizontal plane are generally available, at least for industrialized countries. However, irradiation on tilted surfaces must still be computed. The user has two solutions:

- To use $G_m(\alpha, \beta)$ (computed) values from existing solar radiation atlases, such as the European Solar Radiation Atlas /4.17/ for Europe, which includes data for more than 100 stations and 9 surfaces, as well as maps. Similarly, national atlases are also available e.g., for France /4.18/ /4.19/.

- To compute $G_m(\alpha, \beta)$ for the plane of interest from values of monthly global irradiation on a horizontal plane (or from sunshine duration). Methods and charts are indicated in section 4.1.5.

Data sources are numerous and are summarized in Table 4.3.3.

4.3.4 - Maximal solar irradiance (I_{max})

Maximal solar irradiance is assumed to be observed on clear days, at solar noon for planes facing due south.

As indicated previously (see section 4.1.3), maximal solar irradiance can be computed from the Linke Turbidity factor T_L.

Computation steps are summarized in Table 4.3.4 as a combination of Kasten's (direct, Eq. 4.1.19) and Perrin de Brichambaut's (global, Eq. 4.1.20) equations with Willmott's model for diffuse anisotropy (Eq.s 4.1.24 and 4.1.25).

Computations are done twice a month for every month (8.5 days after the beginning and 8.5 days before the end of the month); the final value to be taken as a maximal irradiance is the larger of the two. Practically speaking, a single calculation would be sufficient at the day for which $\cos(\varphi - \delta - \beta)$ is maximal. Corresponding solar declination values are given in Table 4.3.5.

If incidence angles are large, that is

$$\cos(\varphi - \delta - \beta) / \cos(\varphi - \delta) < (1 + \cos\beta)/2$$

then this method can no longer be used due to a small amount of clear day irradiance on the plane considered: maximal irradiance is assumed to be obtained with a higher diffuse fraction (D/G = 0.3) considered as isotropic.

For large climatic areas, average values of Linke's Turbidity Factor can be adopted with periodic annual variations as indicated in Table 4.3.6.

Imax-charts drawn from the previous method are given in Figures 4.3.4 and 4.3.5. Values found in these figures need to be corrected by the sun-earth distance corrective factor, r (Table 4.3.7).

Errors introduced by this method are small and generally within 5% (Table 4.3.8).

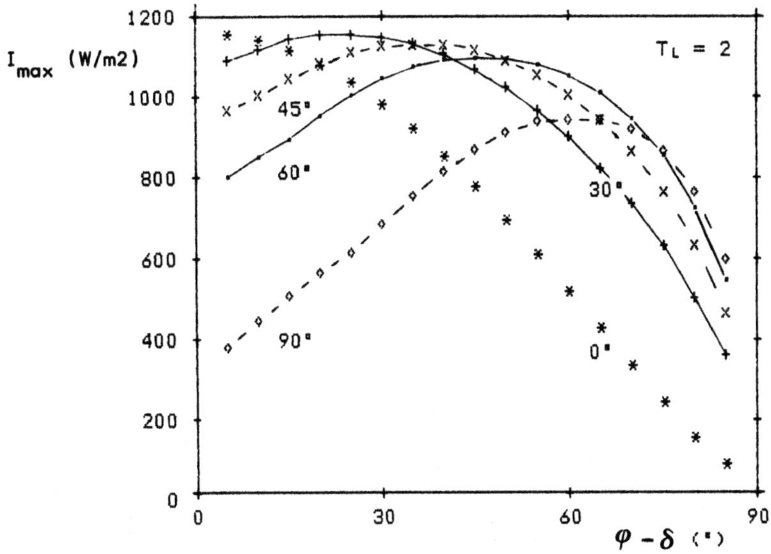

Fig. 4.3.4: Maximal global irradiance I_{max} versus ($\varphi - \delta$) for several tilt angles (surfaces oriented towards the South). Linke turbidity factor T_L = 2.

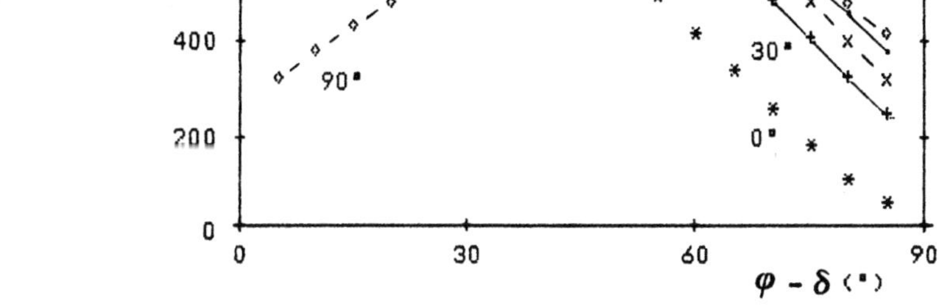

Fig. 4.3.5: Idem Fig. 4.3.4. T_L = 5.

Area or Country	Reference	Year	Data G	Data S	Data H_t	Presentation Maps	Presentation Tables	Nr. of Stations
World	/29/ WMO	1980	*			*		
World	/30/ CLINO	1971		*			*	590
Europe + Middle-East	/23/ CEC 1	1984	*	*		*	*	340
Europe	/17/ CEC 2	1984			*	*	*	102
France	/18/ Claux	1982	*		*		*	39
France	/19/ Tric.	1978	*		*	*		
France	/26/ Garn.	1978		*		*	*	100
Switzerland	/27/ Valko	1982	*		*	*	*	60
United Kingdom	/41/ Page	1986	*		*		*	
Belgium	/42/ Dogn.	1979	*	*	*	*	*	
Northern America	/1/ D/B	1980	*				*	260
USA	/28/ SERI	1980	*			*		
Canada	/43/ McKay	1985	*		*	*	*	129

Table 4.3.3: Sources of solar radiation data: G (mean monthly global irradiation on a horizontal plane or mean monthly clearness index), S (mean monthly duration of sunshine), and H_t (Mean monthly global irradiation on tilted planes).

Angles of incidence: $\sin\gamma = \cos(\varphi - \delta)$

$\cos\theta = \cos(\varphi - \delta - \beta)$

Earth-Sun distance correction:

$$r = 1 + 0{,}034 \cdot \cos(0{,}986 \cdot J - 2°)$$

Global irradiance on a horizontal plane:

$$G_c = r \cdot (1270 - 56 \cdot T_L) \cdot (\sin\gamma)^{(T_L+36)/33}$$

Air mass: $AM = (1 - 0{,}1 \cdot z) / \sin\gamma$

(z: local altitude in km)

Direct irradiance at normal incidence

$$B_c = B_0 \cdot r \cdot \mathrm{EXP}\,[\,-AM \cdot T_L / (0{,}9 \cdot AM + 9{,}4)\,]$$

Diffuse irradiance on a horizontal plane $D_c = G_c - B_c.\sin\gamma$

Isotropic diffuse fraction $F = 1 - B_c / (B_0 \cdot r)$

Conversion factor of isotropic diffuse: $C_s = (1 + \cos\beta) / 2$

Diffuse irradiance on tilted plane

$$D_t = D_c \cdot [\,F \cdot C_s + (1 - F) \cdot \cos\theta / \sin\gamma\,]$$

Maximal irradiance on tilted plane

$$I_{max} = \mathrm{MAX}\begin{cases} B_c \cdot \cos\theta + D_t + \rho_g \cdot (1 - \cos\beta) / 2 \cdot G_c \\ G_c.[\,0.7 \cos\theta/\sin\gamma + 0.3\, C_s + \rho_g.(1-\cos\beta)/2\,] \end{cases}$$

Table 4.3.4: Computation of the maximal irradiance I_{max} at solar noon on clear days on a tilted plane facing due south (T_L: Linke turbidity factor).

Month	Jan.	Feb.	Mar.	Apr.	May	June	July	Aug.	Sep.	Oct.	Nov.	Dec.
δ_1	-22.2	-14.8	-4.6	7.5	17.3	22.9	22.4	16.0	5.4	-6.2	-16.8	-22.8
δ_2	-19.6	-11.0	0.9	12.1	20.5	23.4	20.2	11.6	0.4	-11.3	-20.1	-23.4

Table 4.3.5: Values of the solar declination to be used for computing I_{MAX} (9TH day after the beginning and 9TH day before the end of the month). The value which leads to the highest I_{MAX} (from the Eq.s used in Table 4.3.4) should be adopted.

Area		T_0	a	b
Mediterranean	(coastline)	3.25	-1.1	-0.15
	(hinterland)	3.15	-0.5	-0.05
Atlantic		3.25	-0.7	-0.15
Continental		3.75	-0.2	-0.05
Large urban area		4.05	-0.5	-0.10

Table 4.3.6: Mean yearly variations of Linke's Turbidity factor T_L (at sea level) in Europe according to Eq. (from /4.13/)

$$T_L = T_0 + a.\cos(0.986.J) + b.\sin(0.986.J)$$

T_L decreases by about 0.35 per 1000 m of altitude.

Month	Jan.	Feb.	Mar.	Apr.	May	June	July	Aug.	Sep.	Oct.	Nov.	Dec.
r	1.033	1.026	1.011	0.994	0.978	0.968	0.966	0.974	0.989	1.007	1.023	1.032

Table 4.3.7: Mean monthly correction factor, r, for sun-earth distance (to be applied to I_{MAX} read from Fig. 4.3.4 and 4.3.5).

Carpentras Tilt angle: 0° T_L = 3.15

Month	Jan.	Feb.	Mar.	Apr.	May	June	July	Aug.	Sep.	Oct.	Nov.	Dec.
CFC	454	586	763	880	942	961	954	898	793	644	480	381
EST.	446	591	763	884	943	952	942	895	794	646	484	385

Carpentras Tilt angle: 45° T_L = 3.15

Month	Jan.	Feb.	Mar.	Apr.	May	June	July	Aug.	Sep.	Oct.	Nov.	Dec.
CFC	926	982	1052	1056	1018	980	996	1016	1028	1001	925	872
EST.	927	1010	1055	1049	1004	958	968	1007	1020	1003	934	870

Carpentras Tilt angle: 90° T_L = 3.15

Month	Jan.	Feb.	Mar.	Apr.	May	June	July	Aug.	Sep.	Oct.	Nov.	Dec.
CFC	909	883	846	732	602	510	551	660	790	869	893	891
EST.	900	896	835	701	559	496	520	627	760	854	879	881

Trappes Tilt angle: 0° T_L = 3.60

Month	Jan.	Feb.	Mar.	Apr.	May	June	July	Aug.	Sep.	Oct.	Nov.	Dec.
CFC	353	494	669	804	861	892	874	798	718	545	374	285
EST.	347	493	674	808	878	891	879	821	707	550	384	286

Trappes Tilt angle: 45° T_L = 3.60

Month	Jan.	Feb.	Mar.	Apr.	May	June	July	Aug.	Sep.	Oct.	Nov.	Dec.
CFC	795	912	992	1028	987	976	981	983	998	917	810	740
EST.	800	919	1000	1011	989	954	955	973	969	920	813	721

Trappes Tilt angle: 90° T_L = 3.60

Month	Jan.	Feb.	Mar.	Apr.	May	June	July	Aug.	Sep.	Oct.	Nov.	Dec.
CFC	812	854	832	739	616	556	594	683	787	829	822	801
EST.	818	848	828	727	605	525	553	657	759	809	802	765

Table 4.3.8: Comparison of maximal irradiance I_{MAX}
i) CFC: from reference /4.13/
ii) EST: estimated by the method described in Table 4.3.4 using the average Linke Turbidity factor T_L with monthly variations of table 4.3.6.

4.4 - SOLAR RADIATION DATA: ADDITIONAL ASPECTS

The first additional aspect discussed here concerns the problem of data for concentrating collectors, some of which (such as compound parabolic collectors, CPC) are sometimes used in solar space heating or domestic hot water systems.

The second aspect discussed is the computation of the mean monthly incidence angle modifier.

4.4.1 - Suitable data for concentrating collectors

ESM methods are not limited to flat-plate solar collectors: any type of concentrating collector can be used in solar space heating or domestic hot water systems. High - concentrating devices are unlikely to be used owing to the low temperature level required by SH and DHW systems. However, low-concentrating collectors should not be excluded. Compound parabolic collectors (CPC's) are still under development, but with low concentration ratios. A low-concentration CPC exhibits characteristics quite comparable with evacuated tubular collectors and can thus be included in SH or DHW systems.

Concentrating collectors convert beam radiation and, according to their concentration ratio, a fraction of the diffuse radiation. This type of collector, therefore, requires suitable solar radiation data. CFC data sets for this type of collector have been recently established within the EUFRAT project /4.38-4.40/.

Existing utilizability correlations from COLLARES-PEREIRA and RABL /4.31/ can also be used.

4.4.2 - Mean monthly incidence angle modifier

The incidence angle modifier of a collector has been defined in section 2.1.4. The coefficient, $K(\theta)$, to be applied to the optical efficiency η_0, represents the variation of transmittance by glass and absorptance by the absorber plate with the angle of incidence. This coefficient is a function of the angle of incidence and of the type of collector.

The variation of $K(\theta)$ may be characterized (for flat-plate collectors), according to ASHRAE Standards 93-77, by a single coefficient b_0

$$K(\theta) = 1 - b_0 \cdot (1/\cos\theta - 1) \quad . \qquad (4.4.1)$$

This relationship is valid for instantaneous values and for beam irradiance only. Hence this relationship needs to be extended to define a monthly average incidence angle modifier, which will make it possible to estimate the average optical efficiency of a collector for each month, $\bar{K} . \eta_0$.

The diffuse radiation (diffuse sky radiation + ground reflected radiation) comes from any direction, with all possible angles of incidence; an average effective incidence angle can be defined, according to /4.1/, assuming diffuse radiation to be isotropic. This effective incidence angle is generally close to 60°, thus it justifies the usual rule of thumb leading to an effective incidence angle of 60°.

Therefore, the average incidence angle modifier for diffuse radiation is simply equal to $(1 - b_0)$.

For beam radiation, the monthly average incidence angle modifier is not constant and depends on a number of parameters: solar declination, latitude, orientation and inclination of the surface, ground albedo, and frequency distribution of direct radiation within different hours.

Therefore, the average incidence angle modifier, $\bar{K}$, depends (as well as on the collector characteristic b_0) on these input parameters; furthermore $\bar{K}$ depends on the monthly diffuse fraction Kd

$$Kd = D_m / G_m \quad . \qquad (4.4.2)$$

An analytical method for the estimation of $\bar{K}$ has been developed recently /4.40/ /4.44/. This method was derived from the relationships of COLLARES-PEREIRA and RABL giving the average diurnal distribution of the global and diffuse irradiance /4.31/. $\bar{K}$-charts, derived by this method, are given in Figs 4.4.1 and 4.4.2 for the months of the year versus the quantity $(\varphi - \beta)$ and for two values of the diffuse fraction Kd (a linear interpolation is allowed for $0.30 < Kd < 0.70$). These charts are suitable for Europe (35°N < Latitude < 55°N). For values of b_0 differing from 0.10, a linear correction may easily be made.

The monthly diffuse fraction of solar irradiation on a horizontal surface may be estimated from the monthly clearness index

K_{tm} by a correlation of the type /4.22/

$$Kd = 1.032 - 1.175 \cdot K_{tm} \quad . \qquad (4.4.3)$$

If the monthly average beam irradiations, both on the collector plane $B_m(\alpha, \beta)$ and at normal incidence B_{mn}, are known, $\bar{K}$ may be approximated by

$$\bar{K} = 1 - b_0 \cdot [\ G_m(\alpha,\beta) + B_{mn} - 2 \cdot B_m(\alpha,\beta)\ ] \ / \ G_m(\alpha,\beta) \quad . \qquad (4.4.4)$$

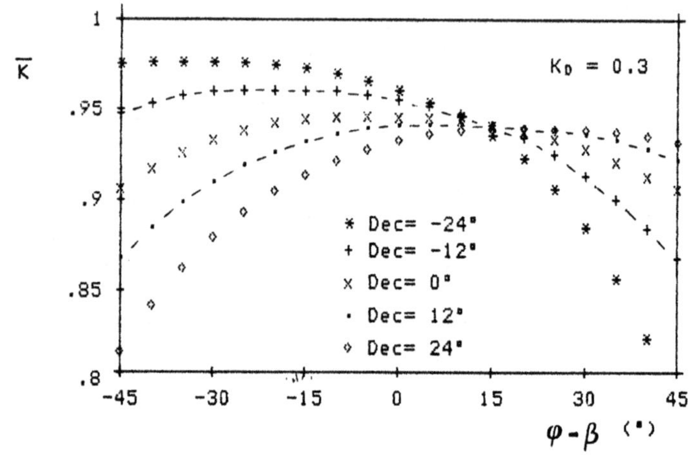

Fig. 4.4.1: K-charts of the monthly average incidence angle modifier for several values of solar declination (Kd = 0.3).

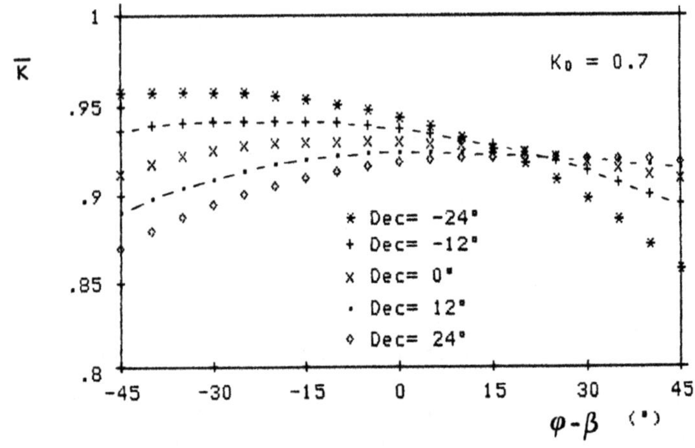

Fig. 4.4.2: Idem Fig.4.4.1 (Kd = 0.7).

REFERENCES OF CHAPTER 4

/4.1/ DUFFIE and BECKMAN. Solar Engineering of Thermal Processes. Wiley, 1980.
/4.2/ Ch.PERRIN DE BRICHAMBAUT et Ch.VAUGE. Le gisement solaire: Evaluation de la ressource énergétique. Tec et doc, Ed. Paris (1982)
/4.3/ A.RABL. Active Solar Collectors and their Application. Oxford Univ. Press, New-York (1985).
/4.4/ HICKEY et al. SOLAR ENERGY, Vol.29, n.2, pp.125-129 (1982).
/4.5/ B.BOURGES. SOLAR ENERGY, Vol.35, n.4, pp.367-369 (1985)
/4.6/ B.BOURGES. LA METEOROLOGIE, 7ème Série, n.11, (1986).
/4.7/ WILLMOTT. SOLAR ENERGY, Vol.28, n.3, pp.205-216 (1982).
/4.8/ PEREZ et al. INTERSOL 85, MONTREAL, Vol.4, pp.2498-2502. PERGAMON (1986).
/4.9/ ERBS et al. SOLAR ENERGY, Vol.28, p.293 (1982).
/4.10/ ORGILL and HOLLANDS. SOLAR ENERGY, Vol.19, n.4, pp.357-359 (1977).
/4.11/ BOURGES and LASNIER. CEC Report EUR 10203 EN (1983).
/4.12/ LUND. CEC Report EUR 9402 EN (1985)
/4.13/ B.BOURGES. Etude des Courbes de Fréquences Cumulées de l'éclairement solaire global reçu par des surfaces inclinées. Préparation d'un atlas pour la France. Rapport ARMINES-AFME 3.350.3168, Paris (1985)
/4.14/ BOURGES. INT. J.of SOLAR ENERGY, Vol.3, pp.241-253 (1985).
/4.15/ B.BOURGES. Les courbes de fréquences cumulées de l'éclairement solaire: analyse statistique et application au calcul des installations solaires. Thèse de Docteur-Ingénieur, Paris 6 (1980).
/4.16/ B.BOURGES and M.GRIECH. INTERSOL 85, Montreal, Vol.4, pp.2444-2448 (1985).
/4.17/ W.PALZ et al. European Solar Radiation Atlas. Volume 2: Inclined Surfaces. Verlag TUV Rheinland, Cologne (1984).
/4.18/ P.CLAUX et al. Atlas solaire français: Energie transmise et calcul de l'effet de masque. PYC Edition, Paris (1982).
/4.19/ J.F.TRICAUD. Atlas énergétique du rayonnement solaire pour la France. PYC Editions, Paris (1978)
/4.20/ KLEIN and THEILACKER. ASME J. of SOLAR ENERGY ENGINEERING, Vol.103, p.29 (1981).
/4.21/ J.K.PAGE. Prediction of Solar Radiation on Inclined Surfaces. REIDEL, Dordrecht, 459p. (1986).
/4.22/ P.BREJON et G.WATREMEZ. CASAMO: Methode de calcul du coefficient B sur micro-ordinateur. XIIIèmes journées d'étude du Comité Français d'Electrothermie, Versailles (1982).
/4.23/ W.PALZ et al. European Solar Radiation Atlas. Volume 1: Horizontal Surfaces. Verlag TUV Rheinland, Cologne (1984).
/4.24/ J.A.BEDEL, J.JAN et V.PERARNAUD. Le gisement solaire en France: Irradiation solaire globale reçue sur une surface horizontale. Direction de la Météorologie, Boulogne (1980)
/4.25/ Direction de la Météorologie. Le gisement solaire en France: Recueil de données statistiques 1971-1980. Données et Statistiques n.6. Boulogne (1984).
/4.26/ M.GARNIER. Durée et fraction d'insolation en France: Période 1951-1970. Direction de la Météorologie, Monographie n.105 (1978)
/4.27/ VALKO. Données Solaires pour la Suisse. Docum. Suisse de Batiment, Blauen (1982).
/4.28/ SERI. Insolation data Manual. Report SERI/SP-755-789 (1980)
/4.29/ WMO (World Meteorological Organization). TN 172 (1981).

/4.30/ WMO. Climatological Normals CLINO 1931-1960. n.117 (1971)
/4.31/ COLLARES-PEREIRA and RABL. SOLAR ENERGY, Vol.23, pp.235-253 (1979).
/4.32/ R.PEREZ et al. SOLAR ENERGY, Vol.39, n.3, pp.221-232 (1987).
/4.33/ G.J.VANDENBRINK. CEC Report ESF-006-80 NL. TPD-TNO, Delft (1982).
/4.34/ Ch.PERRIN DE BRICHAMBAUT et al. Ressources Energétiques Solaires. Cahier COFEDES n.2 (1988)
/4.35/ B.BOURGES. Int. J. Solar Energy, Vol.5, pp.171-184 (1987).
/4.36/ B.BOURGES. Advances in Solar Energy Technology (Congress of the ISES, Hamburg, 1987), Vol.4, pp.3811-3815. Pergamon, Oxford (1988).
/4.37/ C.E.C. DGXII. Test Reference Years TRY. EUR Report 9765 (1985)
/4.38/ B.BOURGES. EUFRAT. Solar Energy Applications to Building and Solar Radiation Data (EC Contractors' Meeting, Brussels, 1987), Vol.4, pp.17-26. Kluwer Ac. Pub., Dordrecht (1988)
/4.39/ B.BOURGES. EUFRAT. EUROFORUM New Energies 88, Sarrebrucken (1988).
/4.40/ B.BOURGES et al. EUFRAT: Climatic data for the design of renewable energy systems in Europe (in press) (1990).
/4.41/ J.K.PAGE and R.LEBENS. Climate in the United-Kingdom. HMSO, London (1986).
/4.42/ R.DOGNIAUX. Distribution de la durée d'Insolation et du rayonnement solaire en Belgique. Institut Royal Météo. de Belgique, Miscel. Série B, n.51 (1979).
/4.43/ D.C.McKAY et al. Solar radiation data for Canada. En56-64/2-1984. Environment Canada, Ottawa (1984).
/4.44/ B.BOURGES. Monthly average incidence angle modifier. EUFRAT technical note /Co-or/10. ARMINES, Paris (1988).
/4.45/ B.BOURGES et al. EUFRAT final scientific report. ARMINES, Paris (1990).

CHAPTER 5

SPACE HEATING AND HOT WATER LOADS.

TEMPERATURE DATA.

Accurate estimations of energy loads, for space heating (SH) as well as domestic hot water (DHW), are the first step in any solar system design method. Energy loads are used, firstly, as a reference for the energy savings brought about by the solar system.

However, they are also essential in the solar computation process in order to determine the base temperature and the recovery factor of active solar gains. A higher space heating load leads to a higher return temperature from the heat emitters, and a smaller utilizability of available solar energy; however, this energy is used more efficiently due to a smaller increase in storage temperature.

The different steps in the calculation of monthly space heating and domestic hot water loads are described in this chapter. Passive solar and free heat gains are accounted for in this process. An estimation method for variable base temperature degree-days is included. Sources of temperature data are indicated.

5.1 - SPACE HEATING LOAD

A number of methods have been published for the computation of space heating loads. Any of these methods can be used to provide an input for ESM, as long as they produce monthly results. Building simulation models, micro-computer simplified models, or manual methods (/5.1-7/) are relevant for this purpose. Some of them have been reviewed in the European Passive Solar Handbook /5.8/. Moreover, national calculation standard methods in European countries are being analysed in order to provide a base for a possible "Eurocode" in this field /5.30/.

A simplified method, to be used manually or as personal computer software is proposed in this Handbook; this method is based on the algorithms of a French buiding energy design tool, CASAMO /5.4/.

Monthly space heating load computations are split into three successive steps: i) net heat loss coefficient of the building; ii) gross space heating load (without passive solar gains); and iii) incident passive solar gains and net space heating load.

The last step is to be included only as a correction: specifically-designed passive solar systems providing high passive solar fractions (e.g., more than 50 % of the yearly space heating load) are not included. In such a case, coupling of active and passive systems introduces particular problems, which are not dealt with by ESM methods but require detailed (hour-by-hour) computer simulation.

5.1.1 - Building heat loss coefficient

All heat loss sources need to be carefully analysed. In many countries, this analysis is made through standard computation rules, sometimes included in official regulations. National references are summarized in Table 5.1.1.

The approach used in this handbook is based on the French rules in this field, developed by the CSTB /5.9-10/. In all countries however, the methods employed are basically similar, with specific aspects adapted to local building technologies.

Each heat loss is characterized by a heat loss coefficient (rate of heat loss per unit of temperature difference between building and exterior). All these coefficients (for walls, roof, window, ventilation, etc.) are added to give the total heat loss coefficient of the building.

Country	Official Body	Title	Ref.
Belgium	BIN-IBN	NBN-B62-301	/27/
Denmark	D.I.F.	DS-418	/24/
France	C.S.T.B.	Règles Th-K et Th-G	/9-10/
F.R.Germany	D.I.N.	DIN 4701	/25/
Netherlands	N.N.I.	NVN 5125	/28/
U.K.	B.R.E.	BREDEM (Dwellings)	/26/
U.K.	C.I.B.S.	Buil.Energy Code (Non-domest.)	/11/
U.S.A.	ASHRAE	Fundamentals	/6/
World	I.S.O.	ISO/TC 163/SC2 85E	/29/

Table 5.1.1: National rules for building heat loss coefficient computation.

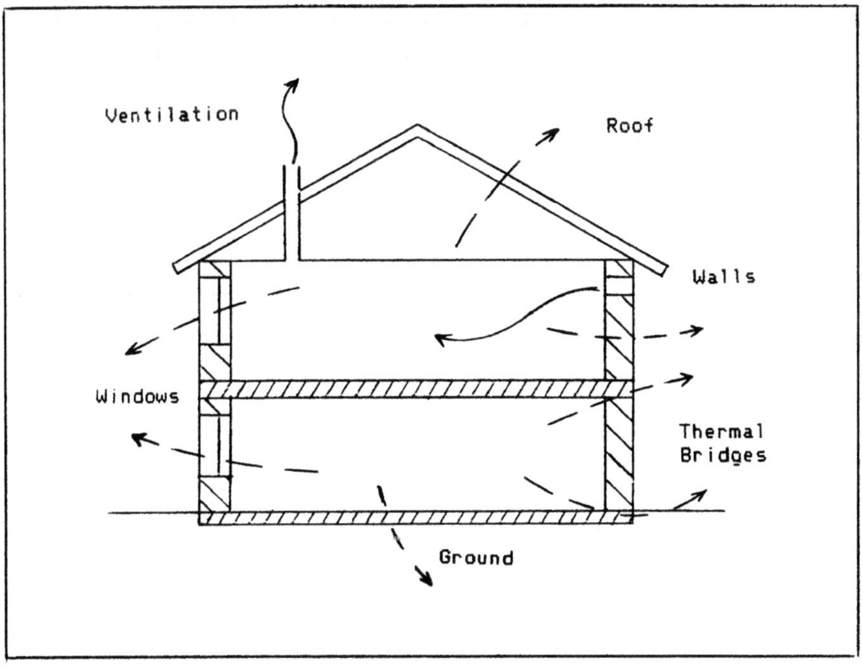

Fig. 5.1.1: Schematic of the different heat loss sources.

5.1.1.1 - Exterior walls and roof

Each element of wall or roof is characterized by a U-value (thermal transmittance as a rate of heat loss per unit area and unit of temperature difference, generally given in W/m2/K). The inverse of this U-value is the thermal resistance of the element and is the sum of resistances r_j for each of the different layers of the wall (numbered with j, ranging from 1 to n)

$$1 / U = R = 1/h_e + \Sigma r_j + 1 / h_i \quad . \qquad (5.1.1)$$

External and internal heat transfer coefficients of the wall, h_e and h_i, are given in Table 5.1.2.

The thermal resistance of a homogeneous material (in the j-th layer) is given as a function of its conductivity k_j and its thickness e_j, by

$$r_j = e_j / k_j \quad . \qquad (5.1.2)$$

Thermal conductivities of common building materials are given in Appendix C.

For inhomogeneous materials, such as bricks or parpens, thermal resistance is sometimes given directly by manufacturers or by Building Research Institutes.

The heat loss coefficient of a wall or roof element is then the product of its U-value and its surface area A (determined from internal dimensions).

	$1 / h_i$	$1 / h_e$
Vertical wall	0.12	0.04
Horizontal wall (Upward flux)	0.10	0.04
Horizontal wall (Downward flux)	0.17	0.04

Table 5.1.2: External and internal surface thermal resistances from ISO /5.30/.

5.1.1.2 - Windows

Calculations for windows are similar to those for exterior walls and roof. A global U-value may be given for the whole window or the calculation may be done separately for the glazing and the frame. Night and day U-values are distinguished, in order to account for night-insulation whose resistance is added to the window resistance during the night. A mean U-value is obtained by weighting according to the relative lengths of day and night. Some typical values are given in Table 5.1.3.

Glazing	Air space thickness	Frame	With night-insulation	Without night-insulation
Single		Wood	3.7	4.2
Single		Metal	4.2	4.8
Double	6 mm	Wood	2.6	2.9
Double	6 mm	Metal	3.1	3.4
Double	10 mm	Wood	2.4	2.7
Double	10 mm	Metal	2.9	3.2

Table 5.1.3: Typical values of average window U-values from CSTB /5.9/ (Unit: W/m2/K).

5.1.1.3 - Thermal bridges

"Thermal bridges" are certain parts of a building which contribute to a large amount of heat loss (e.g., contacts between walls of a different type, edges or places where the building and ground come into contact with each other). Heat loss due to thermal bridges must be computed carefully, generally using some u_L-values (linear heat loss coefficient per unit length) multiplied by the corresponding length. When external wall insulation is fitted, thermal bridges are largely reduced and may be neglected as a first approximation.

Some common thermal bridges and their typical u_L-values are given in Table 5.1.4. This table is by no means exhaustive; other thermal bridges, not given in this table, may be found in other reference documents /5.9/. Approaches to the remedy of thermal bridges differ considerably and are usually dependent on the country concerned /5.30/.

		e (cm)	u_L (W/m/K)
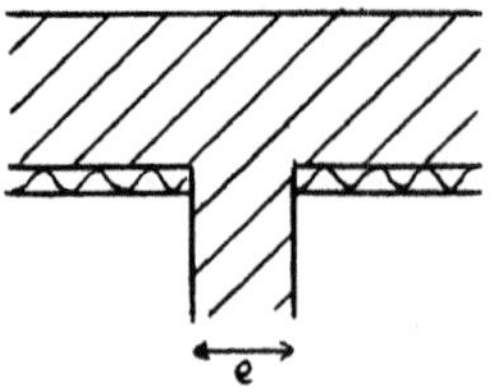	Concrete wall (Internal insulation) + Concrete Floor or partition wall	10 20	0.20 0.35
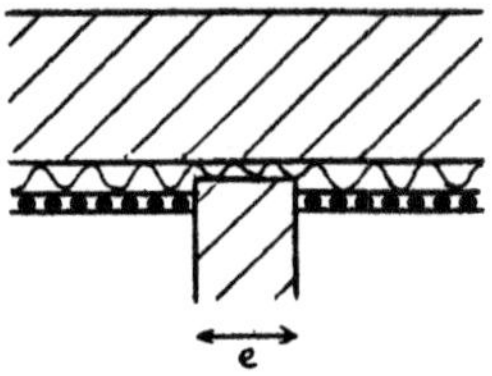	Concrete Wall (Internal insulation + additional internal wall) + Concrete floor or partition wall + Thermal correction (2 cm insulation between wall and partition wall)	10 20	0.05 0.10
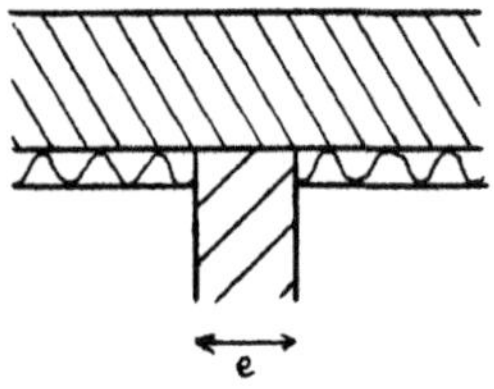	Hollow brick wall (Internal Insulation) + Concrete or masonry partition wall	10 20	0.10 0.20

Table 5.1.4: Some examples of thermal bridges and corresponding values of the linear heat loss coefficient u_L /5.9/.

5.1.1.4 - Ventilation

The ventilation rate depends on the type of ventilation (forced or natural) and infiltration through windows and doors. For a total ventilation flow rate q_f (m3/h), the corresponding heat loss is given by

$$HL_v = 0.34\ q_f \qquad (W/K) \quad . \qquad (5.1.3)$$

The factor 0.34 accounts for the thermal capacity of air and its average moisture content. When a heat recovery system (air/air coil) is incorporated, the ventilation heat loss must be corrected accordingly.

For natural ventilation, the flow rate may be computed from the size of openings, wind exposure, etc. /5.10/.

5.1.1.5 - Buffer spaces

Unheated spaces (such as a greenhouse, storeroom, or garage) in contact with heated spaces are considered as "buffers" and are computed separately. Heat losses (through walls, windows, thermal bridges, and ventilation) between the heated and the unheated spaces are first estimated, giving a heat loss coefficient HL_h. Then heat losses between unheated spaces and outdoors are estimated in the same way, giving a heat loss coefficient HL_u.

Heat losses from the heated space to outdoors through the buffer space are then given by the product $HL_h \cdot \tau$, where

$$\tau = HL_u \;/\; (HL_h + HL_u) \quad . \qquad (5.1.4)$$

Default values are sometimes given for common configurations /5.10/.

5.1.1.6 - Global heat loss coefficient

From all the above-mentioned sources of heat losses, the global heat loss coefficient of the building GV (W/K) may be computed as the sum

$$GV = \Sigma\, U.A + \Sigma\, u_L.l + \Sigma\, \tau.HL_h + HL_v \quad . \qquad (5.1.5)$$

V is the heated volume of the building and G is called the "volumetric heat loss coefficient" (although the product of these two terms is used for space heating load computations).

5.1.2 - Gross space heating load (without passive solar gains)

If T_a is the external ambient temperature and T_c the internal set-point temperature, the instantaneous rate of heat loss from the building (to be compensated by energy from the space heating system) is expressed as

$$P_{sh} = GV \cdot (T_c - T_a) \quad . \qquad (5.1.6)$$

Only positive values of this heat loss are considered, otherwise P_{sh} is taken to be zero. This is denoted by

$$P_{sh} = GV \cdot (T_c - T_a)^+ \quad . \qquad (5.1.7)$$

Furthermore, it is known that in an occupied building "free" heat gains are produced by the inhabitants, appliances, kitchen, etc. with an average power P_{fg}.

Various values have been proposed in the literature and reviewed by the Passive Solar Working Group /5.8/, ranging from 9 to 29 kWh per day for a single-family house. The standard value used in France is 1.5 W per m3 of heated space; this corresponds roughly to the lowest of these daily values.

The rate of heat loss from the building has now to be corrected as

$$P_{sh} = [\ GV \cdot (\ T_c - T_a\ ) - P_{fg}\ ]^+ \quad . \tag{5.1.8}$$

As before, only positive values are considered. The "No-Heating Temperature" (or "temperature of no-more heating") of the building may be defined as

$$T_{nh} = T_c - P_{fg} / GV \quad . \tag{5.1.9}$$

and the rate of heat loss becomes

$$P_{sh} = GV \cdot (\ T_{nh} - T_a\ )^+ \quad . \tag{5.1.10}$$

T_{nh} is also the space heating reference temperature, i.e. the value of the external temperature at which P_{sh} is zero.

This equation enables the calculation of the maximum rate of heat loss (and then the sizing of the space heating system, generator and heat emitters) using the lowest measured external temperature at the site, T_{min}.

The space heating load for one particular day (taking the external ambient temperature at its average daily value) is given by

$$L' = t_h \cdot GV \cdot (\ T_{nh} - T_a\ )^+ \tag{5.1.11}$$

with $t_h = 0.024$ if L' is expressed in kWh
$t_h = 0.0864$ if L' is expressed in MJ.

Over a long period (a month or a heating season), the gross space heating load may be written as

$$L_{0g} = t_h \cdot GV \cdot DD(T_{nh}) \tag{5.1.12}$$

where $DD(T_{nh})$ are the degree-days at base temperature T_{nh} for the site, computed from the sum

$$DD(T_{nh}) = \sum_{days} (T_{nh} - Ta)^{+} \quad . \qquad (5.1.13)$$

Degree-days at the base temperature T_{nh} are the sum of the difference between T_{nh} and the mean daily ambient temperature (positive values only) for all days of the period. They can be computed from the frequency distribution of ambient temperature for the period (see section 5.2).

When different thermostat set-points are used, e.g. during day and night, this method still applies using the two "no-heating temperatures" and their respective degree-days (with a weighting according to their respective durations). This is appropriate only for light-weight buildings (low thermal mass). Energy gains resulting from a lowering of the thermostat setting at night are generally less than computed by this method and may even become negligible for a very massive building. A more complete but still simple method accounting for these effects is given in reference /5.22/. Detailed simulation models of the building are another computation tool suitable for this phenomenon and other types of complex buildings or sophisticated heating controls (e.g., several zones, each with its own thermostat set-point and occupancy scenario).

In the case of ON/OFF controlled space heating (as defined in chapter 2), two set-point temperatures are available (switch-on and switch-off): The set-point temperature considered in equations 5.1.6 to 5.1.9 is their average value.

5.1.3 - Net space heating load (accounting for passive solar gains)

Direct solar gains by windows cannot be neglected as a source of "free" heat gains in a building, even in old houses. It is therefore essential to account for them in the space heating load computation. This cannot reasonably be done with some constant rate (like other free heat gains) because large variations occur according to the season and day. A specific monthly correction has therefore to be included for the estimation of net space heating loads, before any consideration of "active solar systems".

In this section, only methods for direct solar gains are given. Other types of passive solar components like greenhouses or Trombe walls, should be treated by specific methods /5.4-7/. Most of these also apply to direct gains and include more complex computations, such as Method 5000, as selected by the European Passive Solar Working Group /5.8/. Monthly results produced by detailed simulation models may also be used /5.1-3/.

A simple corrective algorithm (from the CASAMO method and also selected by the PSWG /5.8/) is proposed in this Handbook.

The gross monthly incident energy transmitted by glazings is first estimated for each window as

$$E_w = N \cdot H_t \cdot s_n \cdot \bar{s}/s_n \cdot f_s \cdot f_c \cdot A' \qquad (5.1.14)$$

with

- N number of days in the month,
- H_t daily global irradiation incident on the window (kWh/m2)
- s_n glazing solar factor at normal incidence,
- $\bar{s}/s_n$ monthly incidence angle modifier,
- f_c curtain correction factor (typically from 0.85 to 0.93),
- f_s shading factor, and
- A' window glazing surface area excluding frame (smaller than values used for heat loss).

Due to the part of solar radiation absorbed by glazings, which is transferred to the environment by convection and radiation, the solar factor is slightly higher than the transmittance. Typical values of s_n (solar factor at normal incidence) are given in Table 5.1.5.

In the European Solar Radiation Atlas /5.21/, the product $H_t.\bar{s}$ has been directly tabulated for a number of locations and orientations (also for France in /5.12/). Values of $\bar{s}/s_n$ are given in Appendix B for representative European locations.

The shading factor may be estimated using various methods or software /5.4/ /5.12/. Default values ranging from 0.8 to 0.9 may be adopted.

The solar energy transmitted by glazing is not all useful energy: only a part is used as a saved energy. The rest is lost, due to multiple reflexions and, mostly, due to building overheating above its set-point temperature. Direct solar gains are better exploited when they are a small part of the energy demand, if the building thermal mass is high, and if the space heating system is precisely controlled. The first condition, i.e. that direct solar gains should be a small part of the total energy demands, is crucial and may be accounted for (on a monthly basis) using the following equation for the net monthly space heating load

$$L_0 = L_{0g} \cdot EXP [- 0.9 \cdot X \cdot (1 + 0.3 \cdot X)] \qquad (5.1.15)$$

where $X = E_w / L_{0g}$. (5.1.16)

X is the ratio of gross solar incident energy (transmitted through glazings) to gross space heating load (without solar gains).

This formula applies nicely to buildings of moderate to high thermal mass and is only inappropriate for very light-weight buildings.

More accurate computations may be done using other methods, taking thermal mass explicitly into account /5.5//5.7/. The so-called "Un-utilizability" /5.7/ method is noteworthy because it is based on the "utilizability" concept, similar to ESM methods.

Type of glazing	s_n
Single	0.90
Double	0.80
Triple	0.70

Table 5.1.5: Typical solar factors of glazing (values recommended by EUROCODE /5.30/).

5.2 - DEGREE-DAYS AND TEMPERATURE DATA

As indicated in section 5.1.2, monthly degree-days are needed to compute the space heating load. They can be derived from ambient daily temperature statistics (as a frequency distribution), principally characterized by their monthly average value. The monthly average ambient temperature is also needed in ESM methods for the calculation of threshold irradiance.

This section explains how to calculate degree-days at any base temperature, and also gives a list of temperature data sources.

5.2.1 - Degree-days

Degree-days at base temperature T_{nh} are defined, for a given period, as the sum of the differences (positive only) between T_{nh} and ambient temperature for all days of the period

$$DD(T_{nh}) = \sum_{days} (T_{nh} - T_a)^+ \quad . \qquad (5.2.1)$$

If the statistical distribution of temperature is known and characterized by its cumulative frequency distribution Q(T) (that is, the probability of the ambient temperature being smaller than T), degree-days at base temperature T_{nh} are given by the integral (for N days)

$$DD (T_{nh}) = N \cdot \int_{T_{min}}^{T_{nh}} Q (T) \cdot dT \quad . \qquad (5.2.2)$$

Degree-day data are available in many countries for a number of locations, at least for the usual 18°C-base temperature (Tab. 5.2.1). In a few countries, a slightly different standard base temperature is used (e.g., 17°C in Denmark or 15.5°C in U.K.).

If degree-days data are not available, they can be computed with a reasonable accuracy from long-term monthly average ambient temperature and from statistical scattering characteristics. Studies by Erbs /5.13/ and Bourges /5.14/ have shown that the cumulative frequency distribution of daily ambient temperature can be accurately represented by

$$Q (T) = 1 / [1 + EXP (- 2 \cdot a \cdot h)] \qquad (5.2.3)$$

where $$h = (T - T_a) / (s \cdot N^{1/2}) \qquad (5.2.4)$$

T_a is the long-term monthly average ambient temperature,
S the standard deviation of monthly averages over a long period (typically 30 years) for the month considered,
N is the number of days in the month, and
a is an empirically adjusted parameter. Bourges proposed the value

$$a = 2.1 \quad . \qquad (5.2.5)$$

Monthly degree-days at base temperature T are then expressed as

$$DD(T) = S.N^{3/2} \; . \left(\frac{h}{2} + \frac{Ln[\; EXP(-ah) + EXP(ah)\;]}{2.a} \right) . \qquad (5.2.6)$$

The root mean square error obtained for French locations is about 3 degree-days.

The standard deviation of long-term monthly averages, S, can be obtained from weather data statistics or approximated by

$$S = (\; T_{80} - T_{20}\;) \;/\; 1.683 \qquad (5.2.7)$$

where T_{80} and T_{20} are the temperature values corresponding to 80% and 20% probability levels of the cumulative frequency distribution of monthly averages. Typical values of $(T_{80} - T_{20})$ are given in Table 5.2.2 for various geographical zones in Europe. The values given in Appendix B have been calculated with this method.

5.2.2 - Temperature data sources

Monthly average ambient temperatures are needed for the application of ESM methods, not only to determine monthly space heating loads, but also to compute the average efficiency of the system. These data are given in Appendix B for a sample of European locations. The ambient temperature is a widely measured climatic parameter, therefore a number of data sources are available; some of them are summarized in Table 5.2.1.

The ambient temperature varies according to the altitude above sea level of the site considered. A correction of -0.6 K for every 100 m above sea level is generally accepted.

Area or Country	Ref.	Year	Information given: T_a	Degree-days 18°	Degree-days Any Base	No.of Stat.
World	/5.16/	1971	*			1200
Europe	/4.40/	1989	*	*	*	150
France	/5.15/	1984	*	*		117
France	/5.12/	1982	*	*	*	39
U.K.	/5.17/	1986	*	*	*	13
North America	/5.18/	1980	*	*		260
"	/5.6/		*	*		

Table 5.2.1: Data sources of monthly average ambient temperature and degree-days.

Month	Continental mountain	Semi continental	Oceanic North Sea	Atlantic and Mediterranean coast
January	3.0	3.0	2.2	1.4
February	4.6	3.9	3.6	2.6
March	3.4	2.9	2.4	1.6
April	2.9	2.0	1.7	1.3
May	2.1	2.0	2.0	1.3
June	2.1	2.0	1.9	1.7
July	2.1	2.0	1.9	1.7
August	2.0	2.0	1.8	1.7
September	2.3	2.3	2.3	1.7
October	2.7	2.3	2.2	1.9
November	2.5	2.0	2.2	1.7
December	3.4	3.5	3.5	1.7

Table 5.2.2: Typical values of the temperature scattering characteristics (T_{80} - T_{20}) for Europe (Actual values are available in Ref. /5.15/ for France).

5.3 - DOMESTIC HOT WATER LOAD

Domestic hot water loads are easy to compute, but they are complicated by distribution and tank losses. A precise evaluation of the DHW load (or at least a possible range) is required for a good prediction of the solar system performance, but consumption data are not always reliable.

5.3.1 - Domestic hot water demand

The basic (theoretical) domestic hot water monthly load is computed by

$$L_1 = N \, . \, m \, . \, c \, . \, (T_h - T_m) \qquad (5.3.1)$$

where N is the number of days in the month,
m is the daily demand (kg/day),
c is the heat capacity of water (4.18 kJ/kg/K),
T_h is the required hot water temperature, and
T_m is the mains water supply temperature.

Actual daily consumptions vary widely, depending on the occupants' behaviour. There are also weekly and seasonal variations. Values between 30 and 40 litres/day per person are generally accepted /5.19/.

It is sometimes interesting to distinguish "design" and "balance" demand. The "design" demand is the maximum demand (for only a few days in the year) which should be met by the system; it determines the back-up capacity. The "balance" demand is the average demand which determines the auxiliary energy consumption and the performance of the solar sub-system. In this section only balance demand is considered.

Mains water supply temperatures are not well known and depend on the location and season. Typical values for some cities are given in Table 5.3.1. Monthly averages of T_m can be represented by

$$T_m = (T_a + \bar{T}_a) \, / \, 2 \qquad (5.3.2)$$

with no values smaller than 3°C. T_a and $\bar{T}_a$ are respectively the monthly and yearly averages of the external ambient temperature.

If no precise information is available, a default value of 10°C is often adopted for the whole year.

The required hot water temperature generally ranges between 45 and 60°C.

Location	Typical water supply temperature (°C) Jan.	Feb.	Mar.	Apr.	May	June	July	Aug.	Sep.	Oct.	Nov.	Dec.	Year
Marseille	10	10	11	14	16	18	19	19	18	16	14	11	14.7
Lyon	6.5	6.5	7.5	10.5	12.5	14.5	15.5	15.5	14.5	12.5	10.5	7.5	11.2
Paris	6	6	7	10	12	14	15	15	14	12	10	7	10.7
Lille	5	5	6	9	11	13	14	14	13	11	9	6	9.7
Copenhagen	4	3	2	4	8.5	12	14	15	14	12	8	5.5	8.5

Table 5.3.1: Typical monthly mains water supply temperatures for some cities (from /5.23/ and /5.31/).

5.3.2 - Draw-off profile

The daily draw-off profile has an important influence on the performance of active solar water heaters.

Consider the three extreme draw-off profiles represented in Fig. 5.3.1. In profile (a), draw-off is concentrated during the afternoon (from 1400 to 1800); in profile (b), during the night and early morning (from 0200 to 0600), just before the system starts; profile (c) is a diurnal profile from 0800 to 1700, similar to the solar radiation profile. Several studies have shown that, for solar DHW systems with mixed storage tank, profile (a) is the best and (b) the worst with differences in performance within the range 8-15% /5.32-34/. For highly stratified systems (with a very low collector flow-rate), profile (b) remains the worst but (c) becomes the best, with differences still within a similar range /5.34/.

The influence of the draw-off profile is much smaller on combined space heating and domestic hot water systems.

Statistical analyses for a large number of dwellings show concentrated consumption in the morning and in the evening with a secondary peak at noon (Fig. 5.3.2). This information is quite consistent with the normalized profile referred to as the "RAND profile" often used in the USA /5.35/ (Fig. 5.3.3). This type of profile is generally suitable for collective water heating, with a large number of inhabitants using the same system.

For an individual water heater, used by a single family, a more specific profile should be used:

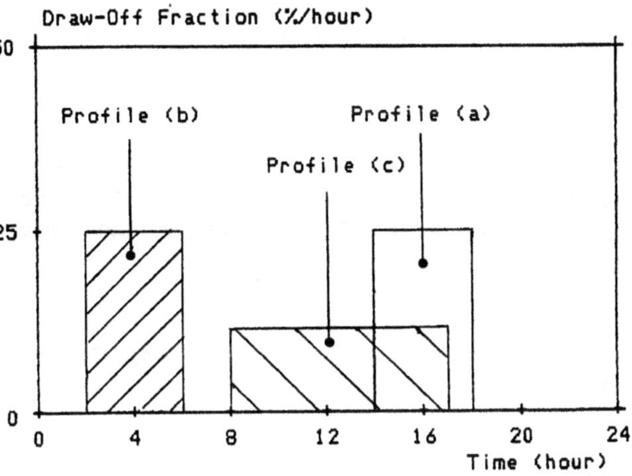

Fig. 5.3.1: Extreme draw-off profiles for solar water heaters.

Fig.5.3.2: Typical daily domestic hot water draw-off profiles (6600 dwellings, weekdays) (from CSTB /5.20/).

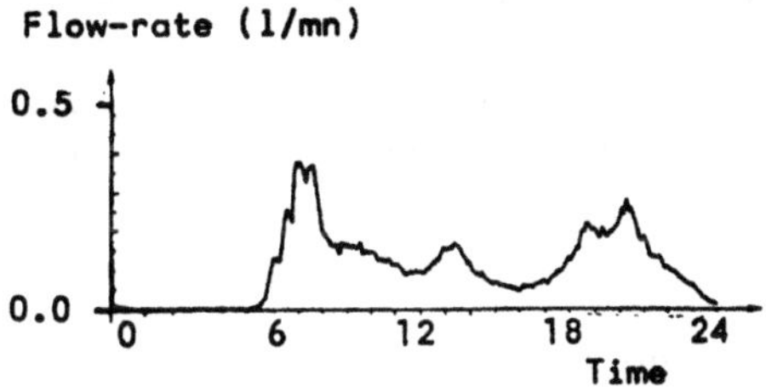

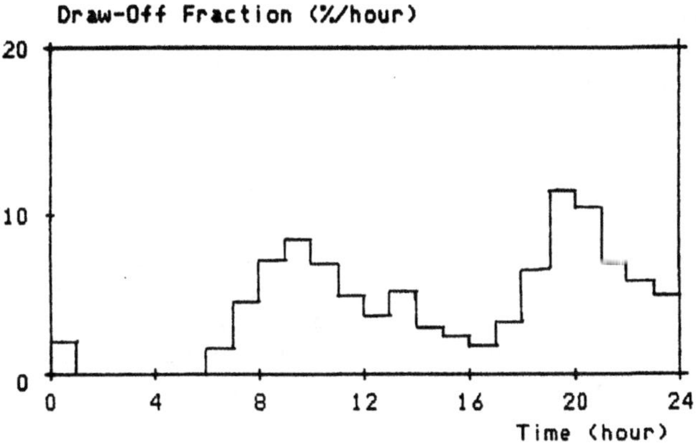

Fig.5.3.3: Normalised DHW draw-off profile (RAND profile) (adapted from Duffie and Beckman /5.18/).

- When both parents are working, 50% of the demand in the morning (0600-0700, True Solar Time) and 50% in the evening (1800-1900).
- For other cases, 40% of the demand in the morning (0600-0700), 20% at noon (1200-1300) and 40% in the evening.
- Evening draw-off (100% from 1800 to 1900) is often a simple, but realistic profile, leading to an intermediate performance.

Typical draw-off profiles for specific applications such as hotels, hospitals, nursing homes, or camping sites, are also available /5.36/.

5.3.3 - Distribution and recirculation losses

The distribution loop and the recirculation loop, as described in section 1.3 (see Fig. 5.3.4) are distinguished in this section. A distribution loop exists in all hot water systems, transferring hot water from the system to the point of use (tap), whereas a recirculation loop is optional.

Monthly heat losses from the distribution circuit are characterized, during draw-off, by the quantity

$$L_d = N \cdot A_dU_d \cdot t_s \cdot (T_h - T_e) \qquad (5.3.3)$$

with the following notations:

- A_dU_d rate of heat loss for distribution pipes (W/K),
- t_s daily draw-off time,
- T_e environmental temperature of the distribution circuit (generally indoor air temperature).

When draw-off stops, the distribution pipes cool down progressively until they reach T_e; when a new draw-off starts, the user has to wait some time before hot water flows. Therefore, an additional demand (volume) takes place because of the distribution losses. The additional demand depends on the capacity of the pipes in the distribution circuit and the number of draw-offs. It is assumed that the corresponding additional demand is included in the hot water demand m.

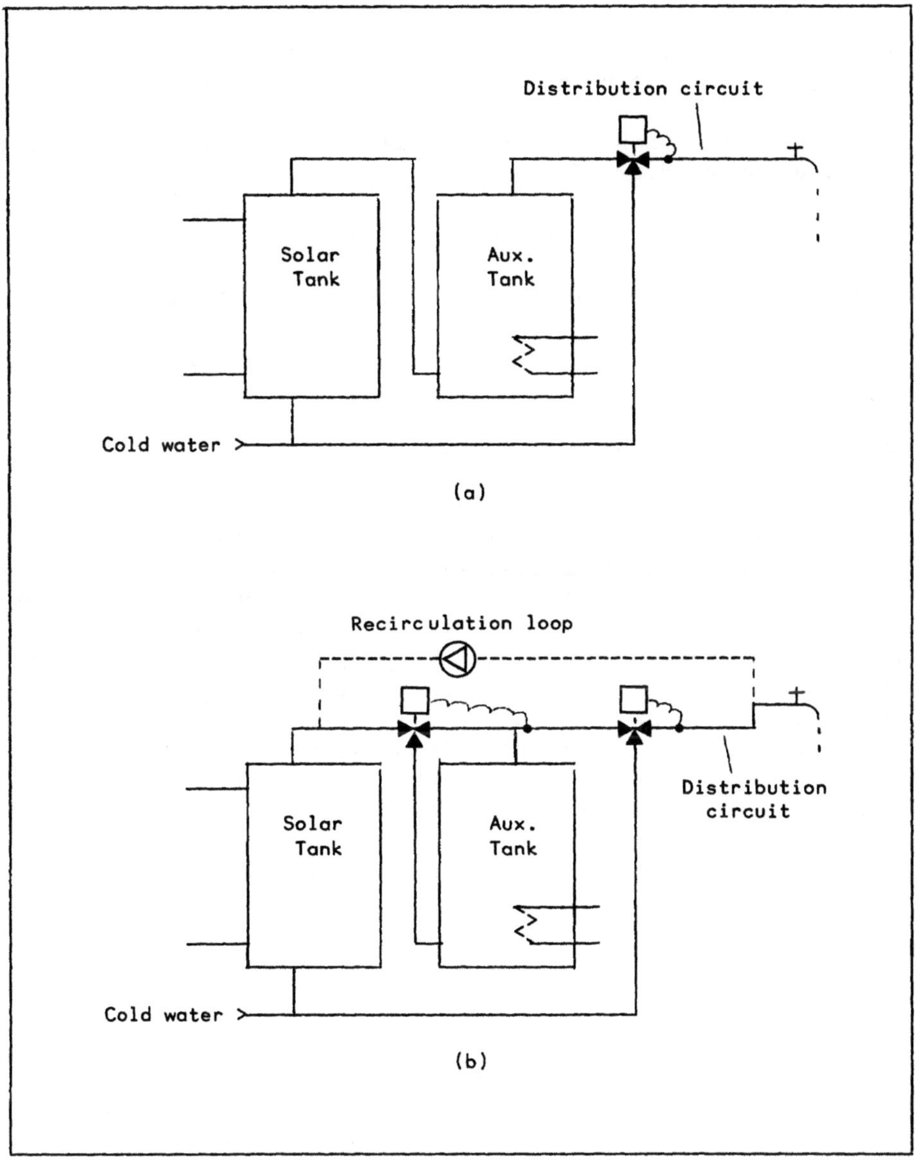

Fig. 5.3.4: DHW distribution circuit:
a) without recirculation loop (series auxiliary tank)
b) with recirculation loop (by-pass of the auxiliary tank with a controlled thee way-valve).

Heat losses from the optional recirculation loop are characterized by:

A_cU_c rate of heat loss for the recirculation loop (W/K).
t_L daily operating time of the recirculation loop ($t_s<t_L$).

We shall distinguish heat losses from the recirculation loop occurring during draw-off

$$L_{c1} = N \cdot A_cU_c \cdot t_s \cdot (T_h - T_e) \qquad (5.3.4)$$

and the supplementary part (losses occurring at other times)

$$L_{c2} = N \cdot A_cU_c \cdot (t_L - t_s) \cdot (T_h - T_e) \quad . \qquad (5.3.5)$$

5.3.4 - Auxiliary tank heat losses

There are two types of back-up for solar water heaters: an auxiliary heater (electric or gas-fired) or an auxiliary storage tank (containing an electric immersion or other indirect heat source). When there is an auxiliary storage tank, heat losses from this tank have to be included in the DHW load (but not the losses from the solar DHW pre-heating tank, since they are included in those from the solar tank). For solar tanks whose upper part is used as an auxiliary tank with back-up, the losses have to be split into two parts: "solar" losses are included in the solar tank loss coefficient, while auxiliary losses are included in the DHW load.

Heat losses of the auxiliary tank can be estimated from the following variables:

A_aU_a rate of heat loss for the auxiliary tank (W/K).
M_a storage capacity of the auxiliary tank (kg).
t_h duration of the day (24 hours).

5.3.4.1 - On/off controlled auxiliary tank

For an on/off controlled auxiliary tank, (back-up operating continuously), two additional set-point temperatures are required:

T_{h1} starting set-point temperature (°C).
T_{h2} stopping set-point temperature (°C).

The auxiliary energy is switched on as soon as the temperature of the auxiliary tank drops below T_{h1} and stops when it rises above T_{h2}. The temperature at the outlet of the tank must be set so that the user will obtain water at the required temperature at the tap; hence the tank temperature must be high enough to cover the distribution losses and half of the recirculation losses. Therefore,

$$T_{h2} > T_{h1} > T_{h\ min}$$

where

$$T_{h\ min} = T_h + t_s.(T_h - T_e).(A_dU_d + A_cU_c/2) / mc . \qquad (5.3.6)$$

The corresponding heat losses are

$$L_a = N . A_aU_a . t_h . [(T_{h1}+T_{h2})/2 - T_e] . \qquad (5.3.7)$$

5.3.4.2 - Off-peak energy auxiliary tank

For an off-peak energy auxiliary tank, the calculations are more complicated (heat losses from the tank depend on the draw-off profile). However, an approximation is proposed in this subsection. In order to approximate heat loss from the tank,

T_{h2} auxiliary stopping set-point temperature (°C), and
t_a auxiliary daily operating time (no draw-off is assumed to take place while the auxiliary is operating)

must be known.

The temperature of the auxiliary tank at the end of the off-peak period (auxiliary operating period) will therefore be T_{h2}. T_{h2} must allow the temperature T_{h1} of the top part of the auxiliary tank at the end of the draw-off period (beginning of the off-peak period) to be higher than $T_{h\ min}$ (as defined by Eq. 5.3.6)

$$T_{h1} = (1 - x) . T_{h2} + x . T_e \qquad (5.3.8)$$

with $$x = (t_h - t_a) . A_aU_a / (M_ac) . \qquad (5.3.9)$$

Whatever the type of control, the time-averaged temperature of the upper part of the auxiliary tank will be approximately $(T_{h1}+T_{h2})/2$.

Heat losses from the auxiliary tank during a 24-hour period will be calculated from its time-averaged temperature. This value is not accurately known, but can be approximated. If a quantity of water m_a at temperature T_{sol} (coming from the solar tank) is drawn off each day through the auxiliary tank, its 24-

hour average temperature can be estimated from

$$T_s = 0.5\ (T_{h1} + T_{h2}) - 0.5\ m_a \cdot (T_{h1} - T_{sol})/M_a \qquad (5.3.10)$$

and the corresponding heat losses will be

$$L_a = N \cdot A_a U_a \cdot t_h \cdot (\ T_s - T_e\ ) \quad . \qquad (5.3.11)$$

By noting

$$L_a' = N \cdot A_a U_a \cdot t_h \cdot [\ (T_{h1}+T_{h2})/2 - T_e\ ] \qquad (5.3.12)$$

heat losses become

$$L_a = L_a' \cdot [\ 1 - \frac{m_a}{2\ M_a} \cdot \frac{T_{h1} - T_{sol}}{(T_{h1}+T_{h2})/2 - T_e}\ ] \quad . \qquad (5.3.13)$$

In all the calculations, it is assumed that if the temperature of the hot water in the tank is too high, the volume of demand will be reduced accordingly (so as to keep a constant energy demand). In practice, this is ensured by the mixing valve included in the distribution circuit.

This assumption, in turn, allows one to derive the following formula for m_a, for a series auxiliary tank:

$$m_a = m \cdot (T_{hmin} - T_m)\ /\ [\ (T_{h1}+T_{h2})/2 - T_m] \qquad (5.3.14)$$

and

$$m_a = m \cdot (T_{hmin} - T_{sol})\ /\ [\ (T_{h1}+T_{h2})/2 - T_{sol}] \qquad (5.3.15)$$

for a series auxiliary tank with a controlled by-pass.

The temperature at the outlet of the solar storage tank T_{sol} is not known a priori and is determined from the solar fraction f:

$$T_{sol} = f \cdot T_{hmin} + (1 - f) \cdot T_m \quad . \qquad (5.3.16)$$

In order to avoid an iterative computation (from a first estimate of the solar fraction without accounting for auxiliary tank losses), a default value 0.5 can be adopted for the solar fraction leading to

$$L_a = L_a' \cdot (\ 1 - 0.25\ m\ /\ M_a) \quad . \qquad (5.3.17)$$

5.3.5 - Effective hot water demand

Due to heat losses from the hot water distribution loop (recirculation loop and auxiliary tank), the actual load is higher than the theoretical load.

It is convenient to define two new parameters:

m_e effective daily hot water demand,
T_{he} effective required temperature of hot water.

The corresponding effective load includes a part of the heat losses defined above (the fraction of those losses which can be met by solar energy)

$$L_1 = N \cdot m_e \cdot c \cdot (T_{he} - T_m) \quad . \qquad (5.3.18)$$

The remaining fraction will be considered as a residual load.

Formulae are given in Table 5.3.2. Distribution losses L_d are integrated in the effective demand, through an increase in the required hot water temperature. The same will occur for recirculation losses but only for the fraction occurring during draw-off L_{cl} (except when the recirculation loop is connected both to the solar tank and to the auxiliary tank with a controlled valve).

Heat losses from the auxiliary tank (series connected) are also included in the effective load, but with a reduced effective demand (the auxiliary tank is generally at a higher temperature than required by the user). On the other hand, in the alternative auxiliary coupling (series-connected auxiliary tank with a controlled three-way valve), heat losses can never be met by solar energy: they are not therefore included in the effective load.

If heat losses from the distribution circuit and auxiliary tank are not precisely known, they can be estimated by a global percentage of the theoretical load (typically 15 to 20%). This percentage will be applied so as to define the effective required temperature, the effective demand remains unchanged.

Table 5.3.2: Distribution and auxiliary heat losses: Effective DHW load. Summary of useful equations.

<u>Theoretical load</u> $L_t = N \cdot m \cdot c \cdot (T_h - T_m)$

<u>Distribution losses</u> $L_d = N \cdot A_d U_d \cdot t_s \cdot (T_h - T_e)$

<u>Recirculation losses</u>

Recirculation loop connected to auxiliary only

$$L_{c1} = N \cdot A_c U_c \cdot t_s \cdot (T_h - T_e)$$
$$L_{c2} = N \cdot A_c U_c \cdot (t_L - t_s) \cdot (T_h - T_e)$$

Recirculation loop connected both to auxiliary and solar tank

$$L_{c1} = N \cdot A_c U_c \cdot t_L \cdot (T_h - T_e) \qquad L_{c2} = 0$$

<u>Minimum required hot water temperature</u>

$$T_{h\,min} = T_h + t_s \cdot (T_h - T_e) \cdot (A_d U_d + A_c U_c / 2) / mc$$

<u>Auxiliary tank losses</u>

$$L_a' = N \cdot A_a U_a \cdot t_h \cdot [(T_{h1} + T_{h2})/2 - T_e]$$

On/off control $L_a = L_a'$

Off-peak energy (series tank) $L_a = L_a' \cdot (1 - 0.25\ m / M_a)$

Off-peak energy (by-pass and controlled three-way valve)
(to be computed after the calculation of solar system performance)

$$L_a = L_a' \cdot \left[1 - \frac{m_a}{2 M_a} * \frac{T_{h1} - T_{sol}}{(T_{h1} + T_{h2})/2 - T_e} \right]$$

$$m_a = m \cdot (T_{hmin} - T_{sol}) / [(T_{h1} + T_{h2})/2 - T_{sol}]$$

$$T_{sol} = f \cdot T_{hmin} + (1 - f) \cdot T_m$$

<u>Effective Demand</u>

Auxiliary tank with by-pass and controlled three-way valve

$$m_e = m \qquad T_{he} = T_m + (L_t + L_d + L_{c1}) / (N . m_e c)$$

Series auxiliary tank

$$m_e = m \cdot (T_{h\,min} - T_m) / [0.5 * (T_{h1} + T_{h2}) - T_m]$$

$$T_{he} = T_m + (L_t + L_d + L_{c1} + L_a) / (N . m_e c)$$

<u>Residual load</u> (cannot be met by solar energy)

Auxiliary tank with by-pass and controlled three-way valve

$$L_r = L_{c2} + L_a$$

Series auxiliary tank $L_r = L_{c2}$

REFERENCES OF CHAPTER 5

/5.1/ ESP. University of Strathclyde, Dpt of Architecture. Glasgow.
/5.2/ BLAST. Lawrence Berkeley Laboratory. Berkeley, California.
/5.3/ CALECO. C.N.R.S., Groupe RAMSES. Orsay, France.
/5.4/ P.BREJON et G.WATREMEZ. CASAMO: Methode de calcul du coefficient B sur micro-ordinateur. XIIIèmes journées d'étude du Comité Français d'Electrothermie, Versailles (1982).
/5.5/ P.CLAUX et al. Methode 5000. PYC Edition, Paris (1982).
/5.6/ ASHRAE. Fundamentals (1985).
/5.7/ W.A.MONSEN et al. Prediction of direct gain solar heating system performance. SOLAR ENERGY, Vol.27, pp.143-147 (1981).
/5.8/ P.ACHARD and R.GICQUEL. European Passive Solar Handbook. CEC DG XII, Brussels, Report EUR 10683 (1986).
/5.9/ C.S.T.B. Règles Th-K 77. Paris (1977)
/5.10/ C.S.T.B. Règles Th-G 77. Paris (1977) (updated in 1982).
/5.11/ C.I.B.S. Building Energy Code: Calculation of energy demands and targets for the design of new buildings and services. London (1981).
/5.12/ P.CLAUX et al. Atlas solaire français: Energie transmise et calcul de l'effet de masque. PYC Edition, Paris (1982).
/5.13/ ERBS et al. Estimation of degree-days. ASHRAE JOURNAL, June 1983.
/5.14/ B.BOURGES. Calcul des degrés-jours. CHAUFFAGE, VENTILATION, CONDITIONNEMENT (AICVF), n.5, Mai 1987 (1987).
/5.15/ Direction de la Météorologie. Normales Climatologiques 1951-1980: Températures. Paris (1983).
/5.16/ WMO. Climatological Normals CLINO 1931-1960. n.117 (1971).
/5.17/ PAGE and LEBENS. Designers'Handbook for U.K. meteo data. (1987).
/5.18/ DUFFIE and BECKMAN. Solar Engineering of Thermal Processes. Wiley, 1980.
/5.19/ J.QUINTON. L'eau chaude sanitaire: Importance énergétique. CHAUFFAGE VENTILATION CONDITIONNEMENT, Octobre 1978.
/5.20/ D.BIENFAIT. Contribution à une meilleure connaissance des consommations d'eau chaude des logements. Cahiers du C.S.T.B., n.205, Cahier 1617 (1979).
/5.21/ W.PALZ et al. European Solar Radiation Atlas. Volume 2: Inclined Planes. TUV Rheinland Verlag, Koln (1984).
/5.22/ G.LEFEBVRE et A.NEVEU. Inertie-Confort-Besoins de Chauffage. BATIMENT-ENERGIE, n.44 (1986).
/5.23/ M.CHATEAUMINOIS et al. Calcul d'installations solaires à eau. PYC-Edition, Paris (1979).
/5.24/ Dansk Standard DS418. Rules for the calculation of heat loss from building. Dansk Ingeniorforening, Copenhagen (1977).
/5.25/ DIN 4701. Regeln fur die Berechnung des Warmebedarfs von Gebauden, Teil 1 und 2. DIN, Benth Verlag, Berlin (1983).
/5.26/ BREDEM-BRE. Domestic Energy Model: Background, philosophy and description. Building Research Establishment, Garston, U.K. (1985).
/5.27/ NBN-B62-301. Isolation Thermique des Batiments. Niveau de l'isolation thermique globale. BIN-IBN, Brussels (1979).
/5.28/ NVN 5125. Energy consumption in Buildings: Heating of Dwellings. Simplified calculation method and target values. NNI, Delft (1984).
/5.29/ ISO. Isolation Thermique: Règles de calcul. Partie 1. ISO/DIS 6946/1.
/5.30/ CEC-DG III. Study for a Eurocode on the rational use of energy in buildings. Doc. 2973/III/86 (1986).

/5.31/ S.Ø.JENSEN (TUD). Personal communication (1987).
/5.32/ W.E.BUCKLES and S.A.KLEIN. SOLAR ENERGY, Vol.25, pp.417 (1979).
/5.33/ J.E.NIELSEN. The influence of the hot water drawoff on the output of solar water heaters. Danish Solar Energy Testing Laboratory (1986).
/5.34/ WUESTLING et al. ASME J. of Sol. Energy Engineering, Vol.107, pp.215 (1985).
/5.35/ J.J.MUTCH. RAND report R1498 (1974) (cited by DUFFIE and BECKMAN, Wiley, 1980).
/5.36/ AICVF. Guide Pratique de l'eau chaude (in press).

CHAPTER 6

ESM1: A CORRELATION FOR COMBINED SPACE HEATING AND DOMESTIC HOT WATER SOLAR SYSTEMS

The calculation method ESM1, which is presented in this chapter, is intended for the performance evaluation of active solar heating systems which provide space heating and, possibly, preheating of domestic hot water.

Beginning with a set of about 30 parameters which describe the system, it is possible to calculate the auxiliary heating requirements and the savings due to solar energy, and in particular the efficiency of the system: that is, the number of kWh or MJ saved per square metre of collector. Given the range of parameters included in the method, sensitivity analyses are possible and permit a true optimisation of the system design. The collector surface is not the only important variable in this respect.

The method is based on two fundamental concepts, described in Chapter 3: a base temperature (below which heat derived from solar energy cannot be used by the emitters) and a threshold radiation (the minimum intensity of solar radiation required to yield energy at a given temperature). As far as climatic data are concerned, the method requires not only an average incident radiation, but a synthesis of the distribution of solar radiation in the form of Cumulative Frequency Curves, or Utilizability Curves, for each month. The derivation of these data has been described in Chapter 4.

The detailed simulation model of solar heating systems, EMGP2 (developed and validated within the framework of the European project OPSYS), has been used as a reference in the present work.

The basic detailed version of ESM1 is presented in this chapter; it is best used with the aid of a microcomputer (how to use the corresponding software is described in Chapter 9). A simplified version which may be used "manually" with slide rule or calculator is presented in Appendix D.

The first section of this chapter contains a detailed description of the systems in question. This should be read carefully in order to determine whether or not a proposed system design is amenable to analysis with ESM1: Certain configurations (or components) are specifically excluded.

The second section illustrates the underlying principles of ESM1, discusses the range of variables which the method takes into account and explains the manner in which the less obvious system parameters are obtained.

The third section is concerned with a detailed examination of the various steps in the calculation, and constitutes a reference to the theoretical principles.

An example is presented in the fourth section. The accuracy of the method is addressed in the last section and results of the comparison with the detailed simulation programme EMGP2 are presented. Limitations of use are summarized.

This chapter has been written so that it may be read independently; some material given elsewhere in the book has therefore been summarized or repeated here.

6.1 - TYPES OF SYSTEM CONSIDERED

ESM1 deals only with a well defined type of solar heating system, which is currently the most common installation. These types of installation are called "active solar heating systems incorporating liquid circulation with short-term energy storage". An installation of this type performs the three essential functions of all solar heating systems: the collection of solar radiation including its conversion into thermal energy and transfer to a useful location; the storage of this energy (sufficient to allow heat collected during the day to be used during the night or possibly the following day); and the distribution of this energy for space heating and hot water.

6.1.1 - Active solar system principles

Active solar systems have been described in detail in Chapter 1. The principal components are:

* The solar collectors, usually of the flat plate type, through which circulates a heat transfer fluid, usually water or water with antifreeze

* The circuit through which energy is transferred to the store, which may incorporate a heat-exchanger (this circuit will be referred to as the "solar loop"). Circulation of the working fluid is by a pump with an appropriate controller.

* A storage tank, usually containing water, providing short-term storage of sensible heat.

* A "distribution circuit" to distribute heat from the store to the emitters - the latter may be one of several common types (radiators, floor-heating panels, fan heaters, etc.). A suitable thermostat and controller are always required.

* A heat-exchanger for the preheating of domestic hot water (in combined systems).

* A control system incorporating separate controls for each of the collection and distribution circuits.

Only essential elements of the system have been listed above; some components, which are necessary for the operation or safety of a system, have been omitted since they do not influence the overall performance.

6.1.2 - Schematic system designs

The possible arrangements of the basic components in active solar systems are described in Chapter 1. Systems which can be assessed with ESM1 are considered as a combination of three subsystems: the solar loop + storage; the space heating distribution loop + auxiliary; and the coupling with domestic hot water. For each subsystem, three options are possible, as described below.

Configurations permitted by ESM1 are constructed by combining three of the options, i.e. one option is selected for each subsystem. Figs 6.1.1 to 6.1.3 present three typical combinations and show all possible subsystems. Other combinations are also permitted (3x3x3 = 27 different schematics). For example, the combination of the solar loop in Fig. 6.1.1 (direct loop) with the heat distribution in Fig.6.1.2 (series-connected auxiliary) and DHW coupling in Fig.6.1.3 (preheating pipe) is a permitted configuration.

6.1.2.1 - Auxiliary heating

Two broad categories of systems may be defined according to the location of the auxiliary heating source.

Supplementary heating can be completely independent of the solar distribution network: this is known as separate or parallel auxiliary (see Fig. 6.1.1). This is the case for electric convector heaters or when there are two completely separate heating networks, one relying on solar energy and the other connected to an auxiliary source. In this case, the operation of the two circuits could be controlled independently, each by its own thermostat.

In contrast, when energy from the solar store and from the auxiliary are distributed by the same network, the auxiliary is said to be in series (Fig. 6.1.2). Here control is more complicated, and a distinction may be made according to whether or not the solar and auxiliary can provide heat simultaneously. If so, the auxiliary is modulated in order to supplement solar heat; if not, the auxiliary operates as "all or nothing", and at any given time, solar energy will make no contribution if the auxiliary is working.

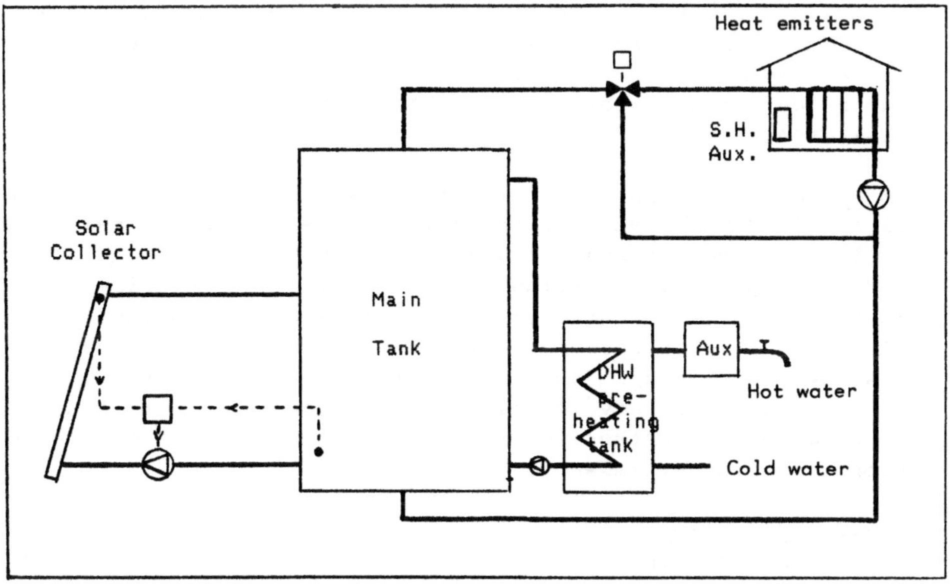

Fig.6.1.1: Typical active solar heating system with parallel auxiliary, direct solar loop, and external D.H.W. pre-heating tank.

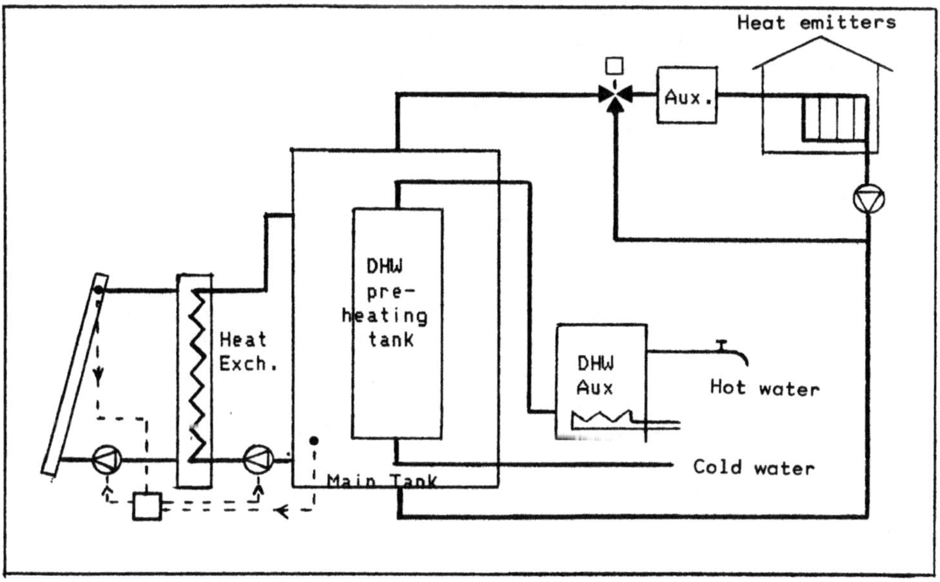

Fig.6.1.2: Typical active solar heating system with series auxiliary, external solar heat-exchanger, and submerged D.H.W. pre-heating tank.

Parallel auxiliary is preferable in terms of energy consumption, but is more expensive; the choice between the two can only be made after optimisation of a particular case. ESM1 handles both configurations and thus permits an assessment of the alternatives.

Warm air heating (in which the auxiliary source is generally positioned "downstream" of the solar heat-exchanger) falls in the first category (parallel). Warm air heating has been introduced as a particular variant, because it can give rise to a rather different schematic (Fig.6.1.3): air circulating in the coil can be either recycled air from the building, fresh air (possibly pre-heated in an air-to-air recuperator), or any mixture of fresh and recycled air. The use of fresh air enables one to obtain a rather low base temperature, and therefore a better collector efficiency.

6.1.2.2 - Solar loop

For the solar loop, three types of sub-systems are allowed. The simplest scheme (fig. 6.1.1) is the direct loop without a heat-exchanger. Since water is used as a heat transfer fluid, this loop requires an automatic drain-down (or drain-back, see Chapter 1) to protect it from freezing. In the second type of subsystem, which is more common in practice, an anti-freeze solution flows in the primary circuit of an external heat-exchanger (plate or tubular heat exchanger) which is inserted between the two parts of the solar loop (Fig. 6.1.2). The third type is the internal heat-exchanger, submerged inside the storage tank (e.g., finned pipe) (Fig. 6.1.3).

6.1.2.3 - Hot water coupling

When domestic hot water is included, even more combinations are possible. Two broad types may be distinguished, depending on whether or not there is a pre-heating tank for the hot water. This tank may be submerged in the solar tank (Fig.6.1.2) or situated externally , with an intermediate heat exchanger in one of the tanks, and connected by forced circulation (Fig. 6.1.1). If there is no pre-heating tank, the DHW supply is preheated in a submerged pipe or heat exchanger inside the main tank (Fig. 6.1.3).

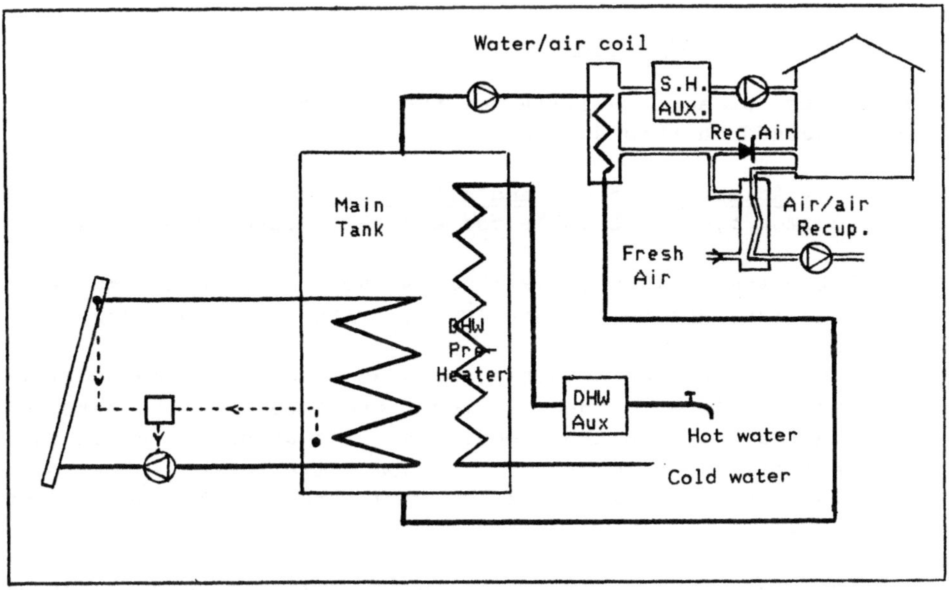

Fig. 6.1.3: Typical active solar heating system incorporating an air distribution loop, an internal solar heat-exchanger, and a D.H.W. pre-heating pipe.

6.1.2.4 - Systems excluded

A whole series of systems is excluded from this analysis: Firstly, those which involve components not described above, for example passive systems (in which collection is an integral part of the building structure and takes place without pumped circulation or control). Certain active systems have components which are not included in this analysis, such as air collectors or heat pumps. System designs which depart greatly from the schematic types described above must be subjected to a detailed analysis according to their particular features and to the characteristics of ESM1. For example, "hybrid" systems which incorporate "active collection" and "passive storage" (into the building structure itself), such as the so-called direct solar under-floor heating (see Chapter 1), are excluded.

It should also be noted that systems which provide domestic hot water only are excluded. The ESM2 method described in Chapter 7 is suitable for systems of this type.

6.1.3 - Which components?

The limits of applicability of ESM1 to each system component are discussed in this subsection.

6.1.3.1 - Solar collectors and solar loop

All common types of collector are taken into account:

- Flat-plate collectors, with or without a selective surface.
- Evacuated tube collectors (ETC), with or without selective surface.
- Concentrating collectors: such as weak (CPC, parabolic), and medium (linear) concentrators, provided that the meteorological data are suitably modified. It should be noted that concentrating collectors are seldom suitable for solar space heating applications, however, the method ESM1 can account for them).

However, solar collectors which may be called "atmospheric" or unglazed, or in which heat transfer with ambient air is dominant are excluded. It should also be noted that these types of components are generally unsuitable for space heating applications, unless a heat pump is incorporated into the system design.

In the solar loop, the working fluid is a liquid: water, water+ antifreeze, or organic. The heat exchanger may be inside or outside the storage tank. The solar loop is regulated by an on-off differential controller.

6.1.3.2 - Storage

The storage is usually a tank of water, designed for the short-term storage of sensible heat. Latent heat stores are not included in the method. The storage capacity should neither be very small ("direct" solar heating is not permitted), nor very large (long-term or seasonal storage is dealt with by other calculation methods). A particular parameter, the characteristic response time of the collector / storage subsystem, should be within certain well-defined limits (see section 6.3).

6.1.3.3 - Distribution network

The distribution network is not subject to particular restrictions, provided that the design is of a type similar to those defined above as "typical schematics" (see section 6.1.2). Underfloorheating, wet radiators, and fan convectors all qualify. It should simply be noted that the lower the temperature at which the distribution network operates, the more efficient the whole system will be, especially with flat-plate solar collectors.

6.1.3.4 - Conclusion

ESM1 is applicable to certain well-defined systems and components. Outside these limits, its validity cannot be guaranteed. That is not to say that the method may never be used on other systems or components, but that each application must be examined carefully.

Other limitations involving several combinations of some parameters are indicated in section 6.5.

6.2 - CALCULATION PRINCIPLES AND BASIC PARAMETERS

This section contains a brief description of the underlying principles of ESM1. The data required by the method are discussed: parameters concerned with the building and its heating system; those concerned with domestic hot water; those concerned with the heating system itself; and, finally, the climatological data.

6.2.1 - Basic principles

The analysis of a solar collector involves the important concept of a threshold irradiance. A solar collector gathers energy according to the incident radiation, but it also loses heat in proportion to the temperature difference between the working fluid and its surroundings. When the incident radiation is too weak (below a certain threshold for a particular fluid temperature) the losses become greater than the gains, and the collector will transfer heat to the atmosphere unless prevented from doing so by the controller.

The energy collected can only be usefully employed for space heating when its temperature exceeds some base temperature. In most cases, this corresponds to the average return fluid temperature in the heating network. This concept and that of threshold irradiance are developed in Chapter 3.

As a result of these two operational constraints, only part of the solar energy incident during a given period can be utilised; this is called the available or collectable energy at the base temperature. When expresssed in a dimensionless form as a fraction of the incident energy (e.g. over a period of a month) it is known as 'Utilizability'. This quantity can be calculated from the frequency distribution of incident solar radiation during the period in question.

Other factors also influence the operation of a solar heating system. For example, the temperature of the storage rises during the day, thus reducing the efficiency of the collectors. As a general rule, it can be said that the larger the solar inputs, the less efficiently are they absorbed by the collectors: their rate of absorption is linked to the capacity of the storage and the size of the heating demand. Other influences include heat loss from the storage or piping, losses due to controller malfunction, and additional optical losses at low angles of solar incidence.

The constraints and influences mentioned above determine the essential steps in a calculation with ESM1. First, the base temperature is estimated according to the distribution network, and the solar energy available at this temperature is calculated. This first estimate is then refined by an iterative process in which the storage capacity and heat demands are introduced. This iterative method has been verified by a range of detailed simulations using the computer program EMGP2 for different system types and climatic conditions.

Other corrections are carried out for secondary sources of heat loss; these may be derived analytically or empirically.

The calculations are made for each month, and a complete evaluation over the heating season allows a conclusion to be drawn regarding the annual performance.

A further stage is necessary in the case of system designs which include the preheating of domestic hot water. Initially the procedure is as above, assuming that the entire load is used for space heating. Domestic hot water is calculated separately by assuming that all the requirements are for hot water. This time, the procedure is analytical: the algorithm is based on the observation that in the summer, energy is used only for hot water, and that the daily demand is much less than the volume of the store. Consequently, the temperature of the storage changes slowly over a given period and approaches an equilibrium value which is a balance between the solar input, the amount of energy used, and the losses. Assuming that every day is the same, the equilibrium state of the system can be calculated and the savings due to the solar heating system can be estimated. Comparisons with the results of detailed simulations have shown that the procedure is satisfactory and can be extrapolated to the winter when space heating requirements are dominant.

The overall performance of the solar heating system is calculated by combining the space heating and hot water figures according to their respective demands.

The various stages in the calculation summarised above are fully described in section 6.3. More detailed information concerning the development of this method may also be found in section 6.5.

6.2.2 - System parameters: Building and heating network.

6.2.2.1 - Space heating load

Heating data are largely concerned with the space heating load. These data should include a deduction for "free" occupancy gains and passive solar contributions. The net heating demand, calculated month-by-month, is a fundamental parameter, since it is used simultaneously to determine the base temperature, the efficiency with which solar energy is collected, and the amount of auxiliary energy required. The heating load can be estimated in the usual way by means of degree-days; in this case, the necessary data are:

(GV) overall rate of heat loss from the structure (W/K), calculated by standard methods (see also section 5.1);

T_c average set-point temperature of the heating system (°C). For an "on/off" controlled auxiliary, this is the half-sum of "auxiliary-on" and "auxiliary-off" set-points (see 6.2.2.2); and

T_{nh} no-heating temperature (°C).

The no-heating temperature is used as a base temperature for the calculation of degree-days $DD(T_{nh})$; it is related to the average rate of occupancy gains, P_{fg} (in W), by the relationship

$$T_{nh} = T_c - P_{fg} / (GV) \quad . \qquad (6.2.1)$$

For each month, the space heating load is computed by the following equation

$$L_0 = t_h \cdot (GV) \cdot DD(T_{nh}) \quad . \qquad (6.2.2)$$

t_h is the length of the daily period (24 hours or 86400 s, according to the units used).

The heating demand can also be estimated by other techniques, particularly if passive solar contributions are considered. In the latter case, specific methods are particularly recommended (see section 5.1.3 for a quick review). Apart from the actual heating demand L_0, only the variable T_c needs to be retained for later stages of the ESM1 calculation.

Detailed information on the space heating load computation is given in Chapter 5: how to estimate the building heat loss coefficient (section 5.1.1), gross space heating load (section 5.1.2), and passive solar gains (section 5.1.3). A simple algorithm for the calculation of monthly degree-days is described and data sources are indicated (section 5.2).

6.2.2.2 - Heating network

The distribution network is characterized by two essential parameters (see also section 2.3):

e_2 effectiveness of heat emitters; and

$\dot{m}_2c$ capacitance rate of the working fluid (W/K). (product of the mass flow rate $\dot{m}_2$ by heat capacity c)

Often, the available data are the heat output under specific conditions (or nominal power), P_n (W), together with the flow and return temperatures (T_f and T_r). These two sets of parameters may be related according to:

$$P_n = \dot{m}_2c \;.\; e_2 \, (T_f - T_c) \tag{6.2.3}$$

and

$$P_n = \dot{m}_2c \;.\; (T_f - T_r) \quad . \tag{6.2.4}$$

This is equivalent to

$$e_2 = (T_f - T_r) \,/\, (T_f - T_c) \tag{6.2.5}$$

$$\dot{m}_2c = P_n \,/\, (T_f - T_r) \quad . \tag{6.2.6}$$

Several types of distribution network are possible, as stated in section 6.1: series auxiliary (with a single distribution network for solar energy and auxiliary energy); parallel auxiliary (with two separate distribution networks); and air distribution loop.

For the air distribution loop, $\dot{m}_2c$ is the water capacitance rate of the water/air fan coil unit and the effectiveness e_2 is estimated with respect to $\dot{m}_2c$ (and not with respect to the smallest capacitance rate). Furthermore, a few additional parameters are required:

q_r recycled air flow-rate (from the building) (m3/h); and

T_{max} maximal air blowing temperature (for safety reasons, the maximum temperature of the air blown into the building is limited).

If the air distribution loop includes fresh air from outdoors, the calculation must account for:

q_f fresh air flow-rate (m3/h); and

e_r air recuperator (heat-exchanger) effectiveness (if such a component is included in the air distribution loop; if not $e_r = 0$).

Of course, the fresh air flow-rate (after correction for the possible heat recuperator) must be integrated in the space heating load.

As far as the heat distribution control strategy is concerned, there is a choice between two possibilities:

- Variable (or modulated) operation of the auxiliary. This refers to a mode of control which allows the solar store to supply heat to the distribution network as long as its temperature is high enough (even if the total solar contribution is insufficient to meet all the demand), with the auxiliary supplying the remainder. Furthermore, the control is assumed to be perfect: the whole system (solar + auxiliary) provides exactly the amount of energy needed at any time (not more, not less).

- 'All-or-nothing' (or on/off) control of the distribution loop. When space heating is required, priority is given to solar heat, which is supplied at the maximal possible rate; if solar energy is insufficient, the auxiliary boiler is then requested to supply energy (also at its maximum power). This can be achieved with a double differential thermostat (with a temperature probe in the living space) controlling the distribution pumps and the auxiliary boiler. Two set-points (on and off) are fixed for the solar heat distribution, and two other set-points for the operation of the auxiliary. The control works in the following manner: if the living space temperature drops below the "solar-on" set-point, distribution of heat starts from the solar tank. If the available amount of solar energy is high enough, the temperature increases in the building and distribution stops as soon as the "solar-off" set-point is reached. If the solar heat is insufficient, the living space temperature decreases and the auxiliary energy source is turned on automatically when the "auxiliary-on" set-point is reached. It is stopped when the building temperature is equal to the "auxiliary-off" set-point.

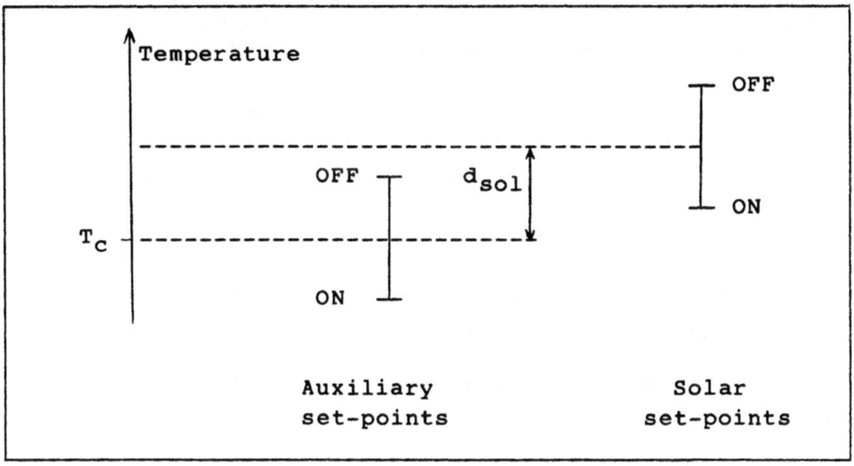

Fig. 6.2.1: Relative temperature levels of the control differentials in the case of an "all-or-nothing" control of heat distribution.

The set-points must be carefully adjusted (Fig. 6.2.1): the "off" set-point is of course always higher than the corresponding "on" set-point; furthermore, the "auxiliary-on" set-point must be lower than the "solar-on" (and the same for "auxiliary-off" with respect to "solar-off"). Otherwise, the system cannot work properly; for example, incorrect adjustment can lead to excessive (or sole) use of the auxiliary.

The following parameter is required by ESM1 when this type of control is incorporated:

d_{sol} differential (in K) between the average solar set-point (half sum of "solar-on" and "solar-off") and the average auxiliary set-point. However, it should be borne in mind that this average auxiliary set-point (half-sum of "auxiliary-on" and "auxiliary-off") is merely the value used for the space heating load estimate (T_c) in this case.

Then, ESM1 allows the following options for the distribution network and its control:

- Type 1: Series auxiliary, modulated control.
- Type 2: Parallel auxiliary, modulated control.
- Type 3: Series auxiliary, all-or-nothing control.
- Type 4: Parallel auxiliary, all-or-nothing control.
- Type 5: Air distribution loop, modulated control.

The base temperature is calculated differently in each of these cases, and therefore it is necessary to identify the correct type of auxiliary heat source and controller in an actual system design.

6.2.3 - System parameters: Domestic Hot Water (see also section 5.3).

The following parameters are used to calculate the hot water demand L_1:

m average daily draw-off (litres/day);

T_m temperature of cold mains water (°C); and

T_h required temperature of hot water (°C).

N number of days in month

c water heat capacity ($kJ.kg^{-1}.K^{-1}$)

$$L_1 = N \cdot m \cdot c \cdot (T_h - T_m) \quad . \qquad (6.2.7)$$

If a part of the heat losses of the DHW auxiliary source can be met by solar energy (i.e. when solar energy is sufficient), m and T_h will be modified accordingly, in order to account for these losses (now included in the D.H.W. load). An effective demand and an effective required temperature of hot water will be defined (see section 5.3.5) (the procedure is the same as for solar hot water only systems).

The capacity of the (optional) pre-heating tank (excluding the auxiliary region) and its heat losses are added to the capacity and losses of the main storage tank (see below).

An additional parameter is required to characterize the quality of heat transfer between the main tank and the DHW production:

e_3 effectiveness of the DHW production

If coupling is done by a specific DHW preheating tank (either external or submerged), this effectiveness is taken equal to unity. If coupling is done by a submerged heat exchanger or a simple pipe, e_3 can be calculated from the average draw-off flow rate $\dot{m}_3 = m / t_s$ (where t_s is the duration of daily draw-off) and the heat transfer rate for the heat exchanger $A_{dh}U_{dh}$ (see section 2.3.4)

$$e_3 = 1 - EXP (- A_{dh}U_{dh} / \dot{m}_3 c) \quad . \quad (6.2.8)$$

As a general rule, $200 < U_{dh} < 400$ W/m2/K. Average draw-off flow rates are between 200 and 600 litres/hour

6.2.4 - <u>System parameters: solar heating system.</u>

The part of a heating system concerned with solar energy comprises the collectors, the connecting plumbing, the optional heat exchanger, and the storage tank.

6.2.4.1 - <u>The collectors</u> (see also section 2.1)

The two basic characteristics of a solar collector are

η_0 optical efficiency for normal incidence (effective transmission - absorption, accounting for the heat transfer factor); and

U_L heat loss coefficient (W/m2/K).

These are usually measured simultaneously according to national standards in approved testing centres. Values are provided by solar collector manufacturers.

The heat loss coefficient may be expressed in terms of the operating temperature of the collector, using the two coefficients U_1 (a constant) and U_2 (a proportional term)

$$U_L = U_1 + U_2 \cdot (T_b - T_a) \quad . \quad (6.2.9)$$

It can be seen from this formula that the higher the fluid temperature, the higher the heat loss coefficient (convective and radiative losses increase faster than linearly with the absorber plate temperature).

Other essential parameters are:

A net surface (or aperture) area of the collector (m2). For flat-plate collectors, this surface area corresponds to the area of the transparent cover. For tubular collectors, it corresponds to the absorber area.

$\dot{m}_0 c_0$ capacitance rate of the fluid passing through the collectors (product of the mass flow-rate and the heat capacity). These values should allow for the presence of any antifreeze (see heat capacities of antifreeze mixtures in Appendix C).

In some countries (e.g., France, USA), values of collector characteristics provided by testing laboratories are related to the gross area. In such a case, these values should be corrected with respect to the net area.

The collectors are also defined by an azimuthal orientation (from south), α ,and an inclination (from the horizontal), β. These parameters are used in the calculation of radiation data.

6.2.4.2 - Optional solar heat-exchanger (see also sect.2.2.2)

The capacitance rate in the collectors (now referred to as "primary" capacitance rate), as well as two other parameters are required:

e_1 heat-exchanger effectiveness; and

$\dot{m}_1 c$ secondary capacitance rate (storage side, between heat-exchanger and tank).

If the heat-exchanger is defined by the quantity of heat P transferred for primary and secondary inlet temperatures T_0 and T_1 respectively,then e_1 can be obtained from the equation

$$P = C_{min} \cdot e_1 \cdot (T_0 - T_1) \qquad (6.2.10)$$

where C_{min} is the smaller of the two values $\dot{m}_0 c_0$ and $\dot{m}_1 c$. In the case of an internal heat-exchanger, the secondary capacity rate is considered to be infinite.

If the heat-exchanger is defined by its overall heat transfer coefficient $A_e U_e$, its effectiveness may be computed from the usual heat exchanger formulas (see section 2.2.2).

When there is no heat-exchanger (direct loop), the following values should be used: $e_1 = 1$ and $\dot{m}_1 c = \dot{m}_0 c_0$.

Other parameters used in the solar loop are:

$A_{in}U_{in}$ total rate of heat loss from external pipes on the inlet side of the collectors (see section 2.2.1).

$A_{ou}U_{ou}$ as above, for the outlet side of the collectors. These two coefficients are in units of W/K. It is advisable to include a further factor for other heat loss paths such as pumps and valves which can easily double the losses from pipes.

d_{on} temperature differential at which the solar loop pump is switched on (typically between 5 and 10K).

d_{off} temperature differential at which collection stops (smaller than the starting differential).

6.2.4.3 - <u>The storage tank</u> (see also section 2.2.3)
The storage tank is defined by

M mass of water in kg (including the domestic hot water pre-heating tank or the part of the tank not affected by the auxiliary heating source).

A_sU_s rate of heat loss (W/K), including losses from the D.H.W. pre-heating tank.

A_pU_p rate of heat loss (W/K) of indoor connecting pipes.

T_e local air temperature surrounding the tank (°C).

6.2.5 - <u>Climatic data</u>

Several climatological parameters are required to use the calculation method. The data for a representative sample of European locations is included in Appendix B.

As far as ambient temperature is concerned, the following parameters are required:

T_a monthly average ambient air temperature (°C).

$DD(T_{nh})$ monthly degree-days for a reference temperature T_{nh}.

Data sources and computation formulae are indicated in section 5.2.

As far as solar radiation is concerned, monthly Cumulative Frequency Curves of solar irradiance (CFC) in the plane of the collectors are the basic input.

These curves make it possible to compute the utilizability $\emptyset$ and collector operating time t_{op}, in terms of the threshold irradiance, I_c. The corresponding definitions and formulae are given in section 4.2.

The CFC's can also be generated for each month with sufficient accuracy from three parameters through the dimensionless quantity

$$v = H_t \;/\; (I_{max} \cdot t_d) \qquad (6.2.11)$$

defined from:

t_d day length (time between sunrise and sunset) (hours or seconds/day);

H_t average total monthly radiation in the plane of the collectors (MJ/m2/day or kWh/m2/day); and

I_{max} monthly maximum solar irradiance in the plane of the collectors (W/m2) (can be determined accurately from atmospheric conditions).

The corresponding charts and data sources are found in section 4.3. A last parameter, related to the solar radiation climate, is required:

$\bar{K}$ monthly average incidence angle modifier

$\bar{K}$ is a correction to be applied to the optical performance of the collectors to allow for the variation in transmission with angle of incidence (monthly average incidence angle modifier). $\bar{K}$ can be obtained from several sources (see section 4.4) or computed. Values obtained for usual glazings can be used as an acceptable approximation for solar collectors. At a given inclination, this factor is constant over fairly large areas, and geographical extrapolation is valid.

There are many possible sources of climatic data, but it is important to verify their consistency before use.

6.3 - DETAILED DESCRIPTION OF ESM1

The different steps in the calculation are discussed below, and are summarised in Table 6.3.1. The steps are grouped in four sections: space heating only, hot water only, monthly evaluations, and annual evaluation.

6.3.1 - Steps for space heating only

Space heating load computations are not specific to ESM1 and have not been included in the present steps: the load is assumed to be an input parameter.

Step 1

The base temperature T_b is calculated in the first step. It is equivalent to the average return temperature of the distribution network (as "observed" by the storage tank, i.e. when the return flow from the emitters goes to the storage). In the "series auxiliary - variable control" case, this temperature is known (that is, the average return temperature), but in all other cases, it depends on the total solar energy distributed by the network, and this quantity can only be known at the end of the monthly calculation. Therefore, an iterative calculation is needed. The first iteration takes the set-point temperature as the base temperature (as if the solar contribution were zero).

This step uses the monthly operating time of the heating system t_e. Where no information is available, it is necessary to take the time corresponding to the number of days on which there was a demand for space heating, or, failing this, simply the number of days in the month, N.

In this first step, the effective space heating load (influencing the solar system) is computed. For an air distribution loop, the effective space heating load which can be covered by solar energy is limited by the maximal air blowing temperature permitted.

Step 2

The collector heat loss coefficient is calculated by assuming that the collectors operate at an average temperature equal to the base temperature T_b.

Step 3

A provisional value for the collector flow-rate correction

factor is calculated. This will be modified after correction for the losses from the plumbing. This factor is not an indication of the collector output; it merely shows the loss of output due to the fact that the working fluid does not extract all the available heat.

Step 4

This step corrects the solar collector data (namely the optical efficiency η_0 and the heat loss coefficient U_L) to allow for the losses from external piping. A correction factor $\bar{K}$ (monthly incidence angle modifier) is introduced to account for the variation in absorption and transmission of the cover with increasing angles of incidence of the solar beam (see section 6.2.5).

Step 5

This step evaluates the efficiency of the complete solar loop F_r', taking the optional heat-exchanger into account. Where there is no heat-exchanger, $F_r' = F_r^*$ (see section 2.2).

Step 6

The threshold irradiance, a function of the base temperature and the properties of the collectors, will be used to calculate the amount of solar energy available at temperature T_b. It is then reduced according to the maximum radiation I_{max}.

Step 7

The utilizability and the operating time of the collectors can be calculated in several ways. Complete formulas are given in Chapter 4.

Step 8

The correction for stratification in the storage applies only if the solar loop heat exchanger is situated outside the storage tank. With an internal heat exchanger, the tank can be considered fully mixed ($K_s = 1$).

Step 9

Supplementary losses due to control differentials are estimated differently, according to whether or not thermal cycling takes place; that is if (see section 2.2.3):

$$d_{on} > d_{off} \; / \; [\; 1 - \exp\; (\; - \; AU_L/\dot{m}_0 c_0\;)\;]\; .$$

Step 10

The ratio Y (available energy / demand) is the relationship between the total energy per month available at temperature T_b

(reduced by the differential losses) and the total heating load L.

Steps 11 and 12

The storage correction is the empirical part of the method. This allows a provisional estimate to be made of the fractional solar contribution f from the ratio Y.

The term t_r is the response time of a reference collector/storage system for which $C_s = 1$:

$$t_r = 75850 \text{ sec} \quad \text{(approximately 21.07 hours).}$$

N.B.: The formula in step 12 is valid only if C_s is within the range $0.25 < C_s < 5$.

Step 13

This step is part of the iteration required to calculate the base temperature and the effective space heating load for types 2 to 5 (auxiliary coupling and control). The value of the solar contribution calculated in the previous step is used. Steps 2 to 12 must now be repeated with these new values of T_b and L_0'.

Step 14

The correction for losses from the storage is an analytical expression, modified by an empirical formula based on a reference heat loss coefficient of 10 W/K.

Assuming the solar loop works about 8 hours per day (one third of the time), heat losses from internal plumbing (inside the building) are included in the solar storage heat losses with a weighting coefficient of 1/3.

Step 15

This final step gives the monthly solar contribution for space heating only, f_{sh} (with respect to the total actual load L_0 and not only the "effective" load L_0').

6.3.2 - <u>Steps for hot water only</u>

The calculation for domestic hot water does not contain any empirical corrections; it is an analytical procedure based on a periodic 24-hour cycle. This steady state, characterised by the temperature at the beginning and end of the collection period, is calculated by two iterations followed by a linear interpolation.

Step 1

T_h' is the temperature of the storage which would allow the entire hot water demand to be met by solar energy. With a perfect heat exchanger ($e_3 = 1$), this equals the demand temperature of hot water T_h.

The initial temperature at the beginning of the collection period (first iteration) (T_b) is taken as the average of T_h' and the temperature of the cold water in the network. The notation T_b has been retained since, as in space heating, it is a base temperature.

Step 2

This step calculates the characteristics of the solar loop, using the same formulae as for space heating (steps 2 to 5 of "Space heating only").

Step 3

The base temperature is raised to a maximum, which must not exceed the stagnation temperature of the collectors, nor 95°C (due to the risk of boiling in the storage tank). On the other hand, it must not be lower than the average ambient temperature, otherwise the normal utilizability formulae are no longer valid.

Steps 4, 5 and 6

The calculation of the reduced threshold, utilizability, and the correction for stratification are identical to steps 6, 7 and 8 in the space-heating-only calculation.

Step 7

The storage correction is based on Phillips' theory (see Chapter 3), and is particularly useful in the absence of flow at the same time as collection.

Step 8

The temperature at the end of the collection period results from an energy balance between the solar input and the losses from the tank. It is not allowed to exceed 95°C.

Step 9

The demand (hot water and heating) is assumed to be entirely for hot water, which leads to a hypothetical daily draw-off m'. This quantity cannot exceed the total capacity of the storage tank. It is assumed that the amount used is drawn off at the end of the collection period.

Step 10

The temperature after draw-off is estimated from the quantity of energy used.

Step 11

At the end of the 24-hour period the night-time losses from the storage are calculated. The final temperature at the end of this period is T_2.

Step 12

The second iteration uses either T_h' (if $T_2 > T_b$) or T_m (if $T_2 < T_b$) as an initial temperature. If the initial and final values are almost the same, the iteration stops and the value of the preliminary solar contribution obtained in step 9 is taken as the final value for hot water only.

Step 13

In the second iteration steps 2 to 11 are repeated.

Step 14

The temperature at the end of the collection period (T_1'') for the fixed 24-hour cycle is calculated by linear interpolation from the results of the two iterations.

Step 15

The final solar contribution for hot water only is calculated from the temperature T_1'', using the same procedure as in step 9.

6.3.3 - Monthly and annual evaluation

The total monthly solar contribution is calculated from the individual contributions to hot water and space heating, weighted according to the sizes of the two demands.

The annual contribution is calculated in a similar way: it is the average of the total monthly solar contributions, weighted according to the total monthly loads.

Table 6.3.1: ESM1 standard version: Steps in the monthly calculation (S.H.only).

1 - Base Temperature and modified load

Auxiliary	Base Temperature
1 Series; variable control	$T_b = T_c + L_0 \cdot (1-e_2) / (t_e.\dot{m}_2c_2.e_2)$ $L_0' = L_0$
2 Parallel; variable control	$T_b = T_c$ $\quad L_0' = L_0$
3 Series; all-or-nothing	$T_b = T_c$ $\quad L_0' = L_0$
4 Parallel; all-or-nothing	$T_b = T_c$ $\quad L_0' = L_0$
5 Air distribution	$T_{air} = [\ q_f.(1-e_r).T_a + (q_r+e_r.q_f).T_c\ ] / (q_f+q_r)$ $T_b = T_{air}$ $L_0' = MIN\ [\ L_0\ ;\ c_{air}.(q_f+q_r).t_e.(T_{max}-T_{air})\ ]$

2 - Heat loss coefficient for collectors alone

$$U_L = U_1 + U_2 \cdot (\ T_b - T_a\ )$$

3 - Collector heat removal factor (provisional value)

$$F_r = \dot{m}_0c_0 \cdot [\ 1 - EXP\ (\ -A \cdot U_L / \dot{m}_0c_0\ )\ ] / (A.U_L)$$

4 - Correction for piping losses

$$U_L^* = U_L + (\ A_{in}U_{in} + A_{ou}U_{ou}\ ) / A$$

$$F_r^* = \dot{m}_0c_0 \cdot [\ 1 - EXP\ (\ -A \cdot U_L^* / \dot{m}_0c_0\ )\ ] / (A.U_L^*)$$

$$\eta_0^* = \eta_0 \cdot \bar{K} \cdot F_r \cdot EXP\ (\ -A_{ou}U_{ou} / \dot{m}_0c_0\ ) / F_r^*$$

5 - Solar loop correction factor

$$F_r' = F_r^* / [\ 1 + A \cdot F_r^* \cdot U_L^* \cdot (\dot{m}_0c_0/e_1/C_{min} - 1) / \dot{m}_0c_0\ ]$$

$$C_{min} = MIN\ (\ \dot{m}_0c_0\ ;\ \dot{m}_1c_1\ )$$

6 - Reduced threshold irradiance

$$y_c = (\ U_L^* / \eta_0^*\ ) \cdot (\ T_b - T_a\ ) / I_{max}$$

7 - Calculation of utilisability Ø and collector operating time t_{op}

8 - Correction for stratification

Heat-exchanger outside storage (or absent)

$$a = 1 - A \cdot F_r' \cdot U_L^* / \dot{m}_1c_1 \qquad K_s = \frac{Ln\ (\ 1 / a\ )}{(\ 1 - a\ )}$$

ESM1 SPACE HEATING ONLY

K_s = MIN [K_s ; 1 + (K_s - 1) . $\dot{m}_1$. t_{op} / M]

Heat-exchanger inside storage (submerged) $K_s = 1$

9 - Losses due to control differentials

$$d' = \text{MAX} [\ d_{on}\ ;\ d_{off} / (1 - \text{EXP} (- A.U_L^*/\dot{m}_0 c_0))\ ]$$

$$Q_d = - 0.25\ A\ .\ (U_L^*/\eta_0^*)^2\ .\ N\ .\ t_d\ .\ d_{on}\ .\ d'\ /\ I_{max}$$

10 - Ratio Input/Demand

$$Y = (A\ .\ F_r'\ .\ K_s\ .\ \eta_0^*\ .\ N\ .\ \emptyset\ .\ H_t + Q_d) / (L_1 + L_0')$$

11 - Storage Correction factors

$$C_s = M\ .\ c\ / (t_r\ .\ A\ .\ F_r'\ .\ U_L^*)$$

$$C_1 = 1.0205\ .\ [1 - 0.47\ .\ \text{EXP} (- 1.785\ .\ C_s)]$$

$$C_2 = 0.205 + 0.1515\ .\ \text{Ln} (C_s)$$

12 - Solar fraction (provisional value)

$$f = 1 - \text{EXP} [- C_1\ .\ Y\ .\ (1 + C_2\ .\ Y)]$$

13 - Iteration for the base temperature (type of auxiliary: 2 to 5 only)

Type	
3-4	$L_0' = L_0\ .\ DD(T_{nh} + f.d_{sol}) / DD(T_{nh})$
2-3-4	$T_b = T_c + f\ .\ L_0'\ .\ (1-e_2) / (t_e.\dot{m}_2 c_2.e_2) + f\ .\ d_{sol}$
5	$T_b = T_{air} + f\ .\ L_0'\ .\ (1-e_2) / (t_e.\dot{m}_2 c_2.e_2)$

Repeat steps 2 to 12

14 - Correction for storage losses

$$T_s = T_b + f\ .\ L_0' / (\dot{m}_2 c_2 t_e)$$

$$A_s U_s^* = A_s U_s + A_p U_p /3$$

$$Y_1 = \text{MAX} [0 ; 10\ .\ t_e\ .\ (T_s - T_e) / (f.L_0')]$$
$$Y_1 = \text{MIN} (1 ; 0.33\ Y_1^{1/2})$$

$$Y_1 = \begin{cases} Y_1\ .\ A_s U_s^* / 10 & \text{if } A_s U_s^* < 10 \\ Y_1\ .\ (A_s U_s^* / 10)\ .\ \text{EXP} [- Y_1\ .\ \text{Ln} (A_s U_s^*/10)] & \text{if } A_s U_s^* > 10 \end{cases}$$

15 - Monthly Solar Fraction (Space Heating; final value)

$$f' = f\ .\ (1 - Y_1)$$

$$f_{sh} = \begin{cases} 1 - (1 - f')\ .\ (L_0' + L_1) / (L_0 + L_1) & \text{if } L_0' > L_0 \\ f'\ .\ (L_0' + L_1) / (L_0 + L_1) & \text{if } L_0' < L_0 \end{cases}$$

ESM1 SPACE HEATING ONLY

Table 6.3.2: ESM1 standard version: Steps in the monthly calculation (H.W.only).

1 - Base Temperature (First iteration)

$$T_h' = T_m + (T_h - T_m) / e_3$$

$$T_b = (T_m + T_h') / 2$$

2 - Collector heat loss coefficient and effect of piping losses

$$U_L = U_1 + U_2 . (T_b - T_a)$$

$$F_r = \dot{m}_0 c_0 . [1 - EXP (- A . U_L / \dot{m}_0 c_0)] / (A.U_L)$$

$$U_L^* = U_L + (A_{in} U_{in} + A_{ou} U_{ou}) / A$$

$$F_r^* = \dot{m}_0 c_0 . [1 - EXP (- A . U_L^* / \dot{m}_0 c_0)] / (A.U_L^*)$$

$$\eta_0^* = \eta_0 . \bar{K} . F_r . EXP (- A_{ou} U_{ou} / \dot{m}_0 c_0) / F_r^*$$

$$C_{min} = MIN (\dot{m}_0 c_0 ; \dot{m}_1 c_1)$$

$$F_r' = F_r^* / [1 + A . F_r^* . U_L^* . (\dot{m}_0 c_0 / e_1 / C_{min} - 1) / \dot{m}_0 c_0]$$

3 - Limitation of base temperature

$$T_b = MIN (T_b ; 95°C ; T_a - 2 + \eta_0^* . I_{max} / U_L^*)$$

$$T_b = MAX (T_a ; T_b)$$

4 - Reduced threshold irradiance

$$y_c = (U_L^* / \eta_0^*) . (T_b - T_a) / I_{max}$$

5 - Calculation of utilisability Ø and collector operating time t_{op}

6 - Correction for stratification

Heat-exchanger outside storage (or absent)

$$a = 1 - A . F_r' . U_L^* / \dot{m}_1 c_1 \qquad K_s = \frac{Ln (1 / a)}{(1 - a)}$$

$$K_s = MIN [K_s ; 1 + (K_s - 1) . \dot{m}_1 . t_{op} / M]$$

Heat-exchanger inside storage (submerged) $\qquad K_s = 1$

7 - Storage Correction

$$S = M . c / (A . F_r' . K_s . U_L^* + A_s U_s^*) / t_{op}$$

$$G_{cst} = S . [1 - EXP (- 1 / S)]$$

8 - Storage temperature at end of collection period

$$T_1 = T_b + G_{cst}.[A.F_r'.K_s.\eta_0^*.Ø.H_t - A_s U_s^*.t_{op}.(T_b - T_e)] / (M.c)$$

$$T_1 = MIN (T_1 ; 95°C)$$

ESM1 HOT WATER ONLY

9 - Fraction of demand drawn from store

$m' = MIN [M ; m + L_0 / (N . c . (T_h - T_m))]$

$f = MIN [1 ; m' . c . e_3 . (T_1 - T_m) . N / L]$

$f = MAX (f ; 0)$

10 - Storage temperature after draw-off

$T_2 = T_1 - f . L / (N . M . c)$

11 - Storage temperature at end of daily cycle

$t_s = M . c / A_s U_s^*$

$T_2 = T_e + (T_2 - T_e) . EXP [- (t_h - t_{op}) / t_s]$

12 - Check and base temperature for second iteration

$T_0' = T_b$ $T_1' = T_1$ $T_2' = T_2$

If $T_2 > T_b + 0,5K$ $T_b = T_h'$ Go to step 13

If $T_2 < T_b - 0,5K$ $T_b = T_m$ Go to step 13

If $T_b - 0,5K < T_2 < T_b + 0,5K$ end of hot water calculation
$f_{dhw} = f$ Go to step 16

13 - Second iteration

Repeat steps 2 to 11

14 - Linear Interpolation

$T_1'' = T_1 + (T_1' - T_1) . (T_2 - T_b) / [(T_0' - T_b) - (T_2' - T_2)]$

15 - Monthly Solar Fraction (Hot water; final value)

$f = MIN [1 ; m' . c . e_3 . (T_1'' - T_m) . N / L]$

$f_{dhw} = MAX (f ; 0)$

ESM1 HOT WATER ONLY

16 - Monthly Solar Fraction (Space Heating + Hot Water)

$f = (f_{dhw} . L_1 + f_{sh} . L_0) / L$

ESM1 MONTHLY EVALUATION

6.4 - A WORKED EXAMPLE OF ESM1 CALCULATION

An example of system calculation is presented in this section. The base case is first analysed, with a complete review of the required input parameters. A few variants are then computed and sensitivity analyses are carried out.

Results of calculations done with a simplified "manual" version of the method ESM1 are also available; they are presented in Appendix D.

6.4.1 - Base case

Consider a system with a layout similar to the one in Fig. 6.1.2, for the space heating and combined hot water of a single -family house. It incorporates a solar loop with an external heat-exchanger. Solar heat is distributed to the dwelling by a single under-floor heating network (series-connected auxiliary). Domestic hot water is pre-heated in a submerged tank, with back-up tank. The system is located near Carpentras, in the South of France (latitude 44°N).

6.4.1.1 - Space heating and hot water loads

We assume a heat loss coefficient of the house GV = 250 W/K. The rate of free heat gains is assumed to be 375 W. The internal set-point temperature is 19°C. The no-heating temperature is therefore (from Eq. 6.2.1)

$$T_{nh} = 19 - 375 / 250 = 17.5 \text{ °C}.$$

The space heating load is calculated from degree-days at base 17.5°C. Degree-days will be estimated automatically by the programme ESM 1.4 (results can be checked by linear interpolation from the data given in Appendix B: about 1910 degree-days).

The heating season is from October to May (degree-days in summer months will be considered as zero). The space heating load is given in Table 6.4.1 for each month, with an annual load of 41.3 GJ (11480 kWh).

The domestic hot water demand is 150 litres/day at 55°C with cold water at 10°C. This corresponds to a theoretical load of 28.3 MJ/day or 10.3 GJ/year (2870 kWh). The DHW back-up tank is connected in series with the pre-heating tank (without possible by-pass): corresponding heat losses should be included in the DHW load. It is assumed that corresponding heat losses are not

Table 6.4.1: Space heating and domestic hot water load summary.

Month	Degree Days	SH Load (MJ)	DHW Dem. (l/day)	Cold Temp. (°C)	Hot Temp. (°C)	DHW Load (MJ)	Total Load (MJ)
Jan.	366.0	7906	150	10.0	61.8	1008	8914
Feb.	283.8	6130	150	10.0	61.8	910	7041
Mar.	261.3	5644	150	10.0	61.8	1008	6652
Apr.	180.7	3903	150	10.0	61.8	976	4879
May	64.7	1398	150	10.0	61.8	1008	2406
June	0.0	0	150	10.0	61.8	976	976
July	0.0	0	150	10.0	61.8	1008	1008
Aug	0.0	0	150	10.0	61.8	1008	1008
Sep.	0.0	0	150	10.0	61.8	976	976
Oct.	131.2	2835	150	10.0	61.8	1008	3843
Nov.	270.0	5833	150	10.0	61.8	976	6808
Dec.	353.7	7640	150	10.0	61.8	1008	8648
Year	1911.5	41288	150	10.0	61.8	11869	53157

Table 6.4.2: Input parameters of the worked example: Solar system characteristics (reference case).

SOLAR LOOP	A = 20 m2 Tilt angle = 45° η_0 = 0.80 U_1 = 7 W/m2.K U_2 = 0.03 W/m2.K2 $\dot{m}_0 c_0$ / A = 51.6 W/m2.K d_{on} = 5 K d_{off} = 1 K $A_{in}U_{in}$ = 0 $A_{ou}U_{ou}$ = 0
HEAT-EXCHANGER	Type: External e_1 = 0.93 $\dot{m}_1 c$ / A = 69.8 W/m2.K
STORAGE	M = 1100 kg of water $A_s U_s$ = 7 W/K T_e = 20°C
HEAT EMITTERS	$\dot{m}_2 c$ = 1150 W/K e_2 = 0.312 Series connected; variable control (Type: 1)

precisely known, but estimated as 15% of the theoretical load. The effective required hot water temperature is therefore 61.75°C (effective demand unchanged) with a total yearly DHW load of 11.9 GJ.

6.4.1.2 - Solar system

The solar system comprises 20 m2 of ordinary single-glazed solar collectors inclined at 45°. Flow rates on the two sides of the heat-exchanger in the solar loop are 1000 litres/hour (water + ethylene glycol 40%), and 1200 litres/hour (water) respectively. The heat-exchanger is sized to transfer 9.6 kW under these flow rate conditions with inlet temperatures 30°C (water + antifreeze) and 20°C (water).

The following values are taken for collector characteristics: η_0 = 0.80, U_1 = 7 W/m2/K, and U_2 = 0.03 W/m2/K2. Piping heat losses are neglected.

Furthermore, the system incorporates a storage tank of 1000 litres of water; the thermal insulation of the tank consist of a 5 cm thickness of mineral wool (tank diameter 85 cm, height 2 m). The pre-heating cistern for hot water (submerged in the main tank) has a capacity of 100 litres.

The underfloor heating is designed to operate between 35°C (flow temperature) and 30°C (return) at an external temperature of -4°C.

6.4.1.3 - Input parameters

Input parameters required by method ESM1 (and by the corresponding software ESM 1.4) are listed in Table 6.4.2. Some of the parameters are obtained directly from the previous description. Other values must be estimated.

The collector fluid capacitance rate $\dot{m}_0 c_0$ is calculated fromthe data in section 6.4.1.2. A mixture of water and ethylene-glycol at 40% (by weight) has a thermal capacity c_0 = 3.56 kJ/kg/K and a density of 1045 kg/m3 at 30°C (see Appendix C); the mass flow rate corresponding to 1000 litres/hour is then $\dot{m}_0$ = 0.29 kg/s and the capacitance rate is $\dot{m}_0 c_0$ = 1033 W/K.

On the other side of the heat-exchanger, the mass flow rate of water is $\dot{m}_1$ = 0.333 kg/s and the corresponding capacitance rate $\dot{m}_1 c_1$ = 0.333 . 4185 = 1395 W/K.

The smaller fluid capacitance rate is therefore 1033 W/K, and the effectiveness of the heat-exchanger (computed by Eq. 6.2.9 from the transferred power 9.6 kW with a 10K temperature difference) is e_1 = 9600 / (10 . 1033) = 0.93. It should be noted that this value of the effectiveness of the heat-exchanger will be obtained with an overall heat transfer coefficient A_eU_e = 5900 W/K (from Eq. 2.2.23, the mean logarithmic temperature difference is 1.62 K).

Without any additional specification, control differentials are taken as standard values (starting: 5K; stopping: 1K).

The heat loss coefficient of the storage tank is estimated from the surface area of the tank (6.5 m2) and the thermal resistance of the insulation layer (conductivity of mineral wool: 0.04 W/m/K). This coefficient is found to be 1.37 m2.K/W, leading to a global heat transfer coefficient of 4.7W/K. This is increased by 50% in order to account for thermal bridges, and the value A_sU_s = 7 W/K is adopted.

Characteristics of the heating network are given for a maximum power needed to compensate building heat losses for the minimum possible external temperature. This power is computed from GV

$$P_n = GV \,.\, (T_c - T_{min}) = 250 \,.\, (19 - (-4)\,) = 5750 \text{ W}$$

and Eq.s 6.2.5 and 6.2.6 yield

$$\dot{m}_2c_2 = 5750 \,/\, (35 - 30) = 1150 \text{ W/K}$$

$$e_2 = (35 - 30) \,/\, (35 - 19) = 0.312$$

The option chosen for space heating control is type 1 (series auxiliary, modulated control).

6.4.1.4 - Results

Monthly and yearly results obtained from the ESM1.4 software are presented in Table 6.4.3. The useful solar energy output is 26.5 GJ (7350 kWh or 367 kWh/m2 of collector) compared with a total load of 53.2 GJ: this represents a solar contribution of 50%; the residual load to be met by auxiliary energy is 26.7GJ.

Solar energy is sufficient in summer to obtain a solar fraction of 100% when there is only a hot water load. The absolute contribution of the solar system is maximal in March, when both the incident solar energy and the space heating load are high.

Results compare very well with those of the manual version (see Appendix D) which leads to a solar fraction of 49%.

Table 6.4.3: System performance summary.

Month	Solar Irrad. (MJ/m2)	Total Load (MJ)	Auxil. Energy (MJ)	Solar Energy (MJ)	Solar Frac -
Jan.	12.4	8914	6429	2485	0.279
Feb.	15.6	7041	4126	2914	0.414
Mar.	17.7	6652	3136	3516	0.529
Apr.	19.1	4879	1545	3334	0.683
May	20.9	2406	63	2342	0.974
June	23.7	976	0	976	1.000
July	25.0	1008	0	1008	1.000
Aug.	22.7	1008	0	1008	1.000
Sep.	22.4	976	0	976	1.000
Oct.	17.9	3843	665	3178	0.827
Nov.	14.3	6808	3827	2982	0.438
Dec.	9.4	8648	6914	1735	0.201
Year	18.4	53157	26705	26452	0.498

Table 6.4.4: Worked example: Sensitivity analysis.

Modification with respect to reference	Total load GJ	Solar Fraction --	Energy output per m2 of coll. MJ/m2
Reference	53.2	0.498	1323
Selective collector U_1 = 5 W/m2.K U_2 = 0.015 W/m2.K2	53.2	0.560	1488
Parallel connected auxiliary	53.2	0.518	1377
Higher SH set-point T_c = 20°C	58.2	0.461	1340
Higher free heat gains P_{fg} = 750 W	45.9	0.545	1249
Lower DHW demand m = 100 l/day	49.2	0.491	1208
Piping heat losses $A_{ou}U_{ou} = A_{in}U_{in}$ = 15 W/K	53.2	0.460	1223
Location: MACON	66.1	0.273	901

6.4.2 - Sensitivity Analysis

In a system design process, sensitivity analyses are an important step which makes it possible: i) to optimise some of the system parameters, e.g., collector area, collector type, or storage capacity; ii) to know the influence of uncertainties in a given parameter (a few input parameters, such as heat transfer coefficients or flow rates or even the load are not well known in the design process,); and iii) to determine the effect of changes with time of a given parameter (e.g., damage to insulation of collector or piping).

6.4.2.1 - Elementary variations

First let us consider a few simple variations: one input parameter (or one set of parameters) only is changed with respect to the base case shown in Table 6.4.2. Examples of the modifications discussed here are given in Table 6.4.4.

What is the effect of a change in the type of solar collector? Buying selective flat-plate collectors (instead of ordinary collectors) leads to an increase in performance by 13 %. This has to be compared to the corresponding increase in the system cost (if smaller than 13%, selective collectors are better suited).

Another parameter, which is worthwhile to optimize, is the type of distribution. A parallel auxiliary (Option 2) introduces a 4% improvement in performance, as compared with the reference case.

A set-point temperature higher than the design setting (20°C instead of 19°C) leads, of course, to a higher load (+5 GJ, 12% of the SH load) and also makes the base temperature for the solar system higher. These two simultaneous effects lead to a solar energy output that is considered unchanged (+ 1.4%).

The rate of free heat gains is a rather arbitrary parameter: conventional values are proposed by national regulations such as, in France, 1.5 W per cubic metre of habitable volume. Large errors can be introduced by this parameter: what is the consequence of a 100% error (750 W instead of 375W)? It results here in a decrease in the total load by 14% and a decrease in system performance by 6 %.

The hot water demand is also not known precisely (because of the behaviour of the inhabitants): if hot water demand is smaller than expected (100 litres/day), the solar contribution is reduced by 9% (whereas the total load is reduced by 7.5%).

The system performance is more sensitive to a change in the hot water load than in the space heating load.

If piping heat losses are no longer negligible (15W/K on each side), the performance of the system is also decreased (-8%).

And finally, what will happen if we move our system northwards, from Carpentras (44°N) to Mâcon (46°N) for example? The space heating load is enhanced by 31 % and the system performance decreases by 30 %, due to a much smaller amount of solar radiation.

6.4.2.2 - Continuous variations

Optimization of the system requires, at least for some of the parameters, continuous variations within a reasonable range.

The collector area is a basic parameter (the storage capacity is assumed to remain unchanged here). When the collector area is increased, the system performance also increases, but the productivity (solar energy output per unit area of collector) decreases; this means that any additional collector unit produces less energy per unit area (Fig. 6.4.1).

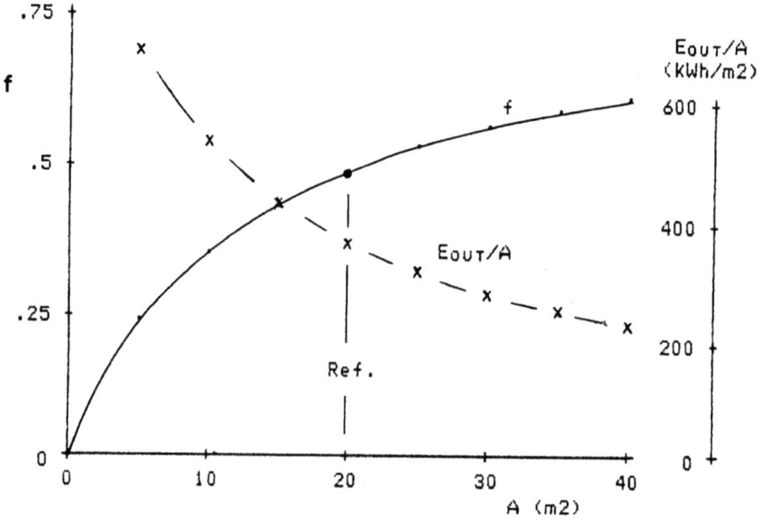

Fig. 6.4.1: Solar fraction and solar energy output per unit area of collector versus collector area A.

———•——— Solar fraction f

x — x — x Solar energy output per unit area of collector Eout/A

The storage capacity is a much less sensitive parameter, provided it remains within a certain range (40 to 100 litres per m2 of collector) (Fig. 6.4.2).

A similar result is observed with the heat-exchanger characteristics (Fig. 6.4.3): for an effectiveness e_1 higher than 0.5, the influence of this parameter is rather small.

As regards the tilt angle of the collector, an optimum is found at 55° (Fig. 6.4.4), but within a range of ± 15°, the difference in performance is smaller than 3%.

On the other hand, the characteristics of the heating distribution network cannot be neglected by the designer (see Fig. 6.4.5, variations with the flow temperature T_f under nominal conditions).

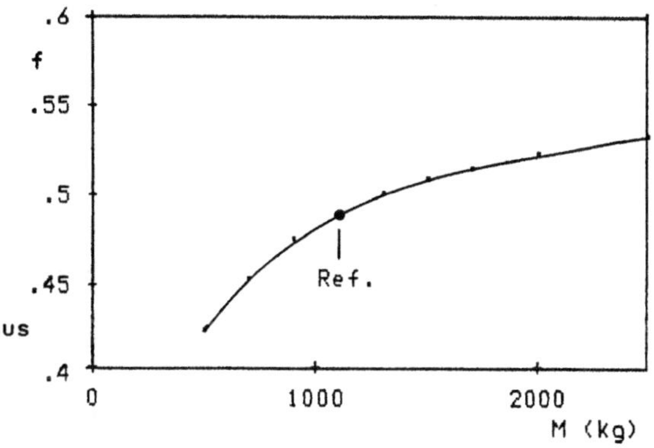

Fig. 6.4.2: Solar fraction versus storage capacity.

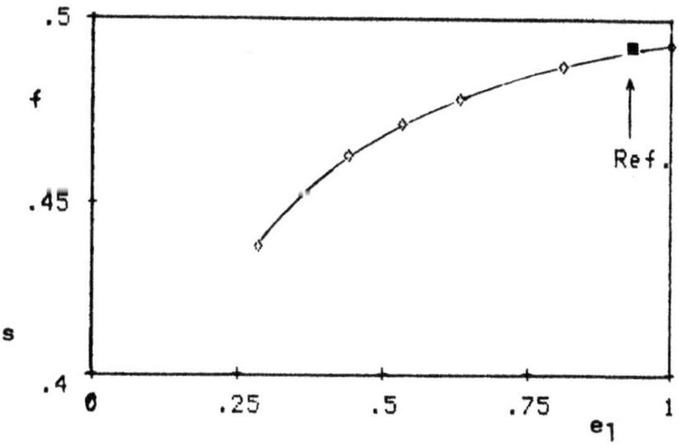

Fig. 6.4.3: Solar fraction versus heat-exchanger effectiveness.

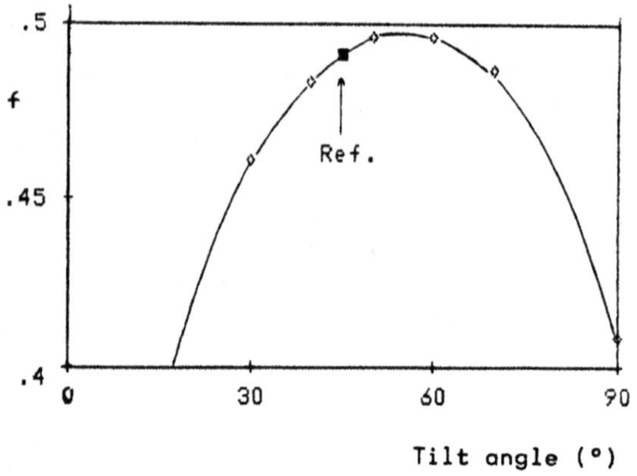

Fig. 6.4.4: Solar fraction versus collector tilt angle.

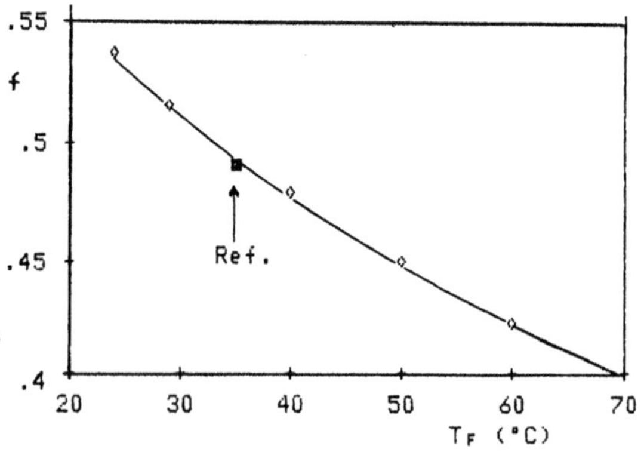

Fig. 6.4.5: Solar fraction versus flow temperature of heat emitters T_f.
($T_r = T_f - 5$ K).

6.5 - VALIDATION OF THE METHOD ESM1

ESM1 has been validated in several steps. The reference used for this purpose was EMGP2 /6.1/, a detailed simulation programme developed by Professor DUTRE at Katholieke Universiteit Leuven within the framework of several European research projects (European Modelling Group from 1981 to 1983; OPSYS from 1986 to 1988). A simple user-friendly version of EMGP2, called EURSOL, has also been used /6.2/.

A preliminary validation step made it possible to derive the basic algorithm of ESM1, involving the base temperature definition and the storage correction factor.

Then a second step made it possible to determine the overall accuracy of ESM1 with respect to EMGP2 under various conditions. A last stage was performed with extreme values of input parameters to define the limitations of the method and the acceptable range for input parameters.

6.5.1 - Deriving ESM1 corrective terms

A first set of EMGP1 (earlier version of EMGP2) runs were used to derive the storage correction factor of ESM1 in the following way. Input parameters are summarised in Table 6.5.1. A very simple (idealized) solar space heating system was considered (no DHW, no storage heat losses).

In the first step, a limited sample of monthly results was used (storage capacity: 65.5 litres per square metre of collector). Plotting the monthly solar fraction versus Y (as defined in step 10, available solar energy divided by load) shows a functional dependence which can be represented by an exponential form (Fig. 6.5.1)

$$f = 1 - EXP [- C_1 * Y * (1 + C_2 * Y)] \quad . \qquad (6.5.1)$$

A least square procedure determined the best values for C_1 and C_2 (0.94 and 0.21 respectively) (Fig. 6.5.2). The same procedure was used for other storage capacities: 133 and 33 l/m2. Interpolation formulae for an interval of storage volume (25 to 150 litres/m2) were then proposed (as stated in step 11).

A first version of ESM1 was published in 1983, including an empirical correction for storage heat losses /6.3/, later an algorithm for combined domestic hot water was completed and added /6.4/. The corresponding software ESM1.2 was then developed /6.5/.

Another set of EMGP2 results was then used to check the general accuracy of the method (Table 6.5.2). The distribution of the ratio f_{ESM}/f_{EMGP2} was studied (ratio of the yearly solar contributions computed respectively with ESM1 and EMGP2). Results are given in Table 6.5.3 in terms of the average and standard deviation of this ratio (see also Fig. 6.5.3): differences are limited to ± 15% in most of the cases /6.4/.

A few improvements were then introduced into the algorithm: the method was extended to on/off controlled distribution and to air distribution. Also, a more accurate corrective term for storage heat losses was developed (steps 14 and 15 for SH). EMGP2 runs were first obtained for several systems at 3 locations, with two values of A_sU_s (0 and 10 W/K, with corresponding monthly solar fractions f_0 and f_1). Monthly heat losses E_L from the storage were estimated at the base temperature T_b

$$E_L = N \cdot t_h \cdot A_sU_s \cdot (T_b - T_e) \qquad (6.5.2)$$

and the relative solar energy decrease $(1-f_1/f_0)$ due to heat losses was plotted against the dimensionless term $x = E_L/(f_0 \cdot L_0)$ (Fig. 6.5.4). A relationship of the following form was found:

$$1 - f_1 / f_0 = 0.33 \quad x^{1/2} \quad . \qquad (6.5.3)$$

For other values of A_sU_s, a linear interpolation was adopted for the range 0 to 10 W/K and a more complicated formula for higher heat loss coefficients (6.5.4)

$$1 - f/f_0 = (1-f_1/f_0) \cdot A_sU_s / 10 \cdot EXP[-(1-f_1/f_0).Ln (A_sU_s/10)]$$

The final version corresponds to the algorithms in Tables 6.3.1 and 6.3.2, as used in the programme ESM 1.4.

6.5.2 - Accuracy of ESM1

The overall accuracy of ESM1 has been tested with a new set of detailed simulation results, using the EURSOL programme (derived from EMGP2 with fixed configurations) /6.2/. The corresponding systems (EURSOL systems 17 and 18) are described in Figure 6.5.5 and their parameters are given in Table 6.5.4.

Examples of results are presented in Table 6.5.5 and in Figures 6.5.6 to 6.5.9 as iso-precision lines. Relative differences (in proportion to the useful solar energy output) between EMGP2 and ESM1 remain, generally, within a limited range (± 10%)

which, in turn, demonstrates the good accuracy of the method. The only exception is observed when systems have a large storage capacity (> 2250 litres) and collector area leading to a high solar contribution (> 50%) under sunny climate conditions: 15% relative differences are then possible.

These results are confirmed by a more limited set of simulations with another system (EURSOL system 20, Fig. 6.5.10) located at Copenhagen (Table 6.5.6, Fig. 6.5.11).

Other limited parametric variations have also been performed in order to check the sensitivity of ESM1 to secondary design parameters.

The sensitivity to DHW load is computed very well by ESM1 (Table 6.5.7). A climatic bias is nevertheless observed (Table 6.5.8) with a slight tendency to underpredict the performance in sunny climates (and to overpredict in cloudy climates).

A last analysis has been performed with an extreme system for the space heating of a highly insulated house, using high performance collectors and low flow-rate heat emitters (Table 6.5.9). This case and its variations (Table 6.5.10) show larger discrepancies with noticeable differences for:

- Low storage capacity (< 20 litres per m2 of collector).
- Low collector flow-rate ($\dot{m}_0 c_0/A$ = 4 W/m2.K).
- High DHW load (200 litres/day represent here about 50% of the SH load).
- High heat losses from internal piping.

On the other hand, the sensitivity of a few parameters is predicted well by ESM1, such as the sensitivity to heat emitter effectiveness. Neglecting the daily draw-off profile for DHW seems justified as differences between extreme profiles (computed by EMGP2) do not exceed 2%.

6.5.3 - <u>ESM1: Limitations of use</u>

A few limitations of use have been derived both from the development process of ESM1 and from the validation exercices reported above; they concern particular input parameters, through minimum and maximum possible values for each of them (the corresponding limits are introduced in the ESM1.4 software, see Appendix E). Furthermore, additional criteria must be respected and are represented by the following sets of input parameters:

DHW load	$L_1 / L_0 < 0.5$
Collector characteristics	$U_L/\eta_0 > 1$ W/m2/K
	$U_L < 10$ W/m2/K
Collector flow-rate	$\dot{m}_0 c_0/A > 5$ W/m2/K
Storage capacity	M/A > 25 litres/m2
Storage heat loss	$A_s U_s/M < .03$ W/K/litres
Storage correction factor	$0.25 < C_s < 5$
Internal piping	$A_p U_p < 1.5$ A U_L
Heat emitters	$\dot{m}_2 c > GV$

As stated earlier, a warning must also be given to users of systems with large storage (> 2250 litres) and high solar fractions (> 50%) in sunny climates (for yearly irradiation on the collector surface greater than 4.2 kWh/m2/day or 15 MJ/m2/day). In such cases, relative differences with EURSOL are larger than 10% (with systematic underprediction by ESM1).

With the exception of these few cases, relative differences with the reference simulation programme are smaller than 10%, which can be considered fully satisfactory with respect to the accuracy of the reference itself or the various uncertainties in the input parameters (climate, system, or behaviour of the inhabitants).

REFERENCES OF CHAPTER 6

/6.1/ W.DUTRE. EMGP2: a European transient simulation model for thermal solar systems. EUR 9777. REIDEL, Dordrecht (1985).

/6.2/ W.DUTRE. EURSOL: an interactive program for the simulation of water based thermal solar systems. Kath. Univ. Leuven (1988).

/6.3/ J.ADNOT et al. Design and validation of simplified methods for sizing active solar space heating systems. CEC contract ESA-M-086 F(G). Centre d'Energétique, ARMINES (1983).

/6.4/ J.ADNOT et al. Modèlisation solaire active. Convention cadre AFME-ARMINES 1984, action 3.2. Centre d'Energétique ARMINES (1985).

/6.5/ B.BOURGES. ESM 1.2: Guide d'utilisation. Centre d'Energétique ARMINES (1985).

/6.6/ J.ADNOT et al. ISES congress, Perth (1983).

/6.7/ J.ADNOT et al. Further development of the European validation of simplified methods for solar heating. ISES congress "INTERSOL 85", Montreal. Pergamon (1986).

Table 6.5.1: Reference system for the development of ESM1 (First version).

Load:	GV = 290 W/K T_c = 20°C No Hot water load
Solar Collectors:	A = 16.5, 31.5 or 46.5 m2 η_0 = 0.775 U_L = 3.83 W/m2.K $\dot{m}_0c_0$ / A = 62.2 W/m2.K
Heat-Exchanger:	External e_1 = 0.44 $\dot{m}_1c_1$ / A = 62.4 W/m2.K
Storage:	M/A = 33, 65.5 or 131 kg/m2 A_sU_s = 0
Distribution:	Series Auxiliary (Variable control) $\dot{m}_2c_2$ = 500 W/K e_2 = 0.36 or 0.58
Locations	Copenhagen, Hamburg, Dublin, London Uccle, Trappes or Carpentras

Table 6.5.2: ESM1 (First version): System used for validation purposes.

Load:	GV = 300 W/K T_c = 20°C m =120 kg/day T_m = 10°C T_h = 50°C
Solar Collectors:	A = 10, 20 or 30 m2 $\dot{m}_0c_0$/A = 29 W/m2.K

Collector	η_0	U_1	U_2
1	0.780	5.70	0.040
2	0.756	3.80	0.022
3	0.622	1.70	0.008

Heat-Exchanger:	External e_1 = 0.60 $\dot{m}_1c_1$ / A = 29 W/m2.K
Storage:	M/A = 37.5 kg/m2 A_sU_s = 3 W/K
Distribution:	Series or parallel Auxiliary (Variable control) $\dot{m}_2c_2$ = 1000 W/K e_2 = 0.00 e_3 = 0.35
Locations	Copenhagen, Uccle, Carpentras

Table 6.5.3: ESM1 (First version): validation results. Mean value and standard deviation of the ratio f_{esm1}/f_{emgp2}.

	Runs	Mean	S.D.
No hot water demand	27	0.96	0.04
DHW (pre-heating pipe)	17	0.99	0.04
DHW (submerged tank)	17	0.90	0.05

Table 6.5.4: Validation of ESM1 (Final version). Reference systems used for EURSOL simulations (Systems 17 and 18; see Fig.6.5.5).

Load:	GV = 300 W/K	T_c = 20°C	T_{nh} = 17°C	
	m = 120 kg/day	T_m = 10°C	T_h = 50°C	
Solar Collectors:	A = 10 to 50 m2	$\dot{m}_0c_0/A$ = 40 W/m2.K		
	Collector	η_0	U_1	U_2
	Non-selective	0.80	5.70	0.040
	Evacuated tube	0.62	1.70	0.008
Heat-Exchanger:	Internal	A_eU_e/A = 36 W/m2.K		
Storage:	M = 500 to 3000 kg	A_sU_s = 15 to 32 W/K		
	M' = 150 kg (DHW pre-heating tank)			
Distribution:	Series or parallel Auxiliary			
	Variable or on/off control			
	$\dot{m}_2c_2$ = 1160 or 870 W/K	e_2 = 0.3 or 0.5		
Locations	Uccle or Carpentras			

Table 6.5.5: Comparison between EURSOL and ESM1 (final version).

SYSTEM 18 SERIES AUXILIARY LOW-TEMPERATURE DISTRIBUTION UCCLE

NON - SELECTIVE COLLECTORS

Area	STORAGE 500 L			STORAGE 1000 L		
	EURSOL	ESM1	DIFF.(%)	EURSOL	ESM1	DIFF.(%)
10 M2	12,00%	11,80%	-1,67%	12,90%	12,80%	-,78%
15 M2	14,60%	14,20%	-2,74%	16,00%	15,90%	-,63%
20 M2	16,40%	15,70%	-4,27%	18,30%	18,00%	-1,64%
25 M2	17,80%	16,90%	-5,06%	19,70%	19,60%	-,51%
30 M2	18,60%	17,80%	-4,30%	20,90%	20,90%	,00%
35 M2	18,90%	18,50%	-2,12%	21,80%	22,00%	,92%
40 M2	19,70%	19,20%	-2,54%	22,60%	22,90%	1,33%
45 M2	20,10%	19,70%	-1,99%	23,30%	23,70%	1,72%
50 M2	20,50%	20,20%	-1,46%	23,90%	24,40%	2,09%

Area	STORAGE 1500 L			STORAGE 2000 L		
	EURSOL	ESM1	DIFF.(%)	EURSOL	ESM1	DIFF.(%)
10 M2	13,30%	13,00%	-2,26%	13,40%	13,00%	-2,99%
15 M2	16,50%	16,50%	,00%	16,80%	16,80%	,00%
20 M2	19,00%	19,10%	,53%	19,50%	19,60%	,51%
25 M2	20,90%	21,00%	,48%	21,70%	21,80%	,46%
30 M2	22,40%	22,50%	,45%	23,50%	23,50%	,00%
35 M2	23,80%	23,80%	,00%	24,80%	24,90%	,40%
40 M2	24,60%	24,90%	1,22%	26,20%	26,50%	1,15%
45 M2	25,30%	25,90%	2,37%	27,10%	27,20%	,37%
50 M2	26,50%	26,70%	,75%	27,80%	28,20%	1,44%

Area	STORAGE 2500 L			STORAGE 3000 L		
	EURSOL	ESM1	DIFF.(%)	EURSOL	ESM1	DIFF.(%)
10 M2	13,40%	12,90%	-3,73%	13,40%	12,80%	-4,48%
15 M2	16,90%	16,80%	-,59%	17,10%	16,80%	-1,75%
20 M2	20,00%	19,80%	-1,00%	20,40%	19,90%	-2,45%
25 M2	22,20%	22,20%	,00%	22,60%	22,40%	-,88%
30 M2	24,10%	24,10%	,00%	24,80%	24,40%	-1,61%
35 M2	25,80%	25,60%	-,78%	26,50%	26,10%	-1,51%
40 M2	27,10%	27,00%	-,37%	27,80%	27,50%	-1,08%
45 M2	28,30%	28,10%	-,71%	29,10%	28,70%	-1,37%
50 M2	29,30%	29,10%	-,68%	30,30%	29,70%	-1,98%

Table 6.5.6: Validation of ESM1 (final version) versus EURSOL. Additional runs for EURSOL system 20 (Location: Copenhagen). System sizes and results.

Load:	GV = 210 W/K T_c = 20°C T_{nh} = 17°C m = 150 kg/day T_m = 10°C T_h = 45°C
Solar Collectors:	A = 10 to 40 m2 $\dot{m}_0 c_0/A$ = 55 W/m2.K η_0= 0.82 U_1= 5.0 W/m2.K U_2= 0.011 W/m2.K2
Heat-Exchanger:	Internal $A_e U_e/A$ = 45 W/m2.K
Storage:	M = 500 to 4000 kg $A_s U_s$ = 8 to 15 W/K M' = 200 kg (DHW submerged pre-heating tank)
Distribution:	Parallel Auxiliary On/off control $\dot{m}_2 c_2$ = 700 W/K e_2 = 0.5
Location	Copenhagen

Run	Description	Solar Fraction		Diff.
		EURSOL	ESM1	
1	A = 10 m2, M = 500 kg (+ 200 kg DHW tank)	0.180	0.183	+1.3%
2	A = 10 m2, M = 1000 kg "	0.186	0.191	+2.6%
3	A = 20 m2, M = 500 kg "	0.255	0.245	-3.9%
4	A = 20 m2, M = 1000 kg " (Reference)	0.270	0.266	-1.8%
5	A = 20 m2, M = 2000 kg "	0.284	0.281	-1.1%
6	A = 30 m2, M = 750 kg "	0.316	0.304	-3.7%
7	A = 30 m2, M = 1500 kg "	0.341	0.332	-2.7%
8	A = 30 m2, M = 3000 kg "	0.361	0.351	-2.9%
9	A = 40 m2, M = 1000 kg "	0.365	0.355	-2.9%
10	A = 40 m2, M = 2000 kg "	0.400	0.386	-3.5%
11	A = 40 m2, M = 4000 kg "	0.426	0.408	-4.3%
12	DHW tank capacity: 150 kg	0.270	0.264	-2.2%
13	DHW tank capacity: 250 kg	0.271	0.267	-1.6%
14	DHW demand: 100 l/day	0.263	0.255	-2.9%
15	DHW demand: 200 l/day	0.278	0.275	-1.3%
16	Another DHW draw-off profile	0.270	0.266	-1.7%
17	Heat loss coef. of the house GV = 130 W/K	0.376	0.369	-1.8%
18	Heat loss coef. of the house GV = 250 W/K	0.241	0.234	-3.1%
19	Effectiveness of heat emitters e_2 = 0.6	0.274	0.268	-2.0%
20	Effectiveness of heat emitters e_2 = 0.4	0.268	0.261	-2.4%

Table 6.5.7: Sensitivities of EURSOL and ESM1 to the Hot Water load.
Reference case: 30 m2 of non-selective collectors
1500 litres of storage (Uccle)
$\dot{m}_2 c_2$ = 1160 W/K e_2 = 0.5 parallel auxiliary (variable).

Daily demand		f_{EURSOL}	f_{ESM1}	Diff.
50 l (50°C)		0.204	0.211	+3.4%
120 l (50°C)	Reference	0.218	0.225	+3.2%
200 l (50°C)		0.230	0.240	+4.3%

Required temperature		f_{EURSOL}	f_{ESM1}	Diff.
40°C (120 l/day)		0.217	0.221	+1.8%
50°C (120 l/day)	Reference	0.218	0.225	+3.2%
60°C (120 l/day)		0.215	0.225	+4.6%

Table 6.5.8: Sensitivities of EURSOL and ESM1 to the climate.
Reference case: idem Table 6.5.7

Location		f_{EURSOL}	f_{ESM1}	Diff.
Copenhagen	(Denmark)	0.226	0.225	-0.4%
Uccle	(Belgium)	0.218	0.225	+3.2%
Trappes	(France, North)	0.276	0.273	-1.1%
Carpentras	(France, South)	0.520	0.490	-5.8%

Table 6.5.9: Validation of ESM1 versus EURSOL. Extreme runs: system size.

Load:	GV = 150 W/K	T_c = 20°C	T_{nh} = 17°C	
	m = 100 kg/day	T_m = 10°C	T_h = 60°C	
Solar Collectors:	A = 10 m2	$\dot{m}_0 c_0/A$ = 41.8 W/m2.K		
	η_0 = 0.642	U_1 = 0.89 W/m2.K	U_2 = 0.007 W/m2.K2	
Heat-Exchanger:	Internal	$A_e U_e/A$ = 50 W/m2.K		
Storage:	M = 650 kg	$A_s U_s$ = 8.4 W/K	$A_p U_p$ = 2.1 W/K	
Distribution:	Parallel Auxiliary	Variable control		
	$\dot{m}_2 c_2$ = 209 W/K	e_2 = 0.7		
	DHW submerged pre-heating pipe	$A_{dh} U_{dh}$ = 250 W/K		
Location	Uccle			

Table 6.5.10: Validation of ESM1 versus EURSOL. Extreme runs: Results. (Solar energy output).

Run	A (m2)	M (kg)	Comments	EURSOL (kWh)	ESM1 (kWh)	Diff. (%)
1	10	400		2836	3151	11.1
2	20	400		3927	4481	14.1
3	40	400		5017	5727	14.2
4	10	800		2931	3149	7.4
5	20	800		4269	4617	8.2
6	40	800		5776	6051	4.8
7	10	1600		2828	3003	6.2
8	20	1600		4386	4492	2.4
9	40	1600		6043	5991	-0.9
10	10	650	<u>Reference</u>	2930	3162	7.9
11	10	650	$\dot{m}_0 c_0/A$ = 20.9 W/m2.K	2933	3154	7.5
12	10	650	$\dot{m}_0 c_0/A$ = 10.4 W/m2.K	2912	3120	7.1
13	10	650	$\dot{m}_0 c_0/A$ = 4.2 W/m2.K	2676	3031	13.3
14	10	650	$A_{dh} U_{dh}$ = 125 W/K	2912	3162	8.6
15	10	650	$A_{dh} U_{dh}$ = 500 W/K	2933	3162	7.8
16	10	650	$A_{dh} U_{dh}$ = 1000 W/K	2913	3162	8.6
17	10	650	e_2 = 0.50	2882	3102	7.6
18	10	650	e_2 = 0.30	2774	2982	7.5
19	10	650	e_2 = 0.90	2961	3199	8.0
20	10	650	m = 150 litres/day	3267	3589	9.9
21	10	650	m = 200 litres/day	3544	3990	12.6
22	10	650	$A_p U_p$ = 4.2 W/K	2823	3128	10.8
23	10	650	$A_p U_p$ = 8.4 W/K	2382	3059	28.4
24	10	650	DHW load: Morning 80%; Eve. 20%	2868	3162	10.3
25	10	650	DHW load: Morning 20%; Eve. 80%	2922	3162	8.2
26	10	650	$\dot{m}_2 c_2$ = 418 W/K	2960	3226	9.0
27	10	650	$\dot{m}_2 c_2$ = 836 W/K	2978	3270	9.8
28	10	650	$\dot{m}_2 c_2$ = 2088 W/K	2988	3308	10.7

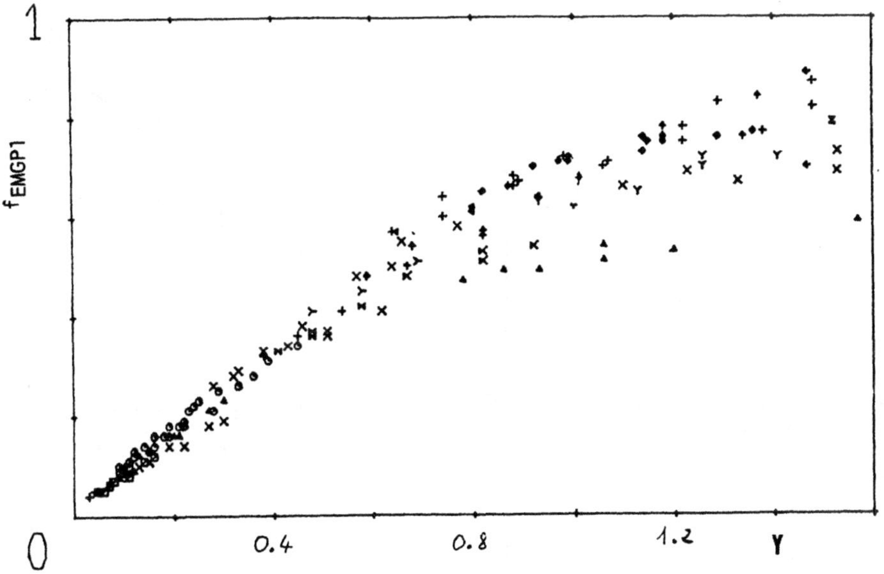

Fig. 6.5.1: Development of ESM1. EMGP1 solar fraction versus available energy (divided by load), Y.

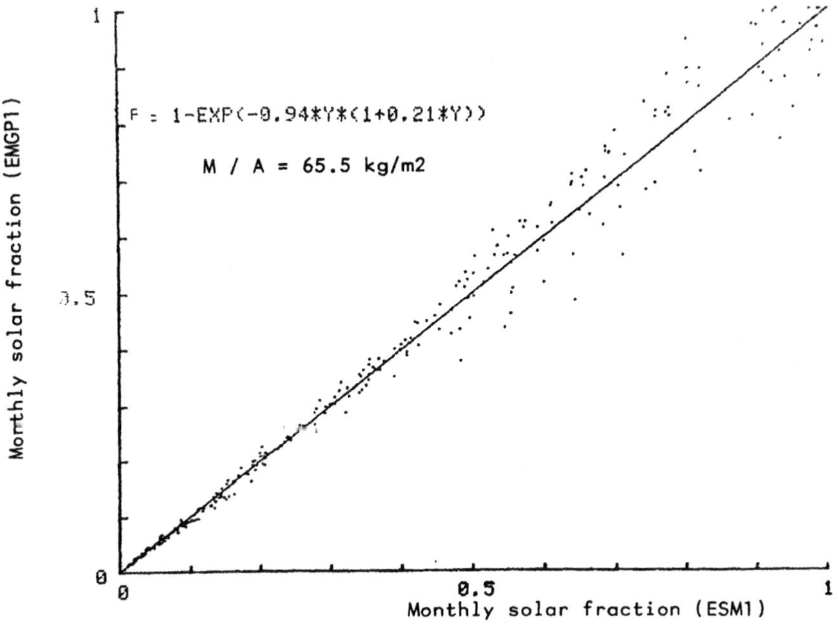

Fig. 6.5.2: Development of ESM1. Accuracy of the storage correction factor.

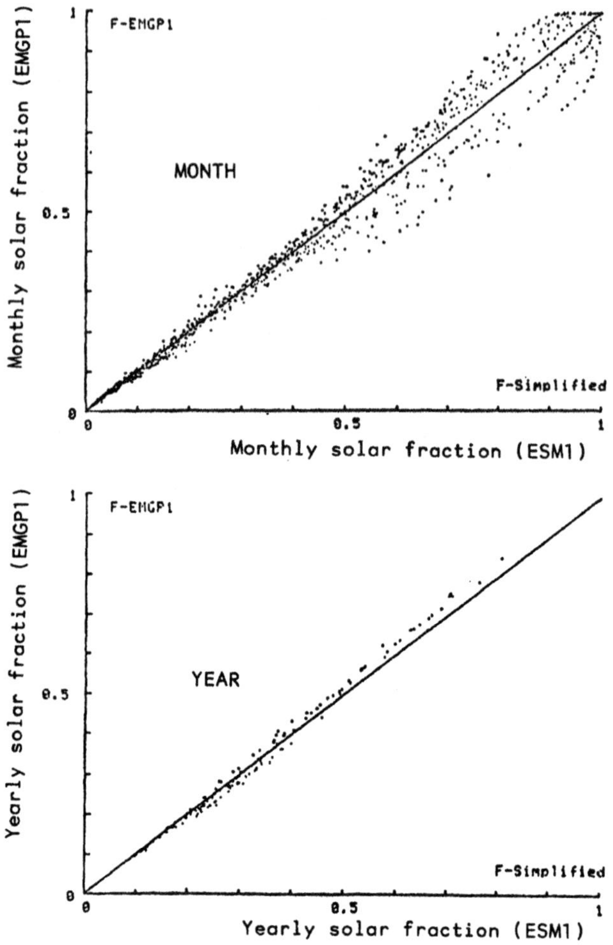

Fig. 6.5.3: ESM1 (first version). Comparison of monthly and yearly results with EMGP1.

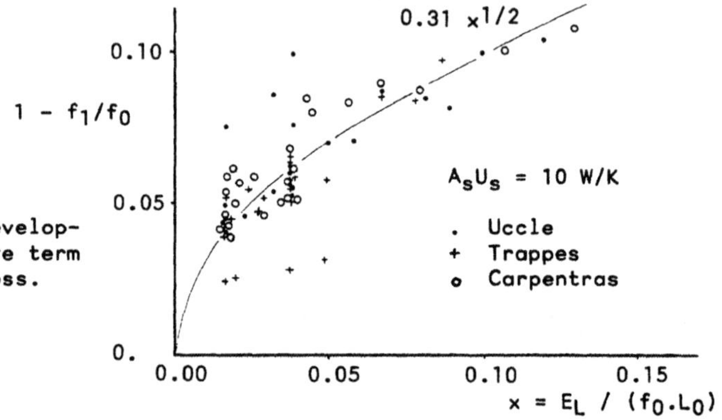

Fig. 6.5.4: ESM1. Development of a corrective term for storage heat loss.

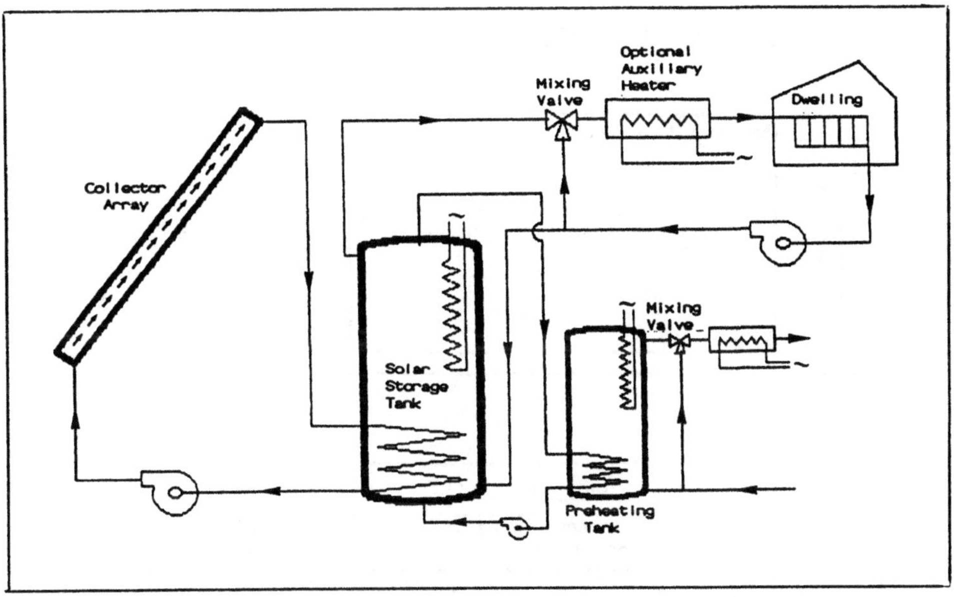

Fig. 6.5.5: ESM1 (final version). Validation with respect to EURSOL. Layout of systems considered (EURSOL systems 17 and 18: internal heat-exchanger; external DHW tank).

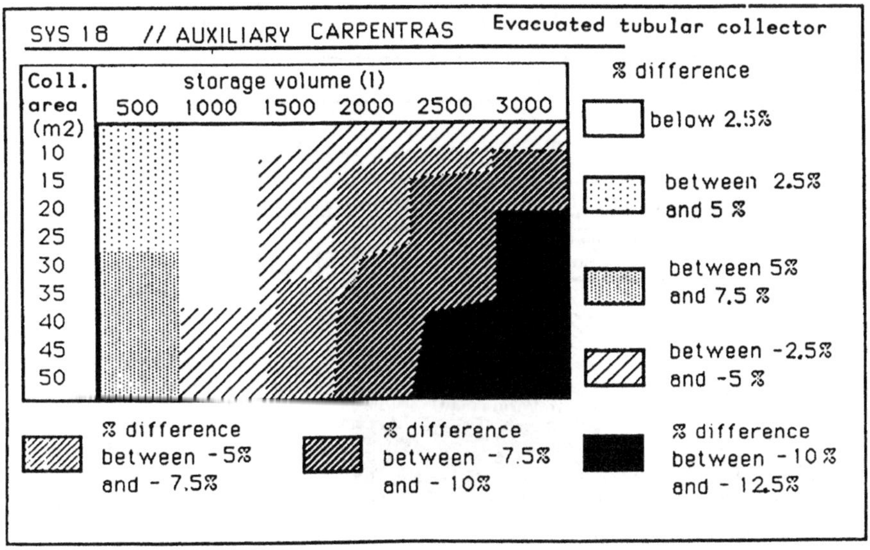

Fig. 6.5.6: Validation of ESM1 (final version). Differences between EURSOL and ESM1.

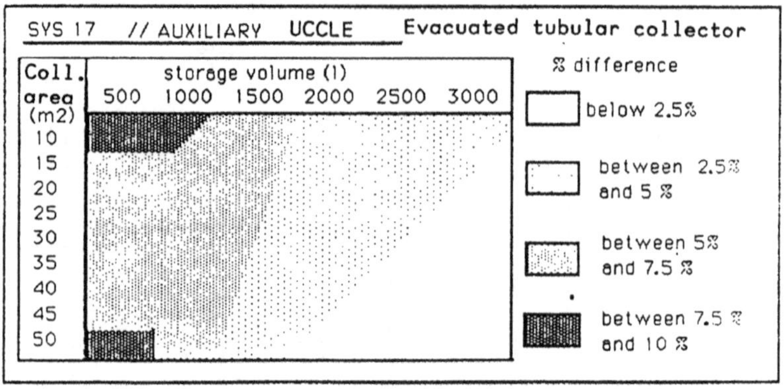

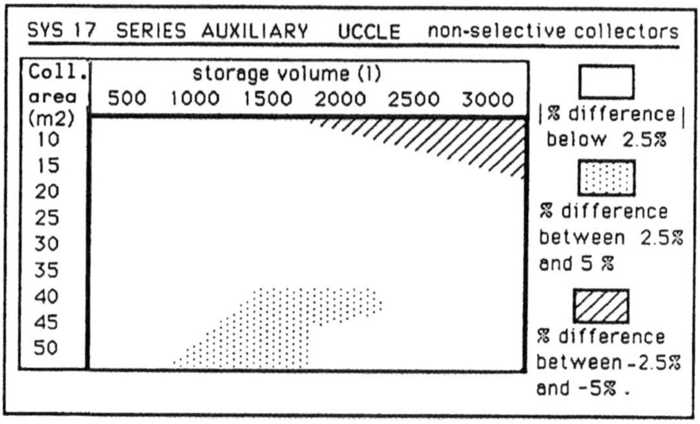

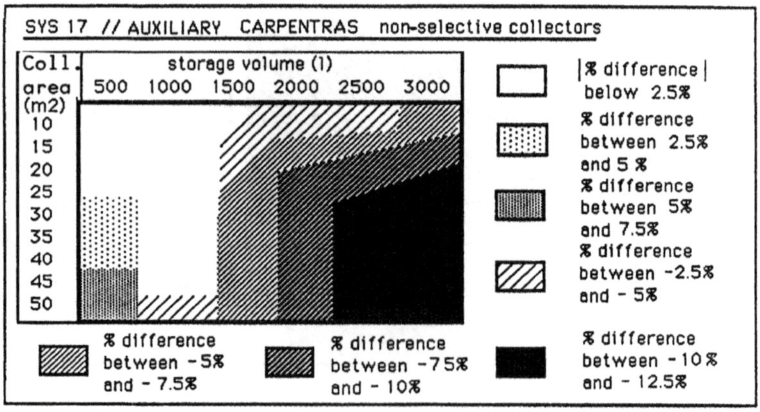

Fig. 6.5.7 to 6.5.9: Differences between EURSOL and ESM1 (final version).

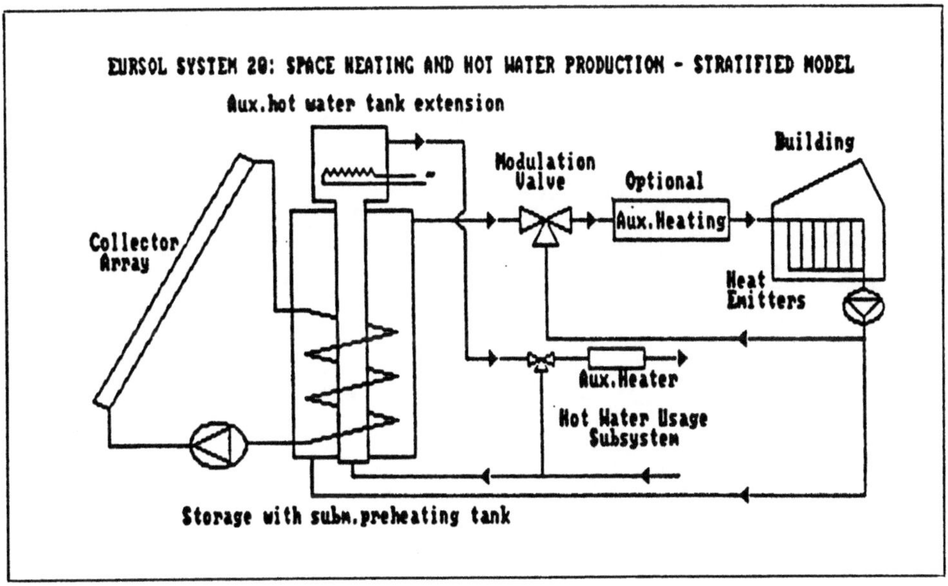

Fig. 6.5.10: Validation of ESM1 versus EURSOL. Additional runs (location: Copenhagen). Layout of system considered (EURSOL no.20: internal heat-exchanger; submerged DHW tank).

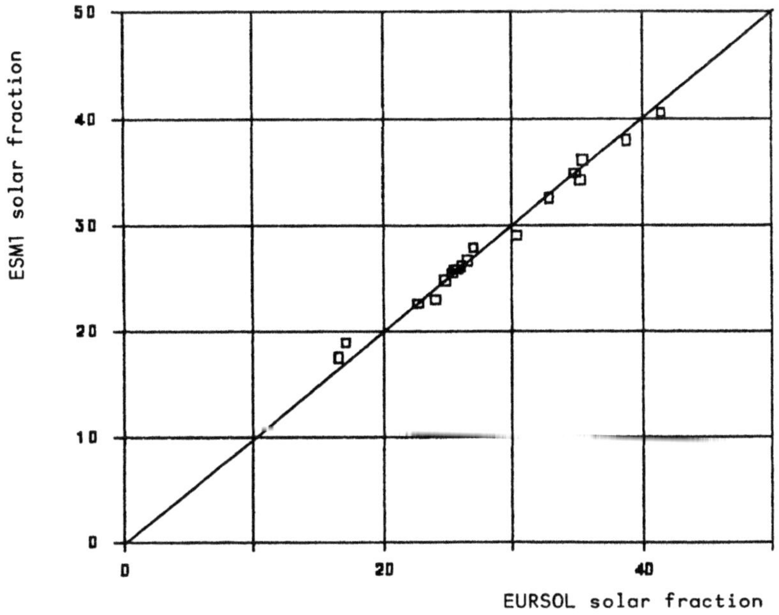

Fig. 6.5.11: Validation of ESM1. Results of additional runs (Copenhagen).

CHAPTER 7

ESM 2: A SIMPLIFIED METHOD FOR SOLAR HOT WATER SYSTEMS

The calculation method ESM2, which is intended for the performance evaluation of solar hot water systems (providing only hot water), is presented in this chapter. The basic principles of ESM2 are similar to the hot water algorithms of ESM1 (for active solar heating systems and preheating of domestic hot water), presented in chapter 6 of this handbook. However, a number of improvements have been incorporated into the ESM2 algorithms.

As in ESM 1, the user begins with a limited set of parameters describing the system. It is possible to calculate the auxiliary heating requirements and the savings due to solar energy, in particular the efficiency of the system: that is, the number of kWh or MJ saved per square metre of collector. Sensitivity analyses and optimisations are possible with respect to the most important parameters.

The method is again based on the fundamental concepts, described in Chapter 3: a base temperature (below which heat derived from solar energy cannot be used by the system, here simply the mains cold water temperature) and a threshold irradiance (the minimum intensity of solar radiation required to yield energy at a given temperature). These concepts are developed within the framework of a one-day repetitive simulation for each month. As far as climatic data are concerned, the method requires not only an average incident radiation, but a synthesis of the distribution of solar radiation in the form of cumulative frequency curves (CFC), or utilizability curves, for each month. The derivation of these data has been described in Chapter 4.

The basic detailed version of ESM2 is presented in this chapter; it requires the use of a microcomputer. The use of the corresponding software, ESM 2.1, is described in Chapter 9. No simplified version which could be used "manually" exists at present.

The first section of this chapter contains a detailed description of the systems considered. This should be read carefully to determine whether or not a proposed system design is amenable to analysis with ESM2: certain configurations (or components) are specifically excluded.

The second section illustrates the underlying principles of ESM2 and discusses the range of variables which the method takes into account. It explains the way in which the less obvious system parameters are obtained.

The third section is concerned with a detailed examination of the various steps in the calculation, and constitutes a reference to the theoretical principles.

An example is presented in the fourth section. The accuracy of the method is addressed in the last section and results of a comparison with the detailed simulation programme EMGP2 are presented. Limitations of use are summarized.

This chapter has been written so that it may be read independently. Some material given elsewhere in the book has therefore been summarized or repeated here.

7.1 - TYPES OF SYSTEMS CONSIDERED

ESM2 is applicable to the most common solar hot water systems, whatever their size, from individual water heaters to large-scale systems for applications such as hospitals, camping sites, hotels, etc. These systems incorporate an active solar collection loop, a short-term storage tank, and an auxiliary energy supply. The storage is needed to allow heat collected during the day to be used at any time during the day or the night. A back-up is generally required, except for over-sized systems in sunny climates.

7.1.1 - Active solar hot water system principles

The principal components are:

* The solar collectors, usually of the flat-plate type, through which circulates a heat transfer fluid, usually water or water with antifreeze.

* The circuit through which energy is transferred to the storage, which may incorporate a heat-exchanger (this circuit will be referred as the "solar loop"). Circulation of the working fluid is by a pump with an appropriate controller.

* A storage water tank providing short-term storage of the hot water.

This solar part is coupled to a conventional hot water production and distribution system which includes:

* An auxiliary energy source which generally heats a specific additional tank. This back-up can be of any type (electricity, oil- or gas-fired boiler) and incorporates its own controller.

* A distribution circuit to the taps.

* For large plants (in which the point of use can be far from the source of heat), a recirculation loop is often included in the plumbing so that hot water may be drawn off as soon as a tap is turned on, even at some large distance from the tanks.

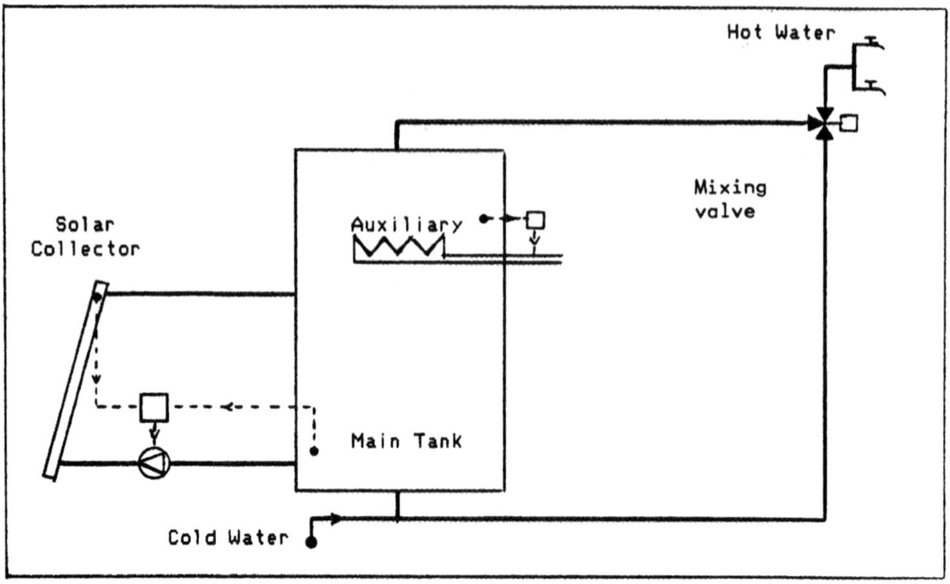

Fig. 7.1.1: Solar water heating system with direct solar loop and auxiliary integrated in the main tank.

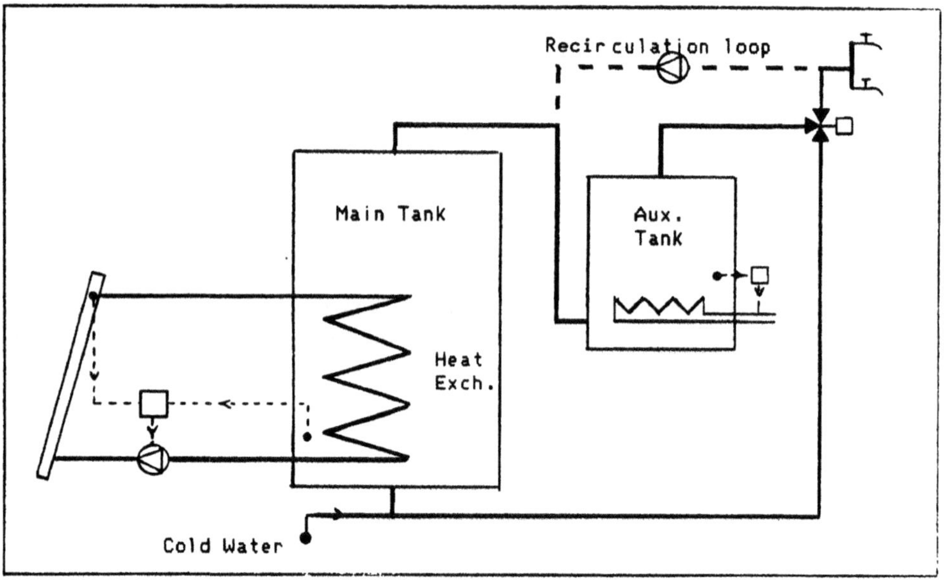

Fig. 7.1.2: Solar water heating system with internal heat-exchanger, series auxiliary tank, and recirculation loop connected to the auxiliary tank (recirculation loop: dashed line).

Only the essential elements of the system have been mentioned above; some components, which are necessary for the operation or safety of a system have been omitted since they do not influence the overall performance.

7.1.2 - Schematic system designs

The possible arrangements of the basic components in DHW solar systems are described in Chapter 1. Systems which can be assessed with ESM2 are considered as a combination of three subsystems: the solar loop+storage; the coupling with back-up; and the recirculation loop. For each subsystem, three options are possible as described below.

Configurations permitted by ESM2 are constructed by combining three options, i.e. one option is selected for each subsystem. Figs.7.1.1 to 7.1.3 present three typical combinations and show all possible subsystems; other combinations are also permitted (3x3x3 = 27 different schematics). For example, the combination of the solar loop in Fig. 7.1.1 (direct loop) with the back-up coupling in Fig 7.1.2 (series-connected auxiliary tank) and recirculation coupling in Fig. 7.1.3 (recirculation loop with a three-way valve) is a permitted configuration.

7.1.2.1 - Solar loop

For the solar loop, three types of subsystems are allowed. The simplest scheme (Fig. 7.1.1) is the direct loop without a heat-exchanger. The loop must be protected from freezing by an automatic drain-down (or drain-back). The direct loop is, of course, a good solution for seasonal applications like hot water for camping sites. In the second scheme, which is more common in practice, an anti-freeze solution flows in the collector circuit and an external heat-exchanger (plate or tubular) is inserted between the two parts of the solar loop (Fig. 7.1.3). The third alternative is the internal heat-exchanger , submerged inside the storage tank (e.g., a finned pipe) (Fig. 7.1.2), or a heating jacket around the wall of the storage tank.

7.1.2.2 - Auxiliary heating

Three categories of systems may be distinguished according to the location of the auxiliary heating source. Supplementary heating can be integrated in the top part of the solar storage tank (Fig. 7.1.1): this is often the case for small systems (individual water heaters). For larger plants, the back-up is

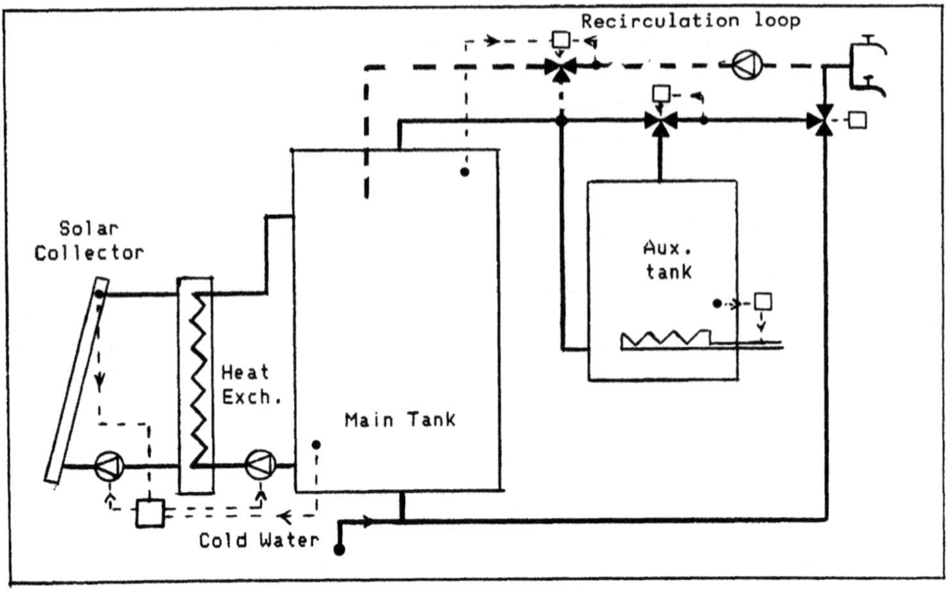

Fig. 7.1.3: Solar water heating system with external heat-exchanger, auxiliary tank (controlled by-pass), and controlled recirculation loop (recirculation loop: dashed line).

generally located in a separate series auxiliary tank (Fig. 7.1.2), which is a good solution if the solar system works as a preheater (the required hot water temperature is seldom obtained in the preheater; the solar fraction remains small). The alternative solution is an auxiliary tank with by-pass (Fig. 7.1.3): if the temperature in the solar tank is sufficient to meet the total load, the auxiliary tank is by-passed. If the temperature is insufficient, the hot water flow-rate from the auxiliary tank is controlled by a three-way valve in order to obtain the required temperature. During very sunny periods, it will be preferable to stop the back-up.

The ESM2 method can account for back-up control in two different ways (whatever the coupling schematic of the auxiliary): i) the usual on/off differential control in which the auxiliary boiler starts if the temperature in the tank drops below a minimum value (starting set-point temperature) and stops if it exceeds a maximum value (stopping set-point); and ii) night charging in which the auxiliary tank is heated during a limited period, until a set-point temperature is reached; off-peak electricity is the usual source of energy in this case.

7.1.2.3 - Recirculation loop

The recirculation loop is required if the water content of the distribution pipes is larger than about 5 litres. Due to heat losses from the recirculation loop, the hot water load is increased.

This optional recirculation loop can be:

- Connected only to the auxiliary tank (Fig. 7.1.2) (in which case the solar energy cannot meet the corresponding additional load).

- Connected either to the solar tank or to the auxiliary tank by a controlled three-way valve (Fig. 7.1.3). If the temperature within the solar tank is sufficient, the return from the recirculation loop flows into this tank; otherwise, it flows into the auxiliary. This is a good solution for systems in which heat losses from the recirculation loop are a significant fraction of the total load and if the solar fraction is rather high.

7.1.2.4 - Systems excluded

A few systems are excluded from this analysis, these are mainly individual water heaters in which no forced circulation takes place. Common examples are thermosiphon systems, integrated collector storage (ICS), or systems incorporating a boiling-condensing fluid for heat transfer (heat pipe) (see section 1.3 for a description of these systems).

System designs which depart greatly from the schematic types described above must be subjected to a detailed analysis according to their particular features and to the characteristics of ESM2.

It should also be noted that systems which simultaneously provide space heating are excluded. The ESM1 method described in Chapter 6 is suitable for such systems.

7.1.3 - Which components?

The limits of applicability of each system component are discussed below.

7.1.3.1 - Solar collectors and solar loop

As well as conventional solar collectors, ESM2 is applicable to "atmospheric" or unglazed collectors (in which there is a significant heat transfer with ambient air). Some examples of the solar collectors which are taken into account with ESM2, are:

- Flat-plate, with or without a selective surface.
- Evacuated tube, with or without a selective surface.
- Unglazed colectors.
- Concentrating: weak (CPC, parabolic), medium (linear concentrators), and even point focus (paraboloid) collectors provided that the meteorological data are suitably modified (direct radiation). Concentrating collectors are suitable only for high temperature applications (> 120°C).

In the collectors, the working fluid is a liquid: water, water+ antifreeze, or organic. The heat exchanger may be inside or outside the storage tank. The solar loop is regulated by an on/off differential controller.

7.1.3.2 - Storage

The storage is an insulated tank of water, designed for short-term storage. Its capacity is generally of the order of magnitude of the daily consumption. The storage capacity should neither be very small ("direct" solar heating is not permitted), nor very large. Precise limitations are indicated in section 7.3.

7.1.3.3 - Distribution network

The distribution network is not subject to particular restrictions, provided that the design is of a type similar to those defined as "Typical schematics" (sect.7.1.2 above).

Hot water distribution subsystems which depart from the schematics or control strategies described in section 7.1.2, must be analysed carefully. The ESM2 method can perhaps be adapted to deal with different systems: for particular auxiliary control strategies, if an effective required hot temperature and an effective hot water demand can be defined (see section 7.2), it may be possible to use the ESM2 method as it is.

Finally, it is important to note that ESM2 is presented as a computation method for solar hot water systems. In fact, ESM2 is applicable to any open loop solar system, provided the system schematic and control are one of the permitted types defined in section 7.1.2. By "open loop", we mean any system in which no feedback occurs in the distribution loop between the cold water (inlet) and the hot water (outlet). In that sense, space heating is not an open loop but a "closed loop" because the return temperature from the heat emitters depends on the flow temperature from the solar storage. In an open loop, the load is defined by three parameters (which may be taken as constant for a given month): "cold" temperature; required "hot" temperature; and daily demand. A number of systems, such as solar industrial process heating (IPH) (with storage), can then be computed with the aid of ESM2.

7.2 - CALCULATION PRINCIPLES AND INPUT PARAMETERS

7.2.1 - Calculation principles

The method ESM2 is based on two fundamental assumptions:

- The monthly behaviour is assumed to be represented by the repetitive day concept: an "average" day is characterized (from the solar radiation point of view) by the monthly frequency distribution of solar irradiance (as a cumulative frequency curve, see Chapter 4), and the average external ambient temperature.

  The calculation over an average day is repeated until a periodic state of the system is reached (storage temperature at the end of the day equals the initial temperature); this periodic state (and the corresponding system performance) is assumed to be representative of the monthly average behaviour of the system.

- The daily behaviour is described by the analytical model developed in Chapter 3. The repetitive day concept requires a daily simulation of the system (24 hour period). This simulation is done using a simplified analytical model of the whole system (solar loop + storage). This model enables quick calculations to be made with reasonable assumptions. It accounts for several types of storage (perfectly stratified or fully mixed): a stratified system is represented by an "equivalent mixed system". An iterative procedure determines the time-averaged storage temperature during the collection period, which in turn determines the average threshold irradiance and the collectable solar energy (from the frequency distribution of solar radiation).

7.2.2 - Hot water load

The hot water load is a basic determinant of the system performance. The theoretical load is characterized by the following parameters:

m	daily hot water demand	(kg).
T_m	cold mains water temperature	(°C).
T_h	required hot water temperature	(°C).

These parameters are assumed to be constant for a given month. Typical values and data sources are given in section 5.3.

The corresponding theoretical load (in energy terms) for N days is given by:

$$L_1 = N \cdot m \cdot c \cdot (T_h - T_m) \quad . \qquad (7.2.1)$$

where c is the heat capacity of water.

Furthermore, the daily draw-off profile must be checked (to avoid extreme cases leading to insufficient accuracy); it is assumed to be the same for each day in the month and it is determined by the following parameters:

q_{sh} fraction of daily demand drawn off during hour h (sum of q_{sh} over the day equals 1).

t_s daily draw-off time (hours or s).

If heat losses from the hot water distribution loop, the recirculation loop, and the auxiliary tank are considered, the actual load is higher than the theoretical load. By adding to the theoretical load, the fraction of those losses which can be met by solar energy, we define an effective DHW load.

The effective DHW load is characterised by two parameters

m_e effective daily hot water demand (kg)

T_{he} effective required temperature of hot water (°C)

$$L_1 = N \cdot m_e \cdot c \cdot (T_{he} - T_m) \quad . \qquad (7.2.2)$$

For example, consider the losses from the auxiliary tank (series-connected auxiliary tank). If the temperature in the solar tank is high enough (higher than the required temperature T_h), these losses can be met by solar energy. On the other hand, in the alternative auxiliary coupling (series-connected auxiliary tank with a controlled three-way valve), heat losses can never be met by solar energy: they are not therefore included in the effective load.

Thus, the computation of the effective load depends on the type of system schematic: Equations are given in section 5.3.5 and summarized in Table 5.3.2 (p.174).

This calculation requires a few additional parameters for the hot water distribution circuit, the recirculation loop, and the auxiliary tank:

A_dU_d	rate of heat loss for distribution pipes (W/K).
A_cU_c	rate of heat loss for the recirculation loop (W/K).
t_L	daily operating time of the recirculation loop.
A_aU_a	rate of heat loss for the auxiliary tank (W/K).
M_a	storage capacity of the auxiliary tank (kg).
t_h	duration of the day (24 hours).
T_{h1}	starting set-point temperature (°C) (only for on-off controlled auxiliary).
T_{h2}	stopping set-point temperature (°C).
t_a	auxiliary daily operating time (hours) (only for off-peak energy auxiliary).

The temperature at the outlet of the tank must enable the user to get the required temperature T_h at the tap; it must be high enough to cover the heat losses (e.g., from the distribution circuit). Thus, the following conditions should be achieved:

- for an on/off controlled auxiliary tank

$$T_{h2} > T_{h1} > T_{h\,min} \quad , \text{ and}$$

- for an off-peak energy auxiliary tank

$$T_{h2} > T_{h\,min}'$$

(in this case, energy stored in the auxiliary tank at the end of the off-peak energy period must be sufficient to cover the hot water load plus the whole heat losses)

where

$$T_{h\,min} = T_h + t_s \cdot (T_h - T_e) \cdot (A_dU_d + A_cU_c/2) \,/\, mc \ , \qquad (7.2.3)$$

and

$$T_{h\,min}' = \text{MAX} \begin{cases} (T_{hmin} - x \cdot T_e) \,/\, (1 - x) \\ T_m + m \cdot (T_h - T_m) \,/\, M_a \\ \quad + (T_h - T_e) \cdot (A_cU_c \cdot t_L + A_dU_d \cdot t_s) \,/\, M_a c \\ \quad + A_aU_a \cdot (t_h - t_a) \cdot (T_{h\,min} - T_e) \cdot (1 + x/2) \,/\, M_a c \end{cases} \qquad (7.2.4)$$

with

$$x = (t_h - t_a) \cdot A_aU_a \,/\, (M_a c) \quad . \qquad (7.2.5)$$

If heat losses from the distribution and recirculation loop or from the auxiliary tank cannot be estimated easily, a default value may be adopted as a fraction of the theoretical load (e.g., 20%); the required hot water temperature is thus increased in proportion to obtain T_{he}.

It will be assumed in all the calculations that, if the temperature of the hot water in the tank is too high, the volume of demand will be reduced accordingly (so as to maintain a constant energy demand). This is achieved in practice by the mixing valve included in the distribution circuit.

7.2.3 - Solar system parameters

The part of a system design concerned with the collection and storage of solar energy comprises the collectors, connecting plumbing, the optional heat exchanger, and the storage tank.

7.2.3.1 - The collectors (see also section 2.1)
The two basic characteristics of a solar collector are:

- η_0 optical efficiency for normal incidence (effective transmission-absorption, accounting for heat transfer factor); and

- U_L heat loss coefficient (W/m2/K).

These are usually measured simultaneously according to national standards in approved testing centres. Values are provided by solar collector manufacturers.

The heat loss coefficient may be expressed in terms of the operating temperature of the collector, using the two coefficients U_1 (a constant) and U_2 (a proportional term)

$$U_L = U_1 + U_2 \cdot (T_b - T_a) \qquad (7.2.6)$$

It can be seen from this formula that the higher the fluid temperature, the higher the heat loss coefficient (convective and radiative losses increase non-linearly with the absorber plate temperature).

Other essential parameters are:

- A net surface (or aperture) area of the collector (m2). For flat-plate collectors, this surface area corresponds to the area of the transparent cover. For tubular collectors, it corresponds to the absorber area.

- $\dot{m}_0 c_0$ capacitance rate of the fluid passing through the collectors (W/K) (product of the mass flow-rate and the heat capacity). These values should allow for the presence of any antifreeze (see heat capacities of antifreeze mixtures in Appendix C).

In some countries (e.g., France, USA), values of collector characteristics provided by testing laboratories are related to the gross area. In such a case, these values should be corrected with respect to the net area.

The collectors are also defined by an azimuthal orientation (from south), α ,and an inclination (from the horizontal), β. These parameters are used in the calculation of radiation data.

7.2.3.2 - Optional solar loop heat-exchanger
(see also section 2.2.2)

The capacitance rate in the collectors (now referred to as "primary" capacitance rate), as well as two other parameters are required in order to describe the performance of the solar loop heat exchanger:

e_1 heat-exchanger effectiveness; and

$\dot{m}_1 c$ secondary capacitance rate (storage side, between heat-exchanger and tank) (W/K).

If the heat-exchanger is defined by the quantity of heat P transferred for primary and secondary inlet temperatures T_0 and T_1 respectively, then e_1 can be obtained from the equation

$$P = C_{min} \cdot e_1 \cdot (T_0 - T_1) \quad (7.2.7)$$

where C_{min} is the smaller of the two values $\dot{m}_0 c_0$ and $\dot{m}_1 c$. In the case of an internal heat-exchanger, the secondary capacitance rate is considered to be infinite.

If the heat-exchanger is defined by its overall heat transfer coefficient $A_e U_e$, its effectiveness may be computed from the usual heat-exchanger formulas (see section 2.2.2).

When there is no heat-exchanger (direct loop), the following values should be used:

$$e_1 = 1 \quad \text{and} \quad \dot{m}_1 c = \dot{m}_0 c_0 .$$

Other parameters used in the solar loop are:

$A_{in} U_{in}$ total rate of heat loss from external pipes on the inlet side of the collectors (see section 2.2.1).

$A_{ou} U_{ou}$ as above, for the outlet side of the collectors.

These two coefficients are in units of W/K. It is advisable to include a further factor for other heat loss paths such as pumps and valves which can easily double the losses from pipes.

7.2.3.3 - The storage tank (see also section 2.2.3)

The storage tank is defined by:

M mass of water in kg.

A_sU_s rate of heat loss (W/K).

A_pU_p rate of heat loss (W/K) of indoor connecting pipes.

T_e local air temperature surrounding the tank (°C).

If the auxiliary heater is located in the upper part of the solar tank, the lower part of the tank is not affected by this auxiliary source. Only this lower part is considered in the storage capacity and in the storage heat losses (heat losses from the auxiliary part are included in the load).

7.2.4 - Climatic data

The same climatic parameters are required for ESM2 as for ESM1 (see section 6.2.5), with the exception of degree-days which are no longer necessary.

Data sources and computation algorithms are described in Chapter 4 (Cumulative frequency curves, CFC's, of solar irradiance; incidence angle modifier) and section 5.2 (Temperature). As ESM2 cannot be used manually, but only as a software package, the user does not need to gather climatic data. Climatic data are already included in ESM2 diskettes.

7.3 - DETAILED DESCRIPTION OF THE ESM2 METHOD

The different steps in the calculation are discussed and summarised in Table 7.3.1.

Monthly calculations are grouped into four sections: initial computations, repetitive day simulation, day-time iteration, and monthly and annual evaluation. Estimation of the hot water load (with heat losses from distribution and auxiliary circuit, as described above) is not included in this standard version. The load is assumed to be an input parameter, characterized by an effective daily demand, m_e; a mains cold water temperature, T_m; and an effective required hot water temperature, T_{he} (subscript $_e$ has been omitted in this section to simplify notation).

7.3.1 - Initial steps

Step 1
The collector heat loss coefficient is calculated by assuming that the collectors operate at a temperature which is a weighted average of T_m and T_h.

Step 2
A provisional value for the collector flow-rate correction factor is calculated. This will be modified after correction for the losses from the plumbing. This factor is not an indication of the collector output; it merely shows the decrease in output due to the fact that the working fluid does not extract all the available heat.

Step 3
This step corrects the solar collector data (namely the optical efficiency η_0 and the heat loss coefficient U_L) to allow for the losses from external piping. A correction factor $\overline{K}$ (monthly incidence angle modifier) is introduced to account for the variation in absorption and transmission of the cover with increasing angles of incidence (see section 6.2.5).

Step 4
This step evaluates the efficiency of the complete solar loop F_r', taking the optional heat-exchanger into account.

Where there is no heat-exchanger, $F_r' = F_r^*$ (see section 2.2).

Steps 5 and 6
Initial values (equal to T_m) are given to T_0 and T_{24}, the space -averaged temperatures of the solar storage tank at the beginning and at the end of a 24-hour period.

7.3.2 - Repetitive day simulation

Step 7
An initial value is set for the base temperature for day-time iterations.

Steps 8 to 14
Day-time iterations (see section 7.3.3) are run until convergence is obtained on T_b. These iterations determine the time-averaged storage temperature during the collection period.

Step 15
Energy distributed during the collection period and heat losses from the storage are estimated from the time-averaged storage temperature. The thermal balance of the storage gives an estimate of energy stored at the end of the collection period.

Step 16
Storage temperature at the end of the collection period is then computed.

Steps 17 to 20
Heat losses and energy drawn off from the storage during the night are computed. Draw-off is assumed to operate in a plug-flow mode with two zones of varying volume. The upper zone is hot (step 18) and the lower cold (at about temperature T_m).

Step 21
Iteration on storage end-of-day temperature: steps 7 to 20 are repeated until the storage state at the end of the 24-hour period is about the same as the initial state (when T_{24} does not differ significantly from T_0, a periodic state is reached). For these iterations the storage temperature at the end of the 24-hour period is taken as an initial temperature for the next day.

7.3.3 - Day-time iterations

Step 8
The threshold irradiance, a function of the base temperature and the properties of the collectors, is used to calculate the amount of solar energy available at temperature T_b. It is then reduced according to the maximum irradiance I_{max}.

Step 9
The utilizability and the operating time of the collectors can be calculated in several ways. Complete formulas are given in Chapter 4.

Step 10
The water draw-off is assumed to be constant during the 24-hour period.

Step 11
Stratification coefficients are estimated according to the theory developed in section 3.3. Storage tanks with a submerged heat exchanger are assumed to be fully mixed during collector operation and stratification coefficients are set equal to one.

Step 12
The time-averaged storage temperatures T_s (of the whole tank) and T_{bot} (of the bottom part) are calculated, assuming a constant irradiance during collector operation according to the average storage temperature procedure (described in section 3.3).

The calculation procedure differs, according to whether or not at least one complete storage cycle takes place during the collection period.

Step 13
A corrected estimate of the average collector inlet temperature is made from the time-averaged bottom storage temperature. This new estimate will be used as the base temperature in the next iteration.

Step 14
Steps 7 to 13 are repeated until convergence (when the difference between T_b^* and T_b becomes sufficiently small).

7.3.4 - Monthly and annual evaluation

Step 22
The monthly contribution is computed; the monthly contribution cannot exceed the effective hot water load.

The annual contribution is calculated as the average of the total monthly solar contributions, weighted according to the total monthly loads.

Table 7.3.1:

ESM2 COMPUTATION OF SHWS MONTHLY AVERAGE PERFORMANCES : STANDARD VERSION

INITIALISATION STEPS

1- collector heat loss coefficient :

$$T_B = 0.6.T_M + 0.4.T_H$$

$$U_L = U_1 + U_2 .(T_B - T_A)$$

2- collector heat removal factor

$$F_R = \frac{\dot{m}_o C_o}{A . U_L} . \left[1 - \exp\left(- \frac{A . U_L}{\dot{m}_o C_o}\right)\right]$$

3- corrections for pipe losses

$$U_L^* = U_L + \frac{A_{IN}U_{IN} + A_{OUT}U_{OUT}}{A}$$

$$F_R^* = \frac{\dot{m}_o C_o}{A . U_L^*} . \left[1 - \exp\left(- \frac{A . U_L^*}{\dot{m}_o C_o}\right)\right]$$

$$\eta_o^* = \bar{K} . \eta_o . \frac{F_R}{F_R^*} . \exp\left(- \frac{A_{OUT}U_{OUT}}{\dot{m}_o C_o}\right)$$

4- efficiency of solar loop (with a heat-exchanger)

$$C_{MIN} = MIN (\dot{m}_o C_o , \dot{m}_1 C)$$

$$F_R' = \frac{F_R^*}{1 + \frac{A . F_R^* . U_L^*}{\dot{m}_o C_o} . \left(\frac{\dot{m}_o C_o}{e_1 . C_{MIN}} - 1\right)}$$

5- set $T_{24} = T_M$

6- initial storage temperature : set $T_0 = T_{24}$

REPETITIVE DAY SIMULATION

7- base temperature for collection (first guess)

set $T_B = T_{24}$

DAY TIME ITERATION

8- reduced threshold

$$Y_C = \frac{U_L^*}{\eta_o^*} . \frac{T_B - T_A}{I_{MAX}}$$

9- computation of utilisability (ϕ) and collector operating time (t_{OP}) from cumulative frequency curve of solar radiation

10- draw-off characteristics

- constant hot water draw-off flow rate : $\dot{m}_3 = m / 24$
- quantity of hot water to be drawn continuously
  - during collection : $m_{OP} = \dot{m}_3 . t_{OP}$
  - during night-time : $m_N = \dot{m}_3 . (24 - t_{OP})$

11- stratification coefficients

* external heat exchanger

$$z = 1 - \frac{\dot{m}_3}{\dot{m}_1}; \quad a = 1 - \frac{A.F'_R.U_L}{\dot{m}_1\, C}; \quad a_1 = a.z.\exp(- \frac{A_sU_s}{\dot{m}_1 C\, .z})$$

$$G_{LOSS} = \frac{\dot{m}_1\, C\, .\, z}{A_sU_s} \cdot \left[1 - \exp\,(- \frac{A_sU_s}{\dot{m}_1\, C\, .\, z}\,)\right]$$

$$K_1 = G_{LOSS} \cdot \frac{\ln(1\, /\, a_1)}{(1 - a_1)}; \quad K_2 = K_1 \cdot \frac{\dot{m}_1 C.z}{A_sU_s} \cdot \left[\frac{1 - a_1}{G_{LOSS}} + a.z - 1\right]$$

* submerged heat exchanger : $z = K_1 = K_2 = 1$

* half-mixed storage : $z = 1$

12- storage average state over collection period

$$T_L = T_B + \left[\frac{\eta_0^*}{U_L^*} \cdot \frac{\phi\, .\, H_T}{t_{OP}} \right]$$

$$n = \frac{t_{OP}\, .\, \dot{m}_1\, .\, z}{M}$$

* $n > 1$

$$T_{STR} = \frac{K_2.A_sU_s.T_E + K_1.z.(A.F'_R.U_L^*.T_L + \dot{m}_3 C.a.T_M)}{K_2.A_sU_s + K_1.z.(A.F'_R.U_L^*.z + \dot{m}_3 C\,)}$$

$$t_{STR} = \frac{M\, .\, C}{K_2.A_sU_s + K_1.z.(A.F'_R.U_L^*.z + \dot{m}_3 C\,)}$$

$$G_{STR} = \frac{t_{STR}}{t_{OP}} \cdot \left[1 - \exp\,(\, - \frac{t_{OP}}{t_{STR}}\,)\right]$$

$$T_S = T_O + (T_{STR} - T_O) \cdot (1 - G_{STR})$$

$$T_{BOT} = K_1 \cdot T_S + \frac{a_2}{1-a_1} \cdot (G_{LOSS} - K_1) \cdot T_L + \frac{a_3}{1 - a_1} \cdot (G_{LOSS} - K_1) \cdot T_M$$

$$+ \left[1 - G_{LOSS} + \frac{a_4}{1 - a_1} \cdot (G_{LOSS} - K_1)\right] \cdot T_E$$

with

$a_2 = (1 - a) \cdot G_{LOSS}$; $a_3 = a \cdot (1 - z) \cdot G_{LOSS}$; $a_4 = 1 - a_1 - a_2 - a_3$

* $n \leq 1$

$$T_{BOT} = T_o + (T_E - T_o) \cdot (1 - \frac{t_{LOSS}}{t_{OP}}) \cdot \left[1 - \exp(- \frac{t_{OP}}{t_{LOSS}})\right]$$

13- computation of base temperature

$$T_B^* = z \cdot T_{BOT} + (1 - z) \cdot T_M$$

14- repeat steps 8 to 14 with $T_B = T_B^*$ until $|T_B^* - T_B| \leq \varepsilon_B$

15- energy balance over collector operation period

$$E_C = A \cdot F_R' \cdot U_L^* \cdot t_{OP} \cdot (T_L - T_B^*)$$

$$E_{D,OP} = t_{OP} \cdot (1 - z) \cdot A \cdot F_R' \cdot U_L^* \cdot (T_L - T_B^*) + z \cdot \dot{m}_3 C \cdot (T_{BOT} - T_M)$$

$$E_{L,OP} = t_{OP} \cdot A_S U_S \cdot (T_S - T_E)$$

$$E_{S,OP} = E_{C,OP} - (E_{D,OP} + E_{L,OP})$$

16- average temperature of storage at the end of collection period

$$T_{OP} = \frac{E_{S,OP}}{M \cdot C} + T_o$$

NIGHT-TIME SIMULATION

17- period duration : $t_N = 24 - t_{OP}$

18- storage hot zone evolution

- time-averaged temperature during t_N :

$$T_{S,H} = T_E - \frac{(T_{OP} - T_E)}{(k+1) \cdot t_N \cdot \dot{m}_3} \cdot \left[\frac{(M - \dot{m}_3 \cdot t_N)^{k+1}}{M^k} - M\right]$$

with $k = \dfrac{A_S U_S}{\dot{m}_3 . c}$

- average temperature at hour 24 :

$$T_{24,H} = (T_{OP} - T_E) . (1 - \frac{\dot{m}_3 . t_N}{M})^k + T_E$$

19- storage cold zone evolution

- time-averaged temperature during t_N, and average temperature at hour 24 :

$$T_{S,C} = T_{24,C} = \frac{A_S U_S . T_E + \dot{m}_3 . c . T_M}{A_S U_S + \dot{m}_3 . c}$$

20- storage status at hour 24

	$m_N \geq M$	$m_N < M$
- hot zone mass, m_H :	0	$M - m_N$
- cold zone mass, m_C :	M	m_N

- average storage temperature :

$$T_{24} = \frac{m_H . T_{24,H} + m_C . T_{24,C}}{M}$$

21- repeat steps 7 to 20 with $T_o = T_{24}$ until $|T_{24} - T_o| \leq \varepsilon_{24}$

22- energy balance

	$m_N \geq M$	$m_N < M$
- energy drawn, $E_{D,N}$:	$M.c.(T_{S,H} - T_E)$	$m_N.c.(T_{S,H} - T_E)$

- storage thermal losses :

$$E_{L,N} = A_S U_S . t_N . [(T_{S,H} - T_E) + (T_{S,C} - T_E)]$$

MONTHLY ENERGY BALANCE

$E_L = N . E_C$

$E_D = N . (E_{D,OP} + E_{D,N})$; $E_D = MAX (E_D , L_1)$

$E_L = N . (E_{L,OP} + E_{L,N})$

7.4 - WORKED EXAMPLE OF A SOLAR HOT WATER SYSTEM

The system presented here is for a small hotel located in the south of France near Carpentras. First the load is analysed, then a base case solar system is chosen and its thermal performance computed. Lastly, a few sensitivity analyses are performed.

7.4.1 - Hot water load

The hot water load is assumed to be 800 litres/day at 50°C; the mains cold water temperature is constant and equal to 10°C. The theoretical load is therefore L_t = 133.9 MJ (37.2 kWh) per day.

The system considered here is shown in Fig. 7.1.3. It has the following features:

- Auxiliary tank with controlled by-pass.
- Auxiliary energy: off-peak electricity.
- Recirculation loop connected to solar and auxiliary tank.
- Solar loop: external heat exchanger.

Characteristics of the distribution circuit and auxiliary tank are given in Table 7.4.1.

Distribution losses are (from Eq. 5.3.3) L_d = 4.3 MJ/day (1.2 kWh). Recirculation loop losses (from Eq.s 5.3.4 and 5.3.5) are much higher: L_c = 34.6 MJ/day (9.6 kWh).

We assume that the stopping set-point of the auxiliary T_{h2} is not known a priori, but we shall take for T_{h2} the minimum value $T_{h\ min}'$ given by Eq.7.2.4. $T_{h\ min}'$ is the greater of two terms. The upper term in Eq. 7.2.4 (here it equals 57.1°C) represents the temperature required to obtain $T_{h\ min}$ (here 53.9°C from Eq. 7.2.3) in the top of the tank at the end of the draw-off period (which is also the beginning of the off-peak period). The intermediate result required by the calculation is the value of x = 0.086 (from Eq. 7.2.5).
The lower term in Eq.7.2.4 represents the temperature required in the auxiliary tank to store sufficient energy to meet the hot water load and all heat losses during the draw-off period for a day without any solar contribution (in the case considered here, it is 64.6°C). Thus we derive

$$T_{h\ min}' = \text{MAX}\ (\ 57.1\ ;\ 64.6\ ) = 64.6°C$$

Table 7.4.1: Base case: Characteristics of the distribution circuit and auxiliary tank.

Indoor air temperature:	T_e = 20°C
Recirculation loop:	A_cU_c = 20 W/K t_L = 16 hours = 57.6 10^3 s
Distribution circuit:	A_dU_d = 5 W/K t_s = 8 hours = 28.8 10^3 s
Auxiliary tank:	A_aU_a = 5 W/K t_a = 8 hours = 28.8 10^3 s M_a = 800 kg

Table 7.4.2: Characteristics of the base case solar system.

Collector:	Tilt angle: Latitude Orientation: South Net area A = 20 m2 η_0 = 0.8 $\dot{m}_0c_0$ = 1000 W/K $A_{in}U_{in}$ = $A_{ou}U_{ou}$ = 0	 U_L = 7 W/K/m2
Heat-exchanger	$\dot{m}_1c$ = 1000 W/K A_eU_e = 1500 W/K	
Storage:	M = 800 kg A_sU_s = 8 W/K A_pU_p = 6 W/K	

Nevertheless, a precise knowledge of auxiliary tank heat losses is not required for solar system performance evaluation because they do not influence the effective demand (this would not be the case with a series - connected auxiliary tank without a three-way valve).

The effective load, derived from the equations of Table 5.3.2 (p.174), incorporates, besides the theoretical load, distribution and recirculation losses (all the recirculation losses because of the double connection of the recirculation loop, both to the auxiliary tank and the solar tank)

$$L_e = L_t + L_d + L_c = 172.8 \text{ MJ/day } (48 \text{ kWh}).$$

This corresponds to an effective demand $m_e = m = 800$ kg/day (as above) and an effective required hot water temperature $T_{he} = 61.6$ °C (to be compared with the required hot water temperature at the tap $T_h = 50$°C). This effective load will affect the solar system behaviour (and the theoretical load will not).

The residual load will be the auxiliary tank heat loss, which will be computed after the solar system performance.

7.4.2 - Base case solar system performance

Input parameters of the base case solar system are given in Table 7.4.2. Monthly results from ESM2 are reproduced in Table 7.4.3. Useful solar energy is equal to 43.4 GJ (compared with a yearly effective load of 63.1 GJ), yielding a solar fraction of 0.68.

Auxiliary tank heat losses can then be computed from an estimate of the average solar tank outlet temperature T_{sol}

$$T_{sol} = T_m + f.L_e/m_e c = f.T_{he} + (1-f).T_m = 45°\text{C}.$$

This yields (from Eq.s 5.3.13 and 5.3.15) auxiliary tank heat losses of about 6.1 GJ (with T_{h2}= 64.6°C) and a total auxiliary energy consumption of 25.8 GJ. The actual solar fraction (accounting for the total system load, and not only for the effective load) is then 0.63.

SYSTEM PERFORMANCE SUMMARY

Month	Solar Irrad. (MJ/m2)	Ambient Temp. (°C)	Total Load (MJ)	Auxiliary Energy (MJ)	Solar Energy (MJ)	Solar Fraction -	Recirc. Loop Losses (MJ)
Jan.	12.4	5.7	5812	3203	2609	0.449	1071
Feb.	15.6	7.4	5281	2421	2860	0.542	968
Mar.	17.7	9.1	5862	2455	3406	0.581	1071
Apr.	19.2	11.5	5681	2246	3435	0.605	1037
May	21.0	15.9	5894	1957	3937	0.668	1071
June	23.9	19.6	5735	1378	4357	0.760	1037
July	25.1	22.3	5942	1064	4878	0.821	1071
Aug.	22.8	22.1	5933	1286	4647	0.783	1071
Sep.	22.4	18.4	5741	1259	4482	0.781	1037
Oct.	17.9	13.4	5886	2078	3808	0.647	1071
Nov.	14.3	8.5	5652	2691	2961	0.524	1037
Dec.	9.3	6.1	5776	3725	2051	0.355	1071
Year	18.5	13.4	69194	25762	43432	0.628	12614

Table 7.4.3: ESM2 results (base case system).

Modification with respect to base case	Effective load (GJ)	Residual load (GJ)	Solar fraction (--)	Energy output per m2 of coll. (MJ/m2)
Base case	63.1	6.1	0.628	2170
Selective collector U_1=5 W/m2/K U_2=0.015 W/m2/K2	63.1	6.8	0.714	2493
Piping heat losses $A_{in}U_{in} = A_{ou}U_{ou}$ = 25 W/K	63.1	5.8	0.550	1895
Poor heat-exchanger A_eU_e / A = 30 W/m2/K	63.1	6.0	0.590	2037
Lower demand (m = 600 l/day)	50.9	6.4	0.628	1796
Higher demand (m =1000 l/day)	75.3	5.8	0.611	2474
Series auxiliary (without controlled by-pass)	68.5	0.0	0.634	2170

Table 7.4.4: ESM2 results (variants of the base case system).

7.4.3 - Sensitivity analysis and optimization

The base case system, whose performance has been studied above, is compared with several other variants in Table 7.4.4.

Figure 7.4.1 shows the influence of variation of the collector area, as estimated from ESM2 (other system parameters remain unchanged).

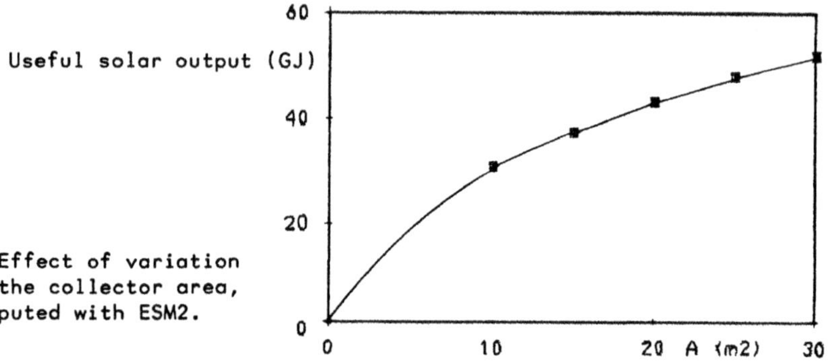

Fig.7.4.1: Effect of variation of the collector area, computed with ESM2.

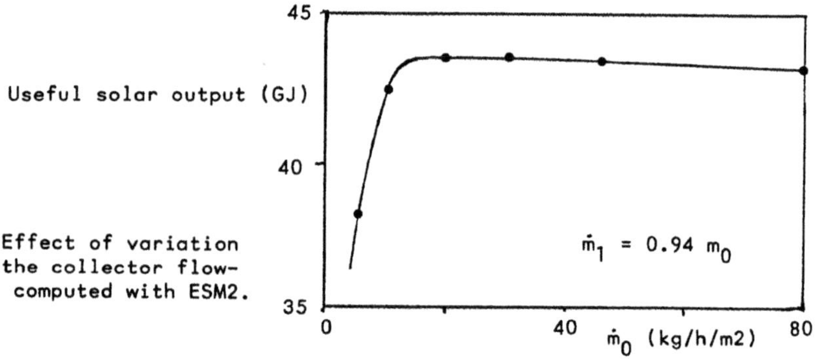

Fig.7.4.2: Effect of variation of the collector flow-rate, computed with ESM2.

Figure 7.4.2 shows the effect of variation of the collector flow-rate at constant heat transfer coefficient (the secondary flow-rate on the other side of the heat exchanger is assumed to be equal to the collector flow-rate).

7.4.4 - Consequences of sizing errors

Method ESM2 can also be used to study the consequences of sizing errors. At the design stage of any system, constraints are sometimes not well defined. For example, the designer must make assumptions about the hot water demand, but what are the consequences of, say, a 25% error?

If the daily demand is 600 litres/day instead of the 800 l/day assumed in the worked example (i.e. a reduction of 25%), the useful solar energy output is reduced by 16% (36.2 GJ instead of 43.4); if the daily demand is 1000 litres/day (+25%), it is increased by 13% (48.7 GJ). From these examples it is easy to see that optimal economic design of a solar hot water system depends greatly on a good provisional estimate of the demand.

It should be noted that the sizing approach is not identical for the auxiliary energy system and for the solar system: a safety coefficient must be applied to the auxiliary part, which on some days must be able to meet a greater demand than expected on average. The solar system however must be sized and optimized for its average performance and not for extreme situations.

Other problems, which are often encountered between design and actual system operation, are related to defective installation. What happens, for example, if pipe losses in the collector loop are greater than anticipated due to poor insulation (e.g., 50 W/K)? These problems can be considered at an early design stage using a computation method such as ESM2.

7.5 - ESM2 VALIDATION

It has been established from the algorithms described in section 7.3 that ESM2 is a purely analytical procedure which does not include "empirical" corrective terms derived from numerous detailed simulations (unlike ESM1 which incorporates such correlations). Nevertheless, it is necessary to try to validate this simplified simulation approach, by comparison with more accurate calculations. This problem is addressed here by comparing a sample of ESM2 results with EMGP2 results. Two levels of validation are distinguished: the first level concerns the "day-time iterations" (or the time - averaged storage temperature procedure), called the daily simulation. The second level of validation concerns the method as a whole, including the "repetitive CFC-day" assumption.

7.5.1 - Validation of daily simulation

In this first approach, a comparison is made between ESM2 and EMGP2 for a set of typical days at different times of the year and starting from various initial temperatures.

The system simulated has the following characteristics:

Collector $A = 20$ m2 $\eta_0 = 0.80$ $U_L = 7$ W/K/m2
heat capacity: 12 kJ/K/m2
$d_{on} = 4K$ $d_{off} = 2K$

No heat-exchanger.

Storage $M = 800$ kg $A_sU_s = 20$ W/K.

No draw-off (comparisons have been made of the energy stored in the tank at the end of the day).

Two variants have been tested: one fully-mixed, with a typical collector flow-rate (0.8 m3/hour); the other, highly stratified, with a low collector flow-rate (0.1 m3/hour).

The comparison shows clearly for both cases (Figs. 7.5.1 and 7.5.2) that no systematic bias appears and that the scattering remains reasonable. This is confirmed by the values of the mean bias error (MBE) and root mean square error (RMSE):

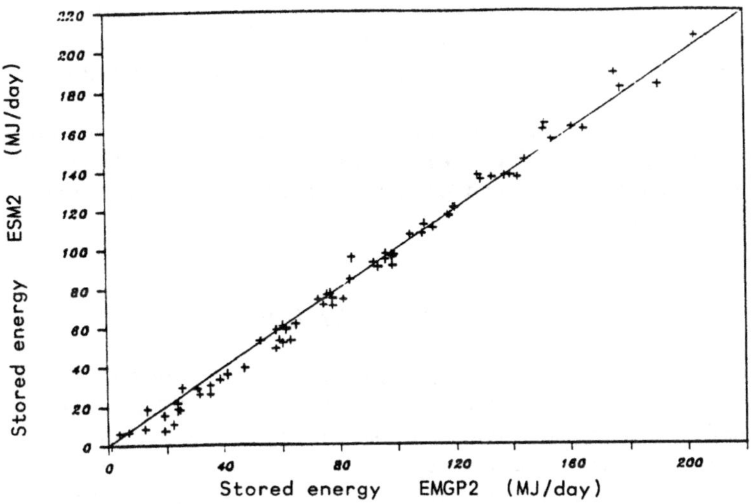

Fig. 7.5.1: Validation of the daily simulation algorithm of ESM2. Stored energy for typical days and several initial temperatures: Comparison of values computed by EMGP2 and ESM2 (mixed tank).

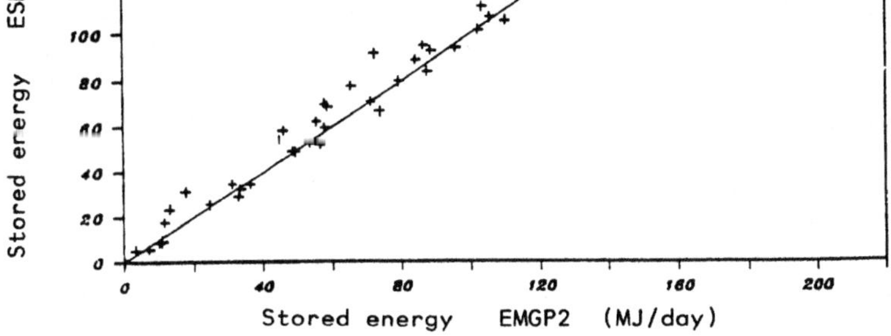

Fig. 7.5.2: Idem Fig. 7.5.1 (perfectly stratified storage tank).

	Mixed case	Stratified case
MBE	1.20 MJ/day	-1.26 MJ/day
RMSE	5.1 MJ/day	7.7 MJ/day

The corresponding average energy output is in both cases about 75 MJ/day: The mean bias error represents roughly 1.5% of this average. Complete results of this comparison have been analysed in reference /7.1/.

7.5.2 - Validation of complete algorithms

The complete set of ESM2 algorithms has been tested with several yearly simulations, still using EMGP2 as a reference programme. The solar system simulated is represented in Figure 7.5.3 with input parameters of the base case. Meteorological data are hourly files of Test Reference Years (TRY) for Trappes.

A first set of results is shown in Table 7.5.1 for the base case and several variants (differing by their storage capacity, daily demand or collector area). The accuracy is very good, differences between ESM2 and EMGP2 being smaller than 5%. Sensitivities of EMGP2 are well reproduced by ESM2; for example, hot water demand multiplied by 2 results in a performance improvement (with respect to the base case) of 34.2% with EMGP2 and 34.3% with ESM2.

Other parametric variations are shown in Tables 7.5.2 to 7.5.6. Sensitivities of the system performance to essential parameters are reproduced very well: storage heat loss coefficient; storage environment temperature; heat exchanger effectiveness; and mains water temperature. The effects of variations in collector surface area are also reproduced well by ESM2 if this area remains not too high (Fig. 7.5.4). However, from 0.04 m2/litre of demand, a systematic overestimate occurs with ESM2.

Sensitivity to storage capacity is correct (Fig. 7.5.5), with a slight tendency to underpredict the performance of systems with a large capacity.

The sensitivity to the collector flow-rate has also been investigated, with constant heat exchanger effectiveness (Fig. 7.5.6) or constant heat transfer coefficient (Fig. 7.5.7). In the last case, the optimum collector flow-rate is indicated well by ESM2.

Finally, results are shown for other types of collectors and other locations (Table 7.5.7).

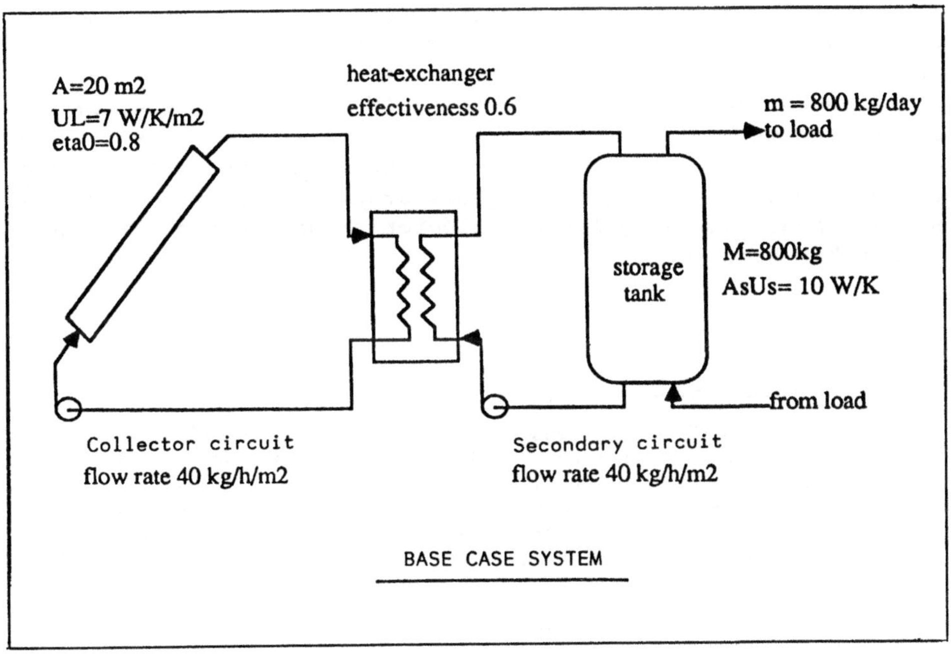

Fig. 7.5.3: Schematic of the hot water system simulated by EMGP2 for ESM2 validation purposes.

Fixed parameters: Load		Demand	m = 800 l/day
		Cold water	T_m = 10°C
		Hot water	T_h = 55°C
		Draw-off:	300 l from 7 to 8 A.M.
			200 l from 1 to 2 P.M.
			300 l from 7 to 8 P.M.
Base case solar system		Collector	A = 20 m2
			η_0 = 0.80
			U_L = 7 W/K/m2
			$\dot{m}_0 c_0$ = 1090 W/K
	External	Heat-exch.	$\dot{m}_1 c$ = 1160 W/K
			e_1 = 0.6
		Storage	M = 800 kg
			$A_s U_s$ = 10 W/K

The main conclusion of this validation exercise is that the accuracy of the simplified method ESM2 with respect to the detailed simulation programme EMGP2 is very satisfactory: Differences are always smaller than 10%, even 5% in most of the cases. The method is suited for common domestic hot water draw-off profiles, but in its present state, it cannot account for extreme profiles (e.g., complete draw-off in the early morning). Except for very large collector areas, no explicit limit of use has been found.

REFERENCES OF CHAPTER 7

/7.1/ L.KADI. Le calcul analytique des performances des systèmes solaires actifs de production d'eau chaude: Modèles et méthode simplifiée. Thèse de Docteur-Ingénieur, Ecole Centrale de Paris (1989).
/7.2/ L.KADI, B.BOURGES et J.ADNOT. Validation de la méthode ESM2 de calcul des performances thermiques des systèmes solaires actifs de production d'eau chaude. Centre d'Energétique, Ecole des Mines de Paris (1989).

Table 7.5.1: Validation of ESM2 versus EMGP2: Comparison of useful solar energy output (MJ) in several cases (Location: Trappes).

Storage capacity		E_{emgp2}	E_{esm2}	Diff.
400 l		25535	26399	+3.4%
600 l		27915	29104	+4.3%
800 l	Base case	28990	29104	+0.4%
1200 l		29710	28569	-3.8%
1600 l		29799	28334	-4.9%

Daily demand		E_{emgp2}	E_{esm2}	Diff.
400 l		18108	18711	+3.3%
800 l	Base case	28990	29104	+0.4%
1600 l		38885	39110	+0.6%

Extreme cases		E_{emgp2}	E_{esm2}	Diff.
A = 10 m2	M = 400 kg	19261	19551	+1.5%
A = 40 m2	M = 1600 kg	37729	38143	+1.1%

Table 7.5.2: ESM2 validation: Sensitivity to storage tank heat loss.

	E_{emgp2}	E_{esm2}	Diff.
A_sU_s = 2 W/K	30238	30602	+1.2%
A_sU_s = 10 W/K (Ref.)	28990	29104	+0.4%
A_sU_s = 20 W/K	27594	27450	-0.5%

Table 7.5.3: ESM2 validation: Sensitivity to tank environment temperature.

	E_{emgp2}	E_{esm2}	Diff.
T_e = 10 °C	27154	26416	-2.7%
T_e = 20 °C (Ref.)	28990	29104	+0.4%
T_e = 30 °C	30824	31886	+3.4%

Table 7.5.4: ESM2 validation: Sensitivity to heat-exchanger effectiveness.

	E_{emgp2}	E_{esm2}	Diff.
e_1 = 0.4	27610	27369	-0.9%
e_1 = 0.6 (Ref.)	28990	29104	+0.4%
e_1 = 0.8	29729	30080	+1.2%

Table 7.5.5: ESM2 validation: Sensitivity to mains cold water temperature.

	E_{emgp2}	E_{esm2}	Diff.
T_m = 10 °C (Ref.)	28990	29104	+0.4%
T_m = 15 °C	25416	25408	-0.0%
T_m = 20 °C	21818	21731	-0.4%

Table 7.5.6: ESM2 validation: Sensitivity to collector surface area.

	E_{emgp2}	E_{esm2}	Diff.
A = 5 m2	13228	13156	-0.5%
A = 10 m2	20314	19797	-2.5%
A = 20 m2 (Ref.)	28990	29104	+0.4%
A = 30 m2	33235	35678	+7.3%
A = 40 m2	35594	39871	+12.0%

Table 7.5.7: ESM2 validation: Sensitivity to collector type and location.

		E_{emgp2}	E_{esm2}	Diff.
Ordinary flat-plate				
Carpentras		43090	44144	+2.4%
Trappes	(Ref.)	28990	29104	+0.4%
Copenhagen		27838	29181	+4.8%
Unglazed collector				
Carpentras		27582	25200	-8.6%
Trappes		17395	17108	-1.6%
Copenhagen		15785	16741	+6.1%
Evacuated tubes				
Carpentras		50408	52361	+3.9%
Trappes		38506	41200	+7.0%
Copenhagen		36846	39881	+8.2%

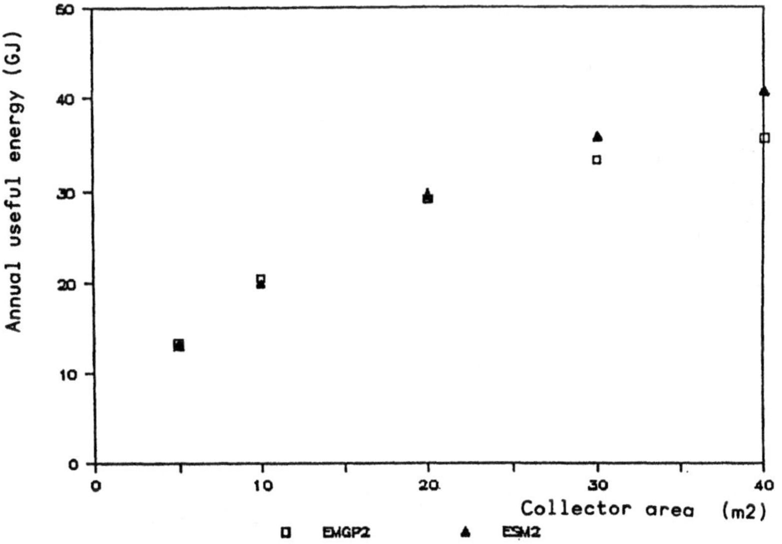

Fig. 7.5.4: Validation of ESM2 versus EMGP2: variation of collector area (ordinary flat-plate collector; location: Trappes).

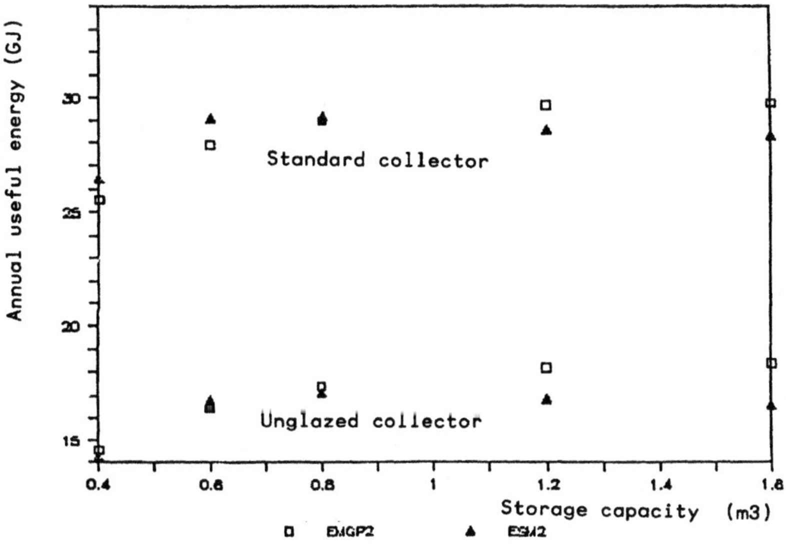

Fig. 7.5.5: Validation of ESM2 versus EMGP2: variation of storage tank capacity (location: Trappes).

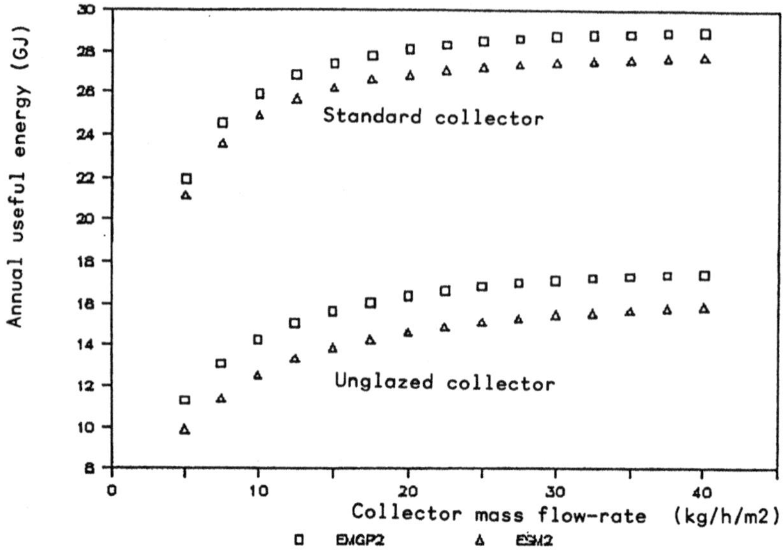

Fig. 7.5.6: Validation of ESM2 versus EMGP2: variation of collector flow-rate (constant heat-exchanger effectiveness).

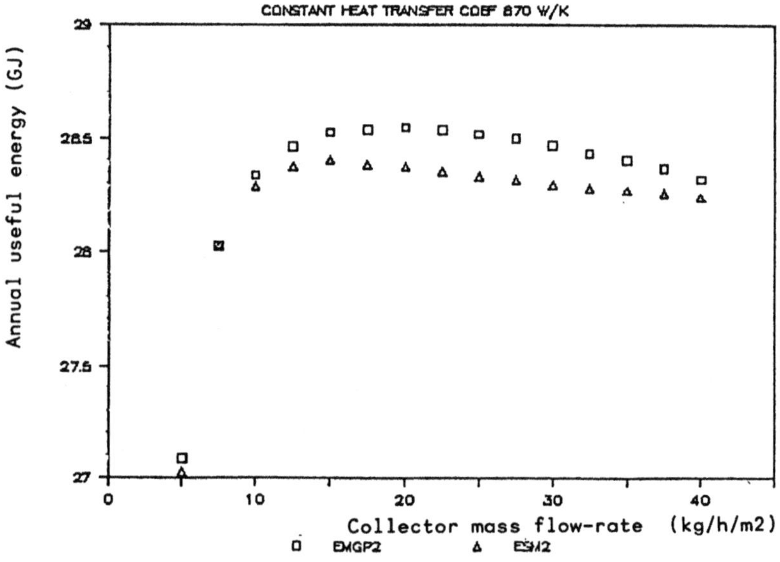

Fig. 7.5.7: Validation of ESM2 versus EMGP2: variation of collector flow-rate (constant heat transfer coefficient).

CHAPTER 8

OPTIMIZATION AND ECONOMIC COMPARISON

An important consequence of the preceding discussions is that no purely technical or energetic optimum can be found for active solar systems (except in a few cases for some particular parameter such as flow-rate). It is evident that system optimization must also take costs into account, and sometimes the economic and energetic environment as well.

This chapter addresses two questions: first, what is the most cost-effective active solar solution - the best combination of components? The answer will be the result of a sizing optimization. Secondly, how does this optimized solar system compare with the cost of other solutions such as a fossil-fuel boiler, electricity, heat pump, passive solar, etc. This second question implies the choice of suitable criteria for an economic comparison. Several of these criteria will be discussed. Finally, a detailed example will be analysed.

The calculations of this optimization procedure can be performed best when using a computer programme, such as the one developed at the Catholic University of Leuven with the help of a spreadsheet.

8.1 - SIZING OPTIMIZATION

The purpose of this first stage in the optimization process is as follows: once a system design and its basic options (reference load, location, and climate; type of heat emitters, auxiliary energy source, type of solar collector) have been chosen, then what is the best combination of these components?

Systems which differ essentially in their initial cost and their yearly thermal performance in terms of energy savings need to be compared. The preliminary choices, regarding system schematic, collector type, and heat emitter type, which have been made, are assumed. Several alternatives can be included; they will be compared later (e.g., two types of solar collectors: selective and non-selective). As the components are identical, system life-time, maintenance costs, etc. are the same. The parameters which are important at this stage, both for the performance and the system cost, are the collector area and the storage capacity (with the first being predominant); both of them can vary continuously within a certain range.

The optimization will consider only the system performance for a given initial cost. It will be seen later how to compare different investments. Calculation of the system performance has already been explained; system cost data are now needed, in order to perform the optimization.

8.1.1 - System cost

Firstly, reliable cost data are needed from manufacturers and installers for a broad range of system sizes. To obtain such data is seldom simple, and data for only a limited number of combinations are available.

It is useful - but not essential and sometimes not possible - to represent the data by a simple linear function of the type

$$Csol = C0 + C1 * A + C3 * M \tag{8.1.1}$$

depending on the collector area A and the storage capacity M.

A quadratic function can sometimes be used; it represents the decreasing marginal cost of collectors with increasing area

$$Csol = C0 + (C1 - C2*A) * A + (C3 - C4*M) * M \tag{8.1.2}$$

The range within which this formula is valid and the type of components included, must be carefully defined.

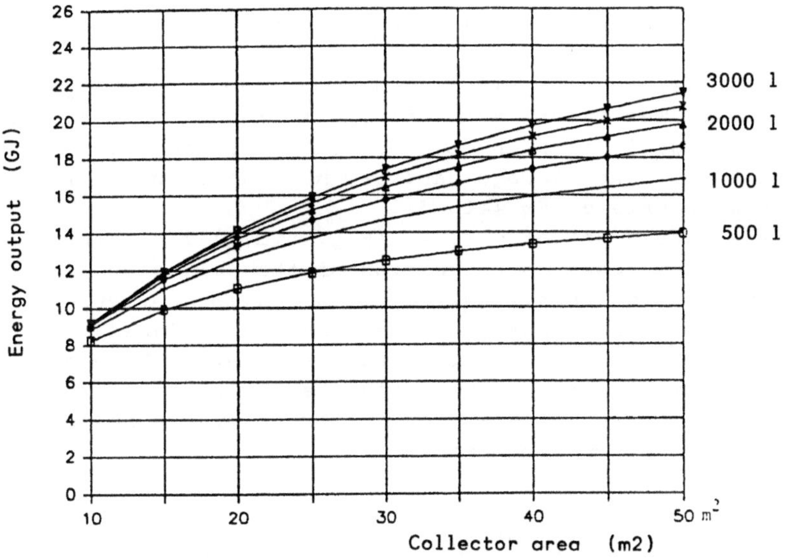

Fig. 8.1.1: Solar energy output versus collector area (parameter: storage capacity).

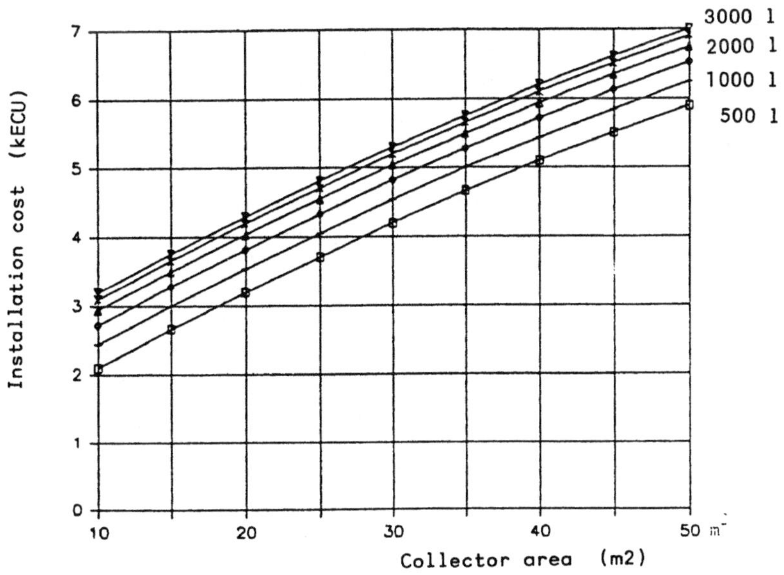

Fig. 8.1.2: Cost function versus collector area (parameter: storage capacity).

8.1.2 - Size optimization for a given initial cost

The available information may be summarized by two sets of curves. The first set (Fig. 8.1.1) gives the yearly energy output Qsol as a function of the collector area A for given values of the storage capacity M. The results are computed by the appropriate ESM method.

The second set of curves (Fig. 8.1.2) represents the cost of the system Csol as a function of the collector area for different storage capacities. This is a simple graphical presentation of the cost data collected by the designer.

From these curves, it is easy to draw a third set of curves (Fig. 8.1.3) which represents the energy output as a function of the collector area for different system costs and clearly shows an optimum in most cases.

8.1.3 - Comparison of alternative investments

The question is now: How to compare different initial costs? A simple criterion is the "thermal performance" of each invested unit of currency. This may be defined as Qsol / Csol, expressed in kWh (or MJ) per year per invested currency unit (Fig. 8.1.4). This ratio is called the specific energy productivity of the investment.

The initial costs may thus be compared by computing this ratio for each optimum-sized system. The different optimization steps are summarized in Fig. 8.1.5 which shows (in the lower left hand corner) the curve of the annual saving per unit capital investment.

Depending on the cost function, this last curve may or may not indicate an optimum. If a linear cost function is chosen (Eq. 8.1.1) an optimum is observed only if the fixed cost C0 is positive; otherwise, if C0 = 0, the optimum is a system with zero collector area (A = 0) due to decreasing system efficiencies (any additional collector array is less productive).

This optimum (in terms of collector area and storage capacity) can then be improved by consideration of additional factors which have a relatively small influence on the system cost and performance: heat-exchanger size, storage insulation, and pump power.

If alternative components are being considered, e.g. selective flat-plate collector versus evacuated tubular collector or under-floor heating versus radiators, their respective optima may now be

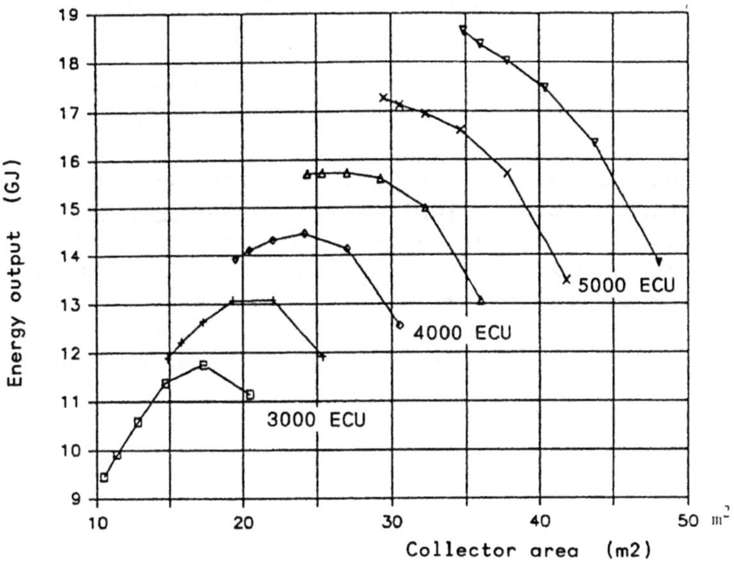

Fig. 8.1.3: Solar energy output versus collector area at several costs.

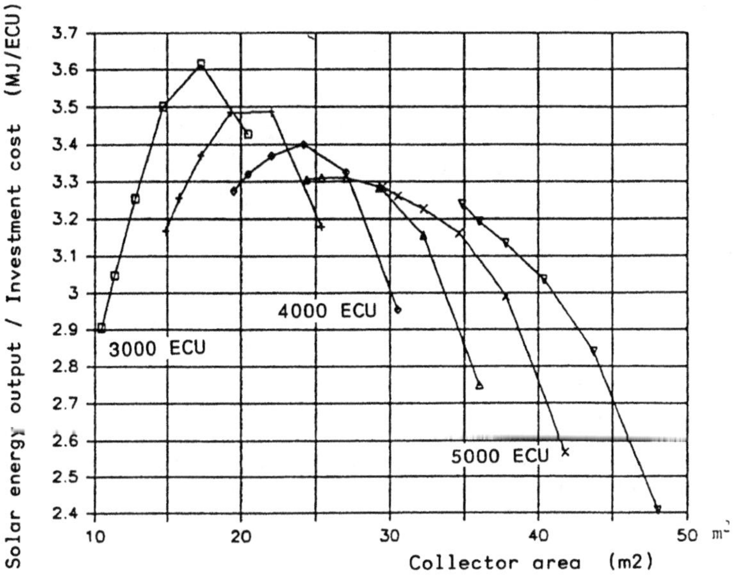

Fig. 8.1.4: Solar energy output per invested ECU versus collector area at constant cost.

compared, still in terms of the specific productivity of the investment.

This procedure leads to "small" optimum systems (e.g., collector areas of 10 to 15 m2 for space heating). The useful information provided by this optimization has limitations which should be emphasized. As will be seen in section 8.2, the optimization is based (implicitly) on the simple payback period or relative net present value ; and hence has the same advantages and drawbacks of this well known simple criterion. Therefore, a detailed system optimization (even one limited to the search for the best solar solution) could require other economic criteria such as those discussed in the next section.

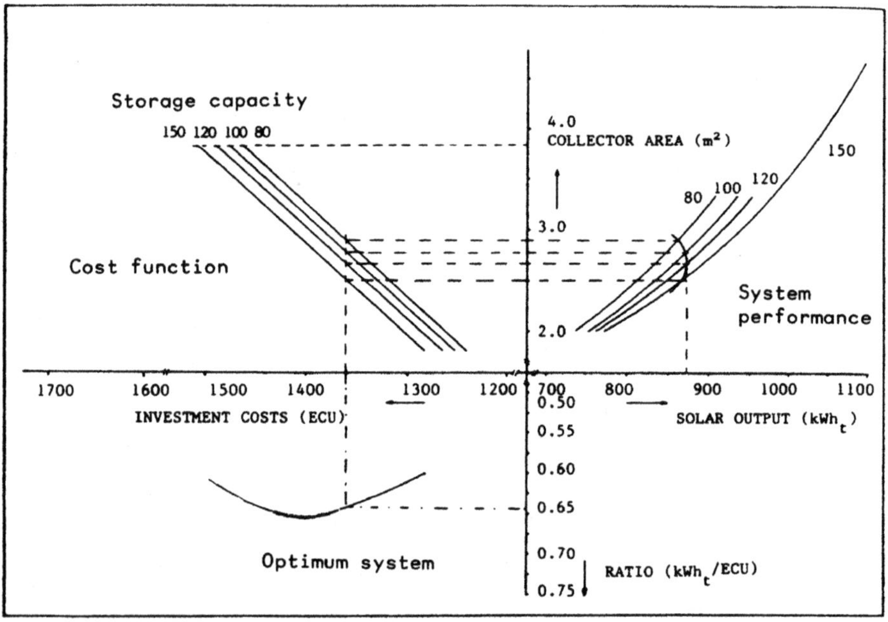

Fig. 8.1.5: Graphical presentation of the optimisation methods (from Ferraro /8.2/). For each investment cost, the optimum system is found (maximum solar output; upper right corner). From this result, the solar energy output per invested ECU is plotted versus the investment cost (lower left corner).

8.2 - CRITERIA FOR ECONOMIC COMPARISON

The method for size optimization presented in the preceding section is simple, but it does not account for present and future economic conditions. Several criteria can be introduced to avoid this drawback. Furthermore, these criteria are essential if one wishes to compare the costs involved in active solar energy systems with those of a conventional or other different heating system.

8.2.1 - Discounting

The first problem to solve is the following: How can expenses and receipts (or savings) occurring at different times be compared?

8.2.1.1 - Definitions

It is known that one ECU (or one franc, pound, mark or any other currency unit) to be received in one year's time is worth less than one ECU possessed today, even in periods without inflation. The difference is accounted for by the interest on investments or loans: if an amount P is borrowed for n years, the amount F to be repaid is (based on the compound interest formula):

$$F = P * (1 + r)^{n} \qquad (8.2.1)$$

where r is the interest rate. Similarly, one can invest an amount P in a savings account today and realize F in n years. It can be said that the present amount P is worth F in n years from now, or that the amount F in n years is worth P today:

$$P = F * (1 + r)^{-n} \quad . \qquad (8.2.2)$$

The ratio P/F is called the present worth factor and indicates the present value of expenses or receipts occurring at different times. This technique is called discounting .

It is usual to eliminate inflation from the calculations and to work in constant currency units (instead of current currency units): expenses and receipts are corrected for inflation. If r' is the market interest rate and i the inflation rate, the real discount rate used will be given by

$$1 + r = (1 + r') / (1 + i) \quad . \qquad (8.2.3)$$

8.2.1.2 - Discount rate

The value to be used for r is not always obvious and is different for different types of investors.

For large public authority investments, a discount rate is sometimes determined by the national Administration (e.g., the Planning Service in France).

For industrial or commercial investors with many investment opportunities, the best solution is probably to take r as the rate of return (see below for the definition of this term) for investments of comparable risk.

Individual investors, such as homeowners, may be in a different situation; if they have enough cash for the investment they intend to make, it seems reasonable to compare the return with the interest rate from a savings account. If they have no cash for the investment, they must borrow from a bank; the discount rate r will then be the interest rate of the loan (generally higher than the interest rate for a savings account). Sometimes, special loans provided by the State in order to promote solar energy applications can lead to a negative real discount rate.

Low discount rates (or even better, negative ones) make long-term investments such as solar systems promising, but high discount rates favour the short-term.

8.2.1.3 - Life-cycle cost

The life-cycle cost of a system (e.g., solar heating) is the discounted amount of all expenses incurred during its life of N years.

The following notation is used:

- C — initial cost (investment), assumed to be paid in cash.
- M — yearly maintenance and operation cost.
- L — yearly energy consumption (auxiliary energy in the case of solar).
- P_e — present unit cost of energy.
- n — average boiler efficiency.
- V_r — salvage value of the system after N years.

It is assumed that energy prices do not follow the general inflation rate but increase at a real rate r_e (r_e is the energy escalation rate and represents fuel inflation).

The system life-cycle cost over N years is computed by Eq.8.2.4

$$LCC = C + \sum_{i=1}^{N} M/(1+r)^i + \sum_{i=1}^{N} L.P_e[(1+r_e)/(1+r)]^i/n - V_r/(1+r).$$

This formula may be modified to account for loan payments, tax reductions, periodic replacement of particular components, etc. Complete formulas can be found in reference /8.1/.

8.2.2 - Life-cycle savings

The basic economic criterion for the comparison of solar heating with a conventional energy system, such as a boiler, as a reference system is defined in this section and referred to as the life-cycle savings (or net present value). Life-cycle savings is simply defined as the difference between the life-cycle costs of the two systems over a period of N years

$$LCS = LCC(ref) - LCC(sol) \quad . \qquad (8.2.5)$$

If maintenance costs and salvage value are neglected (or assumed to be identical in the two cases), and if a reference system identical to the auxiliary circuit used in the solar solution (with the same initial cost and the same energy cost) is considered, the life-cycle savings become

$$LCS = - Csol + \sum_{i=1}^{N} Qsol.P_e.[(1+r_e)/(1+r)]^i / n . \qquad (8.2.6)$$

Csol is the initial cost of the solar part of the system (or solar additional cost). Qsol is the yearly useful solar energy output.

The relative life-cycle savings or relative net present value is sometimes used: it is a measure of the financial efficiency of the investment, defined as

$$RLCS = LCS / Csol \qquad (8.2.7)$$

$$RLCS = (Qsol / Csol) * \left[\sum_{i=1}^{N} P_e . [(1+r_e)/(1+r)]^i / n \right] - 1$$

Size optimization in terms of RLCS does not depend on the economic parameters (r and r_e) but coincides with the maximum of the ratio Qsol/Csol: this is exactly the optimization criterion presented in section 8.1 (with the assumption of constant auxiliary boiler efficiency).

This observation justifies the wide use of the simple size optimization criterion: no assumption is needed for the discount rate or the energy escalation rate, nor even the system lifetime. However, as we are maximizing a relative parameter, we shall get small optimum sizes (sometimes the optimum will be at zero), whereas an absolute criterion such as the LCS would provide a larger optimum collector area.

8.2.3 - Payback period and internal rate of return

The payback period is defined as the period at the end of which the annual savings have paid for the whole initial cost. The simple payback period is the most widely used; it is simply defined as

$$SPP = Csol * n / (Qsol * Pe) \quad . \qquad (8.2.8)$$

This approach has been criticised, because it does not account for the discount rate, the energy escalation rate, system lifetime, etc. However, it does provide some useful information: for any system, the simple payback period must lie within a reasonable range and be less than the system lifetime.

The payback period can be modified by discounting. In this case a discounted payback period is referred to which may also take energy escalation into account. However, the calculation then becomes tedious.

The internal rate of return is the discount rate used in Eq. 8.2.6 for which the life-cycle savings vanish. It is generally calculated iteratively. For any lower value of discount rate, the life-cycle savings are positive. This criterion allows one to classify different investment opportunities: the best one has the highest internal rate of return.

8.2.4 - Which criterion to choose?

Two aspects of the optimization process have been distinguished. The first, called size optimization, is the search for the most cost-efficient solar system in a given class. The second is the economic comparison with conventional energy solutions. A simple size optimization procedure based on the ratio Qsol/Csol, solar energy output per unit of initial cost, has been presented in section 8.1 and certain criteria for an economic comparison based on the discounting technique have been defined in sections 8.2.1-8.2.3. It has been shown that the size optimization procedure proposed as the first step is fully justified, providing the assumptions are checked. It should be borne in mind that some characteristics of the system are independent of the solar system size and performance (namely maintenance and operation cost; auxiliary boiler efficiency; system lifetime, and salvage value). The criterion to be optimized in this first step is the relative life-cycle savings or relative net present value. It is interesting to note that the optimum does not depend on the discount rate.

It can easily be shown that, with the same assumptions, the same optimum size is obtained whether the simple payback period or the internal rate of return is taken as the criterion.

For example, it is obvious from Eq. 8.2.8 that a system optimization based on the minimization of SPP corresponds to our first optimization procedure: a minimum for SPP is obtained with a maximum of Qsol/Csol.

However, even for size optimization, the absolute life-cycle savings could be taken as an optimization criterion and should lead to a different optimum system (varying with discount rate). An example will be shown in the next section.

If one now wishes to compare this optimum solar system with a conventional reference system, the choice of the optimization criterion becomes more important. In this second step, we are no longer interested in the optimal active solar system, but in its economic viability and we must therefore compare it with other energy options. The application of one criterion (e.g. LCS) instead of another (e.g. SPP) can lead to a different investment decision.

8.3 - APPLICATIONS OF ECONOMIC CRITERIA

This section presents a case study in which the criteria defined in section 8.2 are applied. Some general economic results of active solar systems are also considered, and the uncertainties involved in the economic parameters are discussed.

8.3.1 - System optimization and comparison.

Let us consider the active solar space heating (+ DHW) system in Table 8.3.1. The thermal performance is computed by the ESM method and shown in Figure 8.3.1. The collector area is between 5 and 30 m2; the storage capacity is between 0.5 m3 and 3 m3. The total load is 14723 kWh. The cost function (in ECU), derived from the manufacturer's cost data, is

$$Csol = 2250 + 150 * A + 0.3 * M \quad . \qquad (8.3.1)$$

(This is an hypothetical cost function). The reference system is assumed to be an oil-fired boiler, which is also used as an auxiliary in the solar option. The costs of the auxiliary and heat-emitters are not included in Csol.

From these data, we obtain the diagram of solar fraction as a function of collector area at different costs (Fig. 8.3.2). An optimum collector area is obtained for each cost.

It is then possible to compute the energy productivity of the investment, Qsol/Csol (solar energy output per invested currency unit), for each investment cost (at the optimum collector area). This is shown in Figure 8.3.3. The optimum investment appears clearly at 5100 ECU (16 m2 of collector, 1500 litres of storage). Near the maximum the curve is rather flat: within the range 4500 - 6000 ECU, the energy productivity decreases by less than 2%; and from 4000 to 6750, the difference is less than 5%.

These data are summarized in Table 8.3.2. For each investment, several economic criteria have been computed: simple payback period, life-cycle savings, and relative life-cycle savings. These last two parameters have been estimated at different effective discount rates; the effective discount rate corresponds to the real discount rate corrected by the energy escalation rate:

$$1 + r_{eff} = (1 + r) / (1 + r_e) \quad . \qquad (8.3.2)$$

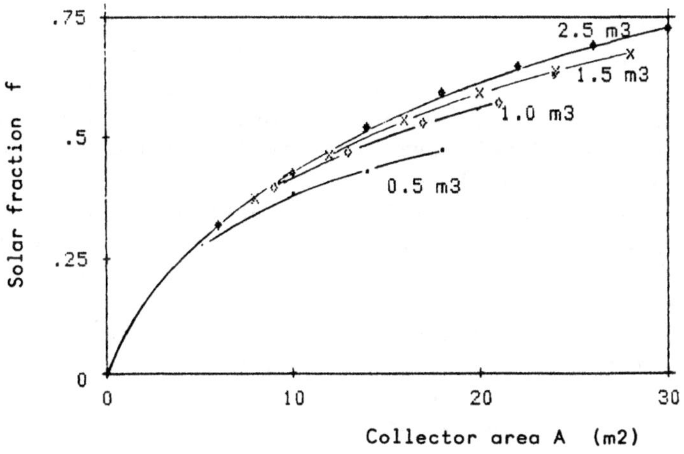

Fig. 8.3.1: Performance of the reference system computed by the ESM1 method. Solar fraction versus collector area at several storage capacities.

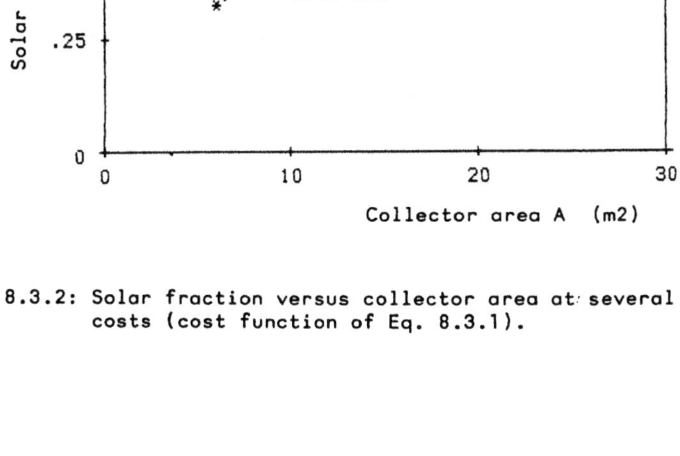

Fig. 8.3.2: Solar fraction versus collector area at several system costs (cost function of Eq. 8.3.1).

The following assumptions have been made:

- The period considered is N = 20 years (salvage value is neglected).
- Unit energy cost P_e / n = 0.045 ECU/kWh.

It can be seen in Table 8.3.2 that the optimum in terms of energy productivity is also the optimum in terms of simple payback period and relative life-cycle savings (whatever the effective discount rate).

The optimum in terms of life-cycle savings is quite different and is very sensitive to the discount rate: with r = 2.5%, the optimum investment is 5100 ECU and it increases to 7500 ECU with r = -2.5%.

For an optimum 16 m2 system, the internal rate of return is 3.4%. For discount rates higher than this limiting value, the system is no longer cost-effective; this can be seen in the negative values of LCS obtained at r = 5%.

The optimum system in terms of relative life-cycle savings does not depend on economic assumptions (discount rate, energy escalation rate, energy cost) although the value of this criterion does: if r increases from 0 to 2.5%, the RLCS decreases from 0.40 to 0.09, as does the "financial efficiency" of the investment. This means that the comparison of a solar system with common non-solar options depends very much on the economic assumptions.

Major uncertainties are involved in these parameters (some values may even be arbitrary); so conclusions must be drawn with caution. One remembers the predictions made in recent years about future oil prices...

As well as the future economic environment, other sources of uncertainty should be considered: system lifetime; climate and system performance; cost of the system. This last point is important, because the optimum system chosen depends on the cost data. Complete and reliable costs are not easy to obtain and must be carefully analysed, particularly when separating fixed and proportional costs. It is preferable to consider a reference solar system with variations around it (with the proportional terms of the cost function being marginal costs). Furthermore, all the costs of the reference system must be considered.

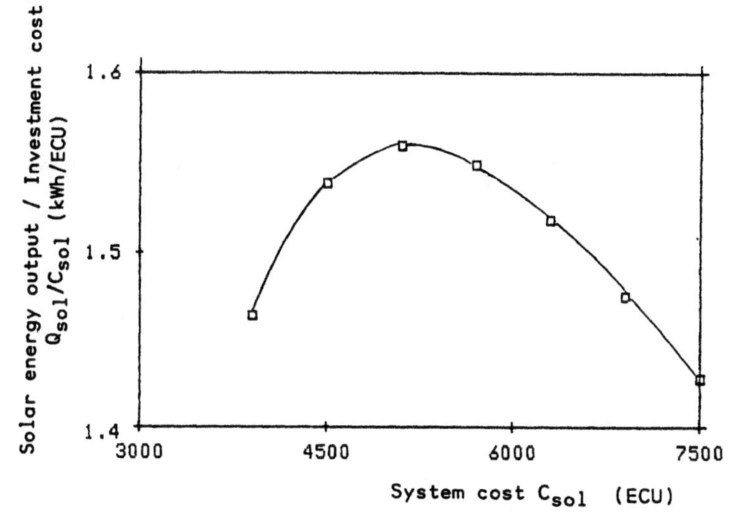

Fig. 8.3.3: Optimal energetic productivity versus system cost.

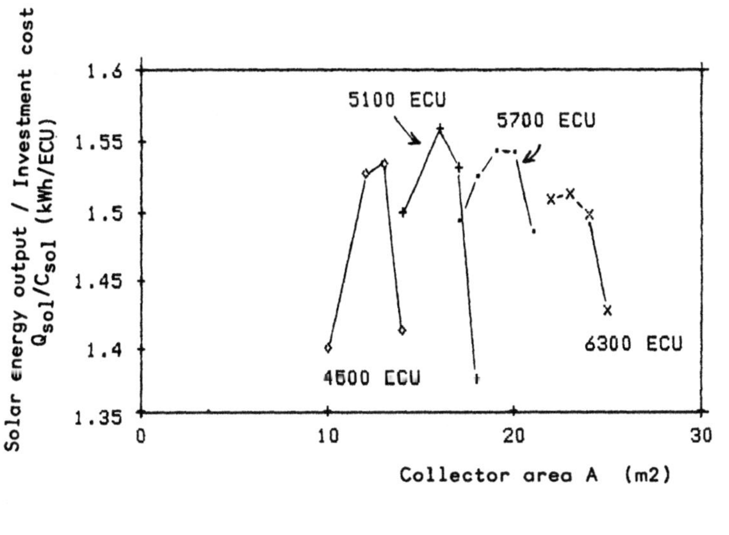

Fig. 8.3.4: Energetic productivity versus collector area at several system costs (cost function of Eq. 8.3.3).

Table 8.3.1: Optimization case study. Solar space heating system characteristics.

Location:	Carpentras (France)		
Space heating load:	Building heat-loss coefficient	GV	= 250 W/K
	Internal set-point temperature	T_c	= 20°C
	No-heating temperature	T_{nh}	= 18°C
Heat emitters:	Capacitance rate	$\dot{m}_2c$	= 750 W/K
	Effectiveness	e_2	= 0.533
	Control and auxiliary coupling	Type	= 2
Domestic hot water:	Daily demand	m	= 150 l/day
	Mains cold water temperature	T_m	= 15°C
	Hot water temperature	T_h	= 55°C
	Heat-exchanger effectiveness	e_3	= 1
Solar collectors:	Tilt angle	INC	= 45°
	Optical efficiency	η_0	= 0.80
	Heat loss coef.: Constant term	U_0	= 4.5 W/K/m2
	Proport. term	U_1	= 0.02 W/K2/m2
	Collector fluid capacitance rate	$\dot{m}_0c_0$	= 60 W/K
	Heat-exchanger fluid capac. rate	$\dot{m}_1c$	= 60 W/K
	Heat-exchanger effectiveness	e_1	= 0.80
	Control differential: Starting	d_{on}	= 5 K
	Stopping	d_{off}	= 0.1 K
	No piping heat losses		
Storage:	Capacity	V	= 500 - 3000 l
	Heat loss coefficient (W/K)	A_sU_s	= 4 + 0.002*V

Coll. area	Csol	Qsol	Qsol/ Csol	SPP	Life-Cycle Savings LCS (ECU)				Relative Life-Cycle Savings		
m2	ECU	kWh	kWh/ECU	Year	r=-2.5%	r= 0.0%	r= 2.5%	r= 5.0%	r=-2.5%	r= 0.0%	r= 2.5%
9.5	3900	5698	1.461	15.2	2862	1228	98	- 705	0.733	0.315	0.025
12.5	4500	6920	1.538	14.5	3712	1728	355	- 620	0.825	0.384	0.079
16.0	5100	7950	1.559	14.3	4334	2055	477	- 642	0.850	0.403	0.094
19.0	5700	8804	1.545	14.4	4747	2224	476	- 764	0.833	0.390	0.084
23.0	6300	9540	1.514	14.7	5021	2286	393	- 951	0.797	0.362	0.062
26.0	6900	10159	1.472	15.1	5155	2243	227	-1204	0.747	0.325	0.033
29.0	7500	10704	1.427	15.6	5202	2134	9	-1498	0.694	0.284	0.001

Table 8.3.2: Optimization case study. Economic criteria.

The collector area is the optimal value at the corresponding cost, which maximizes the solar energy output.
Reference system: fossil-fuel-fired boiler.
Cost function (ECU): Csol = 2250 + 150 * A + 0.3 * V

Csol: Initial investment cost. Qsol: Solar energy output.
SPP: Simple Payback Period.
r: Effective discount rate (corrected for the energy escalation rate).

8.3.2 - Sensitivity analysis

It may be concluded that the sensitivity of the optimum to the principal parameters is as interesting as this optimum itself. The effect of the discount rate has already been shown; in this section, uncertainties in the cost function are analysed.

The same system (see Table 8.3.1) with a modified cost function is considered

$$Csol\ (2) = 2400 + 120 * A + 0.6 * M \quad . \qquad (8.3.3)$$

This formula results in no change in the cost of a 20 m2 / 1.5 m3 solar system (5700 ECU); the collector cost is decreased while the storage cost is doubled. Results are shown in Table 8.3.3 and Fig. 8.3.4. The optimum now has a slightly larger collector array (17 m2, instead of 16 m2) and a smaller storage (1100 litres) but the energy productivity has hardly changed.

A similar conclusion can be drawn with a third cost formula (Table 8.3.4), corresponding to an interchange between fixed cost and collector cost (the price of the reference system 20 m2 / 1.5 m3 is still unchanged):

$$Csol\ (3) = 2850 + 120 * A + 0.3 * M \quad . \qquad (8.3.4)$$

Uncertainties in the cost functions have a small effect, provided that the optimum is near the reference solar system (the cost of which is well known); they generally lead to recalculated optima within the flat region of the optimum curve (Fig. 8.3.3). It seems preferable to define an "optimum range" of investment rather than a fixed optimum system.

The situation is somewhat different if the reference conventional system is modified. Let us compare this solar system (with an electric auxiliary) with an electric space heating system. The cost function is changed because it must account for the cost of heat-emitters:

$$Csol\ (4) = 5250 + 150 * A + 0.3 * M \quad . \qquad (8.3.5)$$

The optimum system is therefore not the same: 25 m2 of collector with 2.5 m3 of storage (total cost: 9750 ECU) with a very different energy productivity (1.027 kWh/ECU). The reference energy cost is also higher (0.0825 ECU/kWh) and leads to life-cycle savings and a payback period quite different from the first scenario (Table 8.3.5).

This illustrates how carefully the reference energy system and the various costs must be analysed.

8.3.3 - Performance of actual systems

The preceding figures are rather theoretical and have been given as an illustration of a methodology and of trends in sensitivity. Actual figures, derived from monitored experiments during the last ten years, have shown that, although it is possible to define optimum-sized solar systems, they are not quite competitive with conventional solutions.

Data collected by the European Performance Monitoring Group in 1982-83 /8.2/ indicated a typical energy productivity per unit investment of 0.56 kWh/ECU for space heating and 0.86 kWh/ECU for hot water systems. More recent figures from France are more encouraging. For example, seasonal hot water production in camping sites /8.3/ (about 1.1 kWh/ECU with a payback period of 11 years with respect to electricity) and even space heating (direct solar underfloor heating for a 90-flat building in the south of France: 1.8 kWh/ECU) /8.4/.

The above analysis is nevertheless useful because it gives information about cost targets which must be achieved in order to make active solar systems cost-effective. Payback periods in the order of ten years could be a reasonable objective. This is attainable for water heating systems (especially in communal applications), and not impossible for space heating, although large cost reductions would be required to compete with present low oil prices (1988).

In conclusion, the economic analysis and optimization of solar systems are both essential and fragile; essential, because these procedures provide guidelines for the improvement of system cost effectiveness and a basis for comparison with other energy options. Fragile, because the apparent exactness and complexity of the mathematical equations may obscure the dramatic uncertainties in the input parameters, particularly those concerned with future economics.

Although the strictly economic considerations cannot be ignored, in the final analysis due weight must be given to the advantages of solar heating, such as protection of the environment, and energy self-sufficiency. In our world of short-term goals, this should not be ignored.

Table 8.3.3: Optimization case study. Sensitivity analysis to cost function:
Csol = 2400 + 120 * A + 0.6 * V
Optimal system: A = 17 m2 V = 1100 l Csol = 5100 ECU.

	Csol = 5100 ECU			Csol = 5700 ECU			Csol = 6300 ECU		
Stor. capa.	900	1100	1300	1100	1300	1500	1300	1500	1700
Coll. area	18	17	16	22	21	20	26	25	24
Sol. Frac.	.532	.536	.533	.593	.598	.596	.645	.649	.649
Qsol	7833	7891	7847	8731	8804	8775	9496	9555	9555
Qsol/Csol	1.536	1.547	1.539	1.532	1.545	1.539	1.507	1.517	1.517

Table 8.3.4: Optimization case study. Sensitivity analysis to cost function:
Csol = 2850 + 120 * A + 0.3 * V
Optimal system: A = 19.5 m2 V = 1700 l Csol = 5700 ECU.

	Csol = 5100 ECU			Csol = 5700 ECU			Csol = 6300 ECU		
Stor. capa.	1100	1300	1500	1500	1700	1900	1900	2100	2300
Coll. area	16.0	15.5	15.0	20.0	19.5	19.0	24.0	23.5	23.0
Sol. Frac.	.522	.525	.524	.596	.597	.595	.655	.656	.655
Qsol	7685	7730	7715	8775	8790	8760	9644	9658	9644
Qsol/Csol	1.507	1.516	1.513	1.539	1.542	1.537	1.531	1.533	1.531

Table 8.3.5: Optimization case study. Economic criteria (idem Tab.8.3.2). Variant with Reference system: electric convectors. The constant term of the cost function accounts for the cost of heat emitters:

$$Csol = 5250 + 150 * A + 0.3 * V$$

Internal rate of return for A = 25m2 (optimum): 5.7%

Csol: Initial investment cost. Qsol: Solar energy output.
SPP: Simple Payback Period.
r: Effective discount rate (corrected for the energy escalation rate).

Coll. area	Csol	Qsol	Qsol/ Csol	SPP	Life-Cycle Sav.		Relat. LCS	
m2	ECU	kWh	kWh/ECU	Year	r=0.0%	r=2.5%	r=0.0%	r=2.5%
12.5	7500	6920	0.923	13.1	3918	1400	0.522	0.187
16.0	8100	7950	0.981	12.3	5018	2125	0.619	0.262
19.0	8700	8804	1.012	12.0	5827	2624	0.670	0.302
23.0	9300	9540	1.026	11.8	6441	2970	0.693	0.320
26.0	9900	10159	1.026	11.8	6862	3166	0.693	0.320
29.0	10500	10704	1.019	11.9	7162	3267	0.682	0.311

REFERENCES OF CHAPTER 8

/8.1/ A.RABL. Active solar collectors and their applications. Oxford Univ. Press, New York (1985).

/8.2/ R.FERRARO and R.GODOY. Solar Heating: Performance of recent systems. CEC-DG XII, Doc. EUR 8947 (1983).

/8.3/ AFME. Performances mesurées des installations solaires à capteurs plans. Colloque Agence Française pour la Maitrise de l'Energie, La Baume les Aix (1985).

/8.4/ T.CABIROL. BATIMENT-ENERGIE, n.38, Mai 1985 (1985).

CHAPTER 9

ESM COMPUTER PROGRAMMES

Most of the methods described in this Handbook have been presented as pieces of software, developed for use on a personal computer.

The ESM 1.4 software deals with the calculation of the long-term performance of active solar space heating and domestic hot water systems. It is based on the ESM1 method described in Chapter 6. Only space heating systems, SHS (with optional domestic hot water production, DHW) are considered by this programme.

Hot water solar systems can be computed with the help of the ESM 2.1 software, based on the ESM2 method, described in Chapter 7.

Both these pieces of software use the same type of weather data as an input (monthly cumulative frequency curves, CFC's, of solar irradiance; monthly average temperature). Such a pre-processed solar radiation data base is provided with the programme diskette. Several ways of obtaining user-specific ESM weather data files are proposed.

This chapter gives practical guidelines for the use of the ESM software with IBM compatible personal computers. The general information in the first section (9.1) must be read carefully before using one of the ESM programmes.

9.1 - GENERAL INFORMATION

Both programmes, ESM 1.4 (S.H.) and ESM 2.1 (H.W.) use a similar structure and similar commands. Likewise, a number of instructions are common to both.

9.1.1 - Getting started

9.1.1.1 - Equipment requirement

The ESM programmes can be run on IBM-PC and compatible personal computers. The following is the minimum configuration needed to run ESM:

- An IBM or fully compatible personal computer (XT or AT).
- Disk Operating System DOS 3.0 or later.
- A supported graphics card.
- A monitor suitable for the graphics card.
- One disk drive (flexible disc 5 ¼").
- 640 K-bytes of RAM memory.
- A supported parallel printer.

Although the instructions below refer to a system with two disk drives, you can run ESM with only one disk drive. The programme will instruct you to swap disks when necessary. ESM can also be run from a hard disk.

9.1.1.2 - ESM diskettes

The ESM software is distributed as single 5 ¼" diskettes (one each for ESM 1.4 or ESM 2.1) including the programme and all required user's routines and files. A meteorological data file, EUROPE-TRY, containing monthly solar radiation and temperature values for 12 locations is included. Other ESM-compatible meteorological data files are also available (see section 9.4).

9.1.1.3 - Keys and commands

You will use the keyboard of your computer to provide the required information. Characters that you have to type on the keyboard are underlined. For example, if "type Return " appears in this document, you should press the R,e,t,u,r, and n keys.

However, if you need to press the return key (the large key to the left of the cursor keys with the ↵ symbol on it), you will be told "type [Return]".

The IBM keyboard has a number of other special keys. For these keys the following symbols are used:

- [Bksp] for the grey backspace key above the return key.
- [Esc] for the "escape" key in the upper left corner.
- [Left], [Right], [Up], [Down], [Home], [End], [PgUp], [PgDn] for the cursor keys in the numeric keypad.
- [Tab] for the grey "tab" key under the [Esc] key.
- [F1] [F2] ... [F10] for the ten function keys.
- [Ctrl-...] for the control codes that are entered by holding down the Ctrl-key and another key simultaneously.
- [Alt-...] for the "alternate" codes that are entered by holding down the Alt-key and another key simultaneously.
- [Grey +], [Grey -], and [Grey *] for the grey +, -, and * keys in the numeric keypad.

Most of these keys correspond to specific commands for ESM. Their effect is described in section 9.1.3.3.
Using the [Grey *] key in conjunction with the [Shift] key corresponds to the [PrtSc] command to dump the contents of the screen to the printer. This rarely results in good ESM output, and a specific ESM command has been included for that purpose.

Variable parts of commands are typed enclosed by angular brackets; e.g. the command

        <u>TYPE</u> <filename> [Return]

When using this command, you must replace the variable part with a specific file name.

9.1.2 - <u>Installation procedure</u>

This section describes how to make backups of the ESM programme, how to install it on floppy disks and hard disks, and how to customize it for different printers and graphic boards.

9.1.2.1 - <u>Overview of the Installation procedure</u>

Before you start the installation, check the file RECENT.INF. It contains recent information not in the printed manual. If there is a change in the installation procedure, it will be noted there, and those instructions override the ones in this chapter. You should make a print-out of that file to view its contents; the procedure for that is as follows:

1 - Insert the disk with the file that you want to print into drive A.

2 - Preview the file on the screen, by typing the command
    TYPE A:RECENT.INF [Return].
    If the file is too long to fit on one screen, you can stop and restart scrolling by pressing [Ctrl-S]. If you have no problem remembering its content, you don't need to print it.
3 - Check that the printer is turned on.
4 - Type COPY A:RECENT.INF PRN: [Return].

9.1.2.2 - Installation on a floppy disk system

You must first make copies of the distribution disk. Then, put the original in a safe place and work ONLY with the copy. Since some of the later steps may involve erasing files on your system disk, it is ESSENTIAL that you do not use the original disk. Here is the procedure to make your working copies:

1 - Put a DOS disk in drive A:.
2 - Type DISKCOPY A: B: [Return].
3 - Put the ESM disk in drive A: and a blank disk in drive B:.
4 - Make sure you put a blank disk in drive B - not a DOS disk, not an ESM disk, etc.
5 - Type [Return].
6 - After the copying is complete, you will be asked whether you want to copy more disks. Answer N (unless you want to make an additional copy of the original ESM disk).
7 - Remove both disks and label the copy ESM 1.4.
8 - Put the original in a safe place.

If you want to prepare an AUTOSTART ESM disk (including the essential DOS functions):

1 - Put a DOS disk into drive A:.
2 - Type FORMAT B:/S [Return].
3 - Make sure you put a blank disk in drive B:, not a DOS disk, not a distribution disk, etc.
4 - Type [Return].
5 - After the formatting is complete, you will be asked whether you want to format more disks. Answer N .
6 - Remove the DOS disk from drive A:.
7 - Put the ESM disk in drive A:.
8 - Type COPY A:*.* B: [Return].
9 - After the copying is complete, remove the ESM disk from drive A:.
10 - Put the DOS disk in drive A:.

11 - Type COPY A:keybxx.EXE B: [Return].
(keybxx.EXE is the national keyboard of your IBM-PC. Refer to your DOS manual).
12 - Type COPY CON:autoexec.bat [Return].
keybxx [Return] [Ctrl-Z] [Return].
13 - Remove both disks and label the copy ESM 1.4.
14 - Put the originals in a safe place.

Now you must prepare one disk for storing your own USER'S ESM DATA FILES:

1 - Put a DOS disk into drive A:.
2 - Type FORMAT B:/S [Return].
3 - Put a blank disk in drive B:.
4 - Make sure you put a blank disk in drive B:, not a DOS disk, not a distribution disk, etc.
5 - Type [Return].
6 - After the formatting is complete, you will be asked whether you want to format more disks. Answer N .
7 - Remove the DOS disk from drive A:.
8 - Remove the disk from drive B: and label it ESM x.x USER'S FILE.

When a ESM USER'S DATA FILE disk is full, you can use this procedure for preparing a new disk.

9.1.2.3 - Installation on a hard disk system

It is suggested that you use a subdirectory \ESM on the root directory of your hard disk to hold the ESM system. If you prefer to place it in a different location, modify the instructions accordingly. It is assumed that your hard disk is on C: - if not, use the appropriate drive letter instead. Here are the instructions:

1 - Insert ESM disk in drive A: and type A: [Return].
2 - Type HARD C:\ESM [Return]. This activates a batch process transferring files onto the hard disk. The directory \ESM is created automatically.
3 - If you want to transfer files from another disk (for example meteo files or another ESM programme) on the directory \ESM of the hard disk, insert the new disk you want to copy into drive A: and type
HARD2 C:\ESM [Return].
4 - Put the original distribution disk in a safe place.

9.1.3 - Using ESM programmes

9.1.3.1 - Running ESM

To start running one of the ESM programmes, e.g. ESM 1.4, first you have to move to the disk directory holding the programme. Instructions given here for ESM 1.4 are also suitable for ESM 2.1 and other ESM software (the name of the programme has to be changed accordingly).

If you have a floppy disk system, insert the ESM 1.4 disk in drive A:, the disk containing your user's data files in drive B: and type

A: [Return].

If you have a hard disk (after installation of the programme on the subdirectory <ESM of the hard disk; see section 9.1.2.3) type

C: [Return]
CD \ESM [Return]

instead.

Then start ESM 1.4 by typing: ESM1_4 [Return].
If you are using meteo files on another disk, you can now remove the ESM disk from drive A: and insert the appropriate disk.

If your ESM disk is an AUTOSTART disk (see section 9.1.2.2), you can start running ESM 1.4 with this simplified procedure:

- If the computer is off, insert the ESM 1.4 diskette into drive A:, close the door and turn the computer on. Running the programme will start automatically.

- If the computer is already on, put the ESM 1.4 diskette into drive A: and press

  [Ctrl - Alt - Del] (The three keys simultaneously).

9.1.3.2 - Software organization

ESM Software is organized in the following structure. At the beginning of the programme, a WELCOME page is displayed with brief information on the programme's purpose. More information can be obtained (as will be the case during any step of the programme) through HELP commands. Then the first step of the programme starts with its successive screen pages. At the end of the first step, the MAIN MENU is displayed: It presents the successive STEPS of the programme. Some of the steps are omitted, depending on earlier choices. At the end of any step, the main menu will reappear, so it is possible to repeat this step (e.g., in order to correct a mistake) or to go to the next step.

Each step includes successive screen PAGES, also sometimes optional. Each page displays information and contains INPUT CELLS. These input cells enable the user to provide the required data to the programme (for example, values of system parameters, numeric input; or comments, sequence of alphanumeric characters). For each data input cell, the current value of the corresponding parameter is displayed (default values are given when starting the programme). This value may be kept or modified by the user. Editing capabilities are provided inside the cell (e.g., with the commands CURSOR TO LEFT, CURSOR TO RIGHT, or DELETE). The new value is then validated by a command (e.g., VALIDATION with the [RETURN] key or NEXT PAGE) (see the section on commands below). During the validation, the new value is checked to ensure that it is within an allowed range; if not, it is rejected. A new value is again requested and a message is displayed.

When a page is run, input cells are reviewed successively. If the new value given as an input in the current cell is correct, after its validation the next data input cell is reviewed.

9.1.3.4 - Commands

Several commands are available to the user. They provide ways of:

- Moving within the programme:
  NEXT PAGE - FORMER PAGE - NEXT CELL - FORMER CELL - RETURN FIRST PAGE
- Editing comments or numeric input data:
  DELETE - RIGHT - LEFT - BACKSPACE - HOME - END - INSERT
- Managing input data diskette files (e.g. weather data files): FILE
- Printing results:
  PRINTING

These commands are obtained by typing special keys or sets of keys. They are summarised in Table 9.1.1

They can be temporarily disabled (e.g., PRINTING as long as complete calculations are not done).

Table 9.1.1: ESM Software Commands.

COMMAND	KEY	FUNCTIONS
NEXT PAGE	[PgUp]	Move to the next page. The current cell is validated; computations and controls corresponding to the current page are done. If the current page is the last page of the current step, we move to the main menu.
FORMER PAGE	[PgDn]	Move to the former page. The current cell is validated; computations and controls corresponding to the current page are done. If the current page is the first page of the current step, it has no effect.
FIRST PAGE	[Ctrl-PgUp]	Move to the first page of the current step.
NEXT CELL	[Down]	Move to the next cell The current cell is validated. If the current cell is the last cell of the page, we move to the first cell.
FORMER CELL	[Up]	Move to the former cell. The current cell is validated. If the current cell is the first cell of the page, we move to the last.
VALIDATION	[Return]	Control and validation of the value displayed in the current cell. Move to the next cell. If the value is within the suitable range, it is accepted as the corresponding parameter. If not, the cell is displayed again with the former correct value; a sound signal warns the user. At the last cell of a page, this command is equivalent to NEXT PAGE.
DELETE	[Del]	Delete the character shown by the cursor in the current cell.

Table 9.1.1 (continued).

COMMAND	KEY	FUNCTIONS
CURSOR TO LEFT	[<--]	Move the cursor one position to the left in the current cell.
CURSOR TO RIGHT	[-->]	Move the cursor one position to the right in the current cell.
INSERT	[Ins]	Insert characters between the cursor and the character to its left. The INSERT mode remains active until the INSERT key is pressed again (even when moving to another cell or page). When the INSERT mode is active, the character I is displayed in the bottom right corner of the screen.
HOME	[Home]	Move the cursor to the most far left position of the current cell.
END	[End]	Move the cursor to the most far right position of the cell
BACKSPACE	[Bksp]	Delete the character on the left side of the cursor and move the cursor one position to the left.
FILE	[F3]	Move to the data file step (reading and updating of diskette files). At the end of this step, go back to the current page. Only one particular type of file can be reached from a given page (e.g. reading meteo data can be made from page 4 in ESM 1.4).
PRINTING	[F4]	Print detailed results.

9.2 - THE ESM 1.4 SOFTWARE FOR ACTIVE SOLAR SPACE HEATING

This sofware is based on the methods described in Chapter 6. Possible system configurations are those given in that chapter, where input parameters are also reviewed. It should be repeated that ESM 1.4 deals with active solar space heating systems with or without coupled domestic hot water.

9.2.1 - Programme steps

There are two main successive steps in the programme (both of which correspond to an option of the MAIN MENU, screen page 1), including several screen pages:

Step 1: Weather data and loads:

* Choice of a weather data file (pages 3, 4 and 24).
* Space heating load input data (pages 5, 6 and 7).
* Domestic hot water (DHW) load input data (pages 8 and 9).
* Space heating and domestic hot water loads: Summary of input data and computation results (pages 10 and 11).

Step 2: Solar system characteristics and performance calculation:

* Project references (page 12).
* Solar collector input data (page 13).
* Solar loop characteristics (pages 14 and 15).
* Storage input data (page 16).
* Space heating and DHW distribution sub-system: input data (pages 17, 18 and 19).
* Data summary and computation results (including warning messages) (pages 20 to 23).

Within each step, all screen pages are reviewed successively. Several commands move from one page to another (NEXT PAGE or FORMER PAGE, see section 9.1.3.4). This general structure is analysed in more detail in the ESM 1.4 AIDE-MEMOIRE (in Appendix E), including a systematic review of input and output parameters.

9.2.2 - Input parameters

Input parameters of the ESM 1.4 software are similar to those described in the standard version of the algorithms (see Chapter 6).

Input parameters are reviewed for each screen page in the AIDE-MEMOIRE, using the same nomenclature as the standard version; it is therefore easy to find complete explanations about each parameter in Chapter 6. The tables of the AIDE-MEMOIRE also contain minimum and maximum values authorized in the programme and default values proposed to the user at the beginning. Blank worksheets for preparing the input data are available in Appendix E.

A few changes with respect to the nomenclature of the standard version have been introduced:

- Space heating loads can be given directly by the user (calculated at a monthly time step with any other computation method), or computed with a degree-day method. In the latter case, monthly degree-days can be either computed by the programme (from the monthly average external temperature) or provided by the user (from any source).

  The correction of degree-days or heating loads for the conventional heating season (e.g., no space heating from May to October) can be done with the following procedure: run the first step normally then (from the MAIN MENU) run it again and choose option 1 or 2 in screen page 5. You can directly correct monthly degree-days or heating loads calculated before by the programme.

- Domestic hot water demand (and also mains cold temperature and required hot water temperature) can be given month by month.

- The concept of "effective" hot water demand and effective required hot water temperature (see Chapter 7) can be used to define hot water loads (accounting for auxiliary heat losses) but heat losses from the auxiliary DHW tank can also be calculated separately by the programme (assuming that the hot water in the tank is controlled at the required temperature 24 hours a day). If one considers the schematics of solar hot water systems defined for ESM2 (Section 7.1), the latter case corresponds to the sub-system 'Series DHW auxiliary with perfect on/off control' (no differential between starting and stopping set-point $T_{h1} = T_{h2}$).

Hence, if the DHW auxiliary tank is in series with the DHW solar tank (or integrated in its upper part) with an on/off control, the actual hot water demand and required temperature will be given as an input in screen pages 8 and 9, and the auxiliary tank heat loss on page 8.

For all other types of DHW auxiliary coupling and control (as described for ESM2), the effective hot water demand and effective required hot water temperature will be given to the programme (integrating auxiliary and possibly distribution heat losses, according to the computation rules described in section 5.3). Auxiliary tank heat losses will be taken as zero when asked for on page 8 of the programme.

- The solar loop heat-exchanger (between solar collector and storage) may be characterised either by its effectiveness e_1 or its heat transfer coefficient A_eU_e (in the latter case, effectiveness is computed assuming a counter-flow heat exchanger).

- Storage capacity M (and also storage heat loss A_sU_s) is divided into two parts: main tank capacity M and solar DHW tank capacity M' (but only their sum is used in the method) (screen page 16).

- The space heating distribution network can be characterized by two sets of data: either water capacitance rate and effectiveness, or flow and return temperature at a nominal power. In the case of an air distribution loop (including a water-to-air coil), only the first set is authorized. The effectiveness e_2 must be given with respect to the <u>water</u> heat capacitance rate $\dot{m}_2c$ (and not, as is usual for a heat-exchanger, with respect to the smaller of the two capacitance rates on both sides of the heat-exchanger). If the air distribution loop admits fresh air (or a mixture of fresh and recycled air), an air-to-air recuperator can be integrated in the circuit, before the water-to-air coil; it will be characterized by its effectiveness e_r (if no recuperator is present, a zero value will be taken for this parameter).

- d_{sol}, thermostat differential between the "solar" set-point and the auxiliary set-point (for space heating distribution, on/off control), see section 6.2.2.2 -, is no longer an input parameter of the programme. Implicit values are assumed:

Series auxiliary: $d_{sol} = 1.5$ K

Parallel auxiliary: $d_{sol} = 0.75$ K

All input parameters can be saved on disk files and used again in another session. They are grouped into several sets of parameters: space heating loads (.LOD), degree-days (.DDJ), domestic hot water load (.DHW) and solar system (.SOL) (see section 9.2.4).

9.2.3 - Output parameters and print-out

There are three types of output parameters:

- Monthly space heating and hot water loads, computed at the end of the first step (screen page 11) after a brief summary of the corresponding input data (screen page 10).

- Monthly and yearly system performance (useful solar energy,auxiliary energy, solar fraction, useful solar energy per square metre of collector) (screen pages 22 and 23) after a complete summary of solar system input parameters (page 20). Energy values are given in Megajoules (MJ); results in kilowatt-hours (kWh) are given only in the yearly summary (page 23). If you prefer kWh, divide results in MJ by 3.6...

- Warning messages are displayed when one or more of the parameters is not within the suitable range (see section 6.5)

  * Yearly hot water load too high
    $L_1 > L_0 / 2.$
  * Internal pipe heat loss too high
    $A_pU_p > 1.5 . A . U_L.$
  * Storage capacity outside the suitable range
    $(M + M') / A < 25$ litres per square metre of collector
    or $C_s < 0.25$ or $C_s > 5.$
  * Storage heat loss too high
    $(A_sU_s + A_sU_s' + A_pU_p/3) / (M + M') > 0.03$ W/K/kg.
  * Solar collector too efficient
    $U_l / \eta_0 < 1$ W/K/m2.
  * Space heating distribution flow-rate too small
    $\dot{m}_2c < GV$
    (i.e. temperature difference in heat emitters is too large, greater than about 25K for maximum power).
  * High solar fraction for system with large storage in sunny climate.
    $f > 0.5$ $H_t > 15$ MJ/day and $M+M' > 2250$ kg of water.

If one of these conditions exists, computations are still permitted, but a message indicates to the user that errors (with respect to EURSOL/EMGP2 reference programme) can be greater than 10%.

Warning messages are indicated briefly on screen pages 20 and 23; they are displayed in detail on a special screen page (page 21).

The print-out of the programme (see Table 9.2.1) contains a summary of all input parameters and complete tables of monthly and yearly performance. It corresponds to the set of the following screen pages: page 12 (project references), page 10 (S.H. load input data), page 11 (DHW load input data; DHW and SH loads computation results), page 20 (summary of solar system input parameters + warning messages), page 22 (monthly system performance), and page 23 (yearly system performance).

9.2.4 - Disk files

Two types of disk files can be used from ESM 1.4:

- Weather data files, which can only be read. Their content is described in section 9.4.
- User's data files which can store the values of a set of input parameters of a solar project. All ESM 1.4 input parameters can thus be saved in several subsets. These files can of course be read or updated from ESM 1.4.

A general rule for ESM file management is that a disk file may be read or updated only from a particular screen page. Particular screen pages are used for file management operations (loading a record, saving a record, creating a new file, deleting a file or record, etc.). Saving data on a disk file is done better at the end of the session (after computations and complete checking have been done). Saving will only affect input parameters validated beforehand (e.g., with the VALIDATION command).

9.2.4.1 - Space heating load files (.LOD)

These files are characterized by the suffix .LOD. Loading or saving data from an .LOD file is allowed from screen page 5 or 6. Each record contains the following information:

- User's comments (entered on page 5 and 6).
- The following parameters: GV, T_c and T_{nh} (entered on page 5), and the 12 monthly values of space heating load (either entered on page 6 by the user or computed by the programme).

9.2.4.2 - Degree-days (.DDJ)

These files can be obtained from screen page 7 and contain the following information:

- User's comments (entered on page 7).
- No-heating temperature T_{nh} (base of degree-days).
- Twelve monthly degree-day values (either entered by the user on page 7 or computed by the programme).

9.2.4.3 - Domestic hot water loads (.DHW)

These files can be read or updated from screen page 9. They are also compatible with ESM 2.1 (hot water). Each record contains:

- User's comments (entered on page 8).
- DHW auxiliary tank heat loss and environment temperature (entered on page 8).
- Twelve monthly values of daily demand m, mains cold water temperature T_m, and required hot water temperature T_h (entered on page 9).

9.2.4.4 - Solar system characteristics (.SOL)

These files are obtained from screen page 12. Each record contains:

- Project references (entered on page 12).
- All input parameters entered from pages 13 to 19, concerning the solar system with the exception of collector orientation and inclination.

9.3 - THE ESM 2.1 SOFTWARE FOR SOLAR HOT WATER SYSTEMS

This sofware is based on the algorithms described in Chapter 7. Possible systems are those given in that chapter, where input parameters are also reviewed. It is emphasized that ESM 2.1 deals only with active solar domestic hot water systems.

9.3.1 - Programme steps

There are two main successive steps in the programme (both of which correspond to an option of the MAIN MENU, screen page 1), including several screen pages:

Step 1: Weather data and loads. Auxiliary and distribution.

* Choice of a weather data file (pages 3, 4, and 5).
* Hot water load input data (screen pages 6, 7, and 8).
* Distribution, recirculation loop and auxiliary input data (type of schematics and heat losses) (screen pages 9 to 14).
* Hot water loads: summary of input data and computation results (screen pages 15 and 16).

Step 2: Solar system characteristics and performance calculation.

* Project references (screen page 17).
* Solar collector input data (screen page 18).
* Solar loop characteristics (screen pages 19 and 20).
* Storage input data (screen page 21).
* Data summary and computation results, including warning messages (screen pages 22 to 25).

Within each step, all screen pages are reviewed successively. Several commands move from one page to another (NEXT PAGE or FORMER PAGE, see section 9.1.3.4). This general structure is analysed in more detail in the ESM 2.1 AIDE-MEMOIRE (in Appendix F), including a systematic review of input and output parameters.

9.3.2 - Input parameters

Input parameters of the ESM 2.1 software are exactly the same as those described in the standard version of the algorithms (see Chapter 7). Input parameters are reviewed for each screen page in the AIDE-MEMOIRE, using the same nomenclature as the standard version; it is therefore easy to find complete explanations about each parameter in Chapter 7. The tables of the AIDE-MEMOIRE also contain minimum and maximum values authorized in the programme and default values proposed to the user at the beginning. Blank worksheets for preparing the input data are available in Appendix F.

A few changes with respect to the nomenclature of the standard version have been introduced:

- Domestic hot water demand (and also mains cold temperature and hot water required temperature) can be given month by month.

- The "effective" hot water demand (see section 7.2 or 5.3 for the definition) is computed by the programme from the information provided by the user. Thus, the hot water demand and hot water temperature asked for by the programme (on screen pages 6 and 7) are the actual demand and temperature, and not the effective ones.

- The solar loop heat-exchanger (between solar collector and storage) may be characterised either by its effectiveness e_1 or its heat transfer coefficient A_eU_e (in the latter case, the effectiveness is computed assuming a counter-flow heat exchanger).

- If there are two or more storage tanks, the storage capacity asked for on screen page 21 is the total capacity (the same for heat loss coefficient), but not including the possible auxiliary energy tank (or the top part of the tank occupied by the auxiliary in some systems).

All input parameters can be saved on disk files and used again in another session. They are grouped into several sets of parameters: hot water load (.DHW), draw-off profile (.HW1), distribution and auxiliary circuit (.HW2), and solar system (.HW3) (see section 9.3.4).

9.3.3 - Output parameters and print-out

There are three types of output parameters:

- Monthly hot water loads, computed at the end of the first step (screen page 16) after a brief summary of corresponding input data (screen page 15).

- Monthly and yearly system performance (useful solar energy, auxiliary energy, solar fraction, useful solar energy per square metre of collector) (screen pages 24 and 25) after a complete summary of solar system input parameters (screen page 22).

  Energy values are given in Megajoules (MJ); results in kilowatt-hours (kWh) are given only in the yearly summary (page 25). If you prefer kWh, divide results in MJ by 3.6...

- Warning messages are displayed when one or more of the parameters is not within the suitable range (see Chapter 7.5). In this case, computation is still permitted, but a message indicates to the user that errors (with respect to EURSOL/EMGP2 reference programme) can be greater than 10%.

  Warning messages are indicated briefly on screen pages 22 and 25; they are displayed in detail on a special screen page (screen page 23).

The print-out of the programme (see Table 9.3.1) contains a summary of all input parameters and complete tables of monthly and yearly performance. It corresponds to the set of the following screen pages: page 17 (project references), page 15 (distribution and auxiliary circuit), page 16 (hot water load input data; HW load computation results), page 22 (summary of solar system input parameters + warning messages), page 24 (monthly system performance), and page 25 (yearly system performance).

9.3.4 - Disk files

Two types of disk files can be used from ESM 2.1:

- Weather data files, which can only be read. Their content is described in section 9.4.
- User's data files which can store the values of a set of input parameters of a solar project. All ESM 2.1 input parameters can thus be saved in several subsets. These files can of course be read or updated from ESM 2.1.

A general rule for ESM file management is that a disk file may be read or updated only from a particular screen page. Particular screen pages are used for file management operations (loading a record, saving a record, creating a new file, deleting a file or record, etc.). Saving data on a disk file is done better at the end of the session (after computations and complete checking have been made).

9.3.4.1 - Hot water loads (.DHW)
These files can be read or updated from screen page 7. They are also compatible with ESM 1.4 (space heating + coupled hot water). Each record contains:

- User's comments (entered on screen page 6).
- Twelve monthly values of daily demand m, mains cold water temperature T_m, and required hot water temperature T_h (entered on screen page 7).

9.3.4.2 - Draw-off profile (.HW1)
These files are characterized by the suffix .HW1. Loading or saving data from a .HW1 file is allowed from screen page 8.
Each record contains the following information:

- User's comments (entered on page 8).
- 24-hourly values of hot water draw-off characteristics (entered on page 8).

9.3.4.3 - Distribution and auxiliary circuit (.HW2)
These files can be obtained from screen page 9 and contain the following information:

- User's comments (entered on screen page 9).
- All input parameters entered from pages 9 to 14, concerning the distribution and auxiliary circuit (schematics, control and heat losses).

9.3.4.4 - Solar system characteristics (.HW3)
These files are obtained from screen page 12. Each record contains:

- Project references (entered on page 17).
- All input parameters entered from pages 18 to 21, concerning the solar system with the exception of collector orientation and inclination.

9.4 - ESM COMPATIBLE WEATHER DATA FILES

ESM programmes for solar space heating systems (ESM 1.4) or hot water (ESM 2.1) require specific climatic information about ambient temperature and solar radiation. Solar radiation data (on the collector surface) must be processed as cumulative frequency curves (CFC) of solar irradiance (see Chapter 4). The CFC enables an estimate to be made of the collectable energy and the collector operating time above any threshold irradiance.

Disk files containing this information are available for several European locations; they can also be created from other programmes. They can be read either from ESM 1.4 or ESM 2.1 (but they cannot be updated or modified from these programmes).

9.4.1 - Content of ESM weather data files

The following information, required by ESM programmes for each month, is contained in the ESM compatible weather data files:

- Temperature data:
  * Average external temperature T_a.
  * A scattering characteristic of the temperature frequency distribution, $(T_{80}-T_{20})$ (see section 5.2; this parameter is used for degree-days calculation).

- Solar radiation data for the collector plane:
  * Global irradiation H_t (MJ/m2/day).
  * Maximum irradiance I_{max} (W/m2).
  * Average incidence angle modifier $\bar{K}$.
  * Number of hours between sunrise and sunset t_d.
  * Fitting parameters of the CFC (see chapter 4) a_i (i=1,...,8).

- Comments about the data: location, altitude, latitude, longitude, albedo, data source, period, orientation, inclination, etc.

Two types of ESM compatible weather data files are available; their filename includes a suffix (.RA1 or .RA5) which distinguishes them.

- Type .RA1: Each record contains general comments, temperature data, and solar radiation data for one particular surface (fixed orientation and inclination). When such a file is used, no change is possible in the collector tilt angle (screen page 13 of ESM 1.4).

- Type .RA5: Each record contains general comments, temperature data, and solar radiation data for a set of surfaces: fixed orientation and five inclinations, 0, 30, 45, 60, and 90 degrees. This type of file permits modification of the collector tilt angle between 0 and 90 degrees (interpolation is made automatically by the programme). Changing collector orientation requires modification of the weather data file and loading the record corresponding to this orientation.

These files can only be read from one specific screen page (e.g., page 4 in ESM 1.4).

9.4.2 - Available ESM weather data files

The following set of files is available with all ESM programme disks. It contains data for 12 European locations (South orientation only, all inclinations):

* Trappes, Carpentras, Mâcon (France).
* Kew, Aberporth, Eskdalemuir (U.K.).
* Hamburg, Wurzburg, Hohenpeissenberg (Germany).
* Uccle (Belgium).
* DeBilt (The Netherlands).
* Copenhagen (Denmark).

These data have been computed from calculations made on test reference year (TRY) files (see section 4.2.3).

Other disk files can also be purchased (Table 9.4.1). Specific computer programmes can also be purchased; they generate suitable ESM weather data files either from hourly radiation files (EUFRAD) or from monthly average solar irradiation on a horizontal surface (CFC-GENER).

Table 9.2.1: Example of ESM 1.4 print-out.

ESM1.4 (Version 05/90) ACTIVE SOLAR SPACE HEATING C ARMINES-B.BOURGES
Step : Load Weather Data File : CARPENTRAS

Page 10 SPACE HEATING LOAD CHARACTERISTICS

COMMENTS : Example Handbook

Building overall Heat Loss Coefficient	GV	(W/K)	:	250
Internal Set-Point Temperature	Tc	(°C)	:	19.0
No-Heating Temperature (Base for Degree-Days)	Tnh	(°C)	:	17.5

Met. file : CARPENTRAS
Monthly Space Heating Loads are
computed by the program with
Degree-Days from the user

DOMESTIC HOT WATER LOAD CHARACTERISTICS

Comments :

DHW back-up tank heat loss coefficient	AsUs	(W/K)	:	0.0
DHW back-up tank environment temperature	Te	(°C)	:	20.0

Page 11 SPACE HEATING AND DOMESTIC HOT WATER LOAD SUMMARY

Month	Degree Day	SH Load (MJ)	DHW Dem. (l/day)	Cold Temp. (°C)	Hot Temp. (°C)	DHW Load (MJ)	Total Load (MJ)
Jan.	366.0	7906	150	10.0	61.8	1008	8914
Feb.	283.8	6130	150	10.0	61.8	910	7041
Mar.	261.3	5644	150	10.0	61.8	1008	6652
Apr.	180.7	3903	150	10.0	61.8	976	4879
May	64.7	1398	150	10.0	61.8	1008	2406
June	0.0	0	150	10.0	61.8	976	976
July	0.0	0	150	10.0	61.8	1008	1008
Aug	0.0	0	150	10.0	61.8	1008	1008
Sep.	0.0	0	150	10.0	61.8	976	976
Oct.	131.2	2835	150	10.0	61.8	1008	3843
Nov.	270.0	5833	150	10.0	61.8	976	6808
Dec.	353.7	7640	150	10.0	61.8	1008	8648
Year	1911.5	41288	150	10.0	61.8	11869	53157

ESM1.4 (Version 05/90) ACTIVE SOLAR SPACE HEATING C ARMINES-B.BOURGES
Weather Data File : CARPENTRAS

Page 12 PROJECT REFERENCES

Project Name : Example Handbook

Location : Carpentras

Date :

Author :

Comments :

ESM1.4 (Version 05/90) ACTIVE SOLAR SPACE HEATING C ARMINES-B.BOURGES
Step : System Weather Data File : CARPENTRAS

Page 13 SOLAR COLLECTOR CHARACTERISTICS

Orientation (0 : South, >0 : West, <0 : East)	(degrees)	0
Surface area	(m2)	20.0
Tilt Angle	(degrees)	45.0
Optical efficiency ETAO	--	0.800
Overall Heat loss coefficient : Constant term	(W/K/m2)	7.00
Overall Heat loss coefficient : Proportion al term	(W/K2/m2)	0.030
Coll. fluid heat capacitance rate per unit area	(W/K/m2)	51.6
Control Starting Differential	(K)	5.0
Control Stopping Differential	(K)	1.0

Page 15_4 SOLAR LOOP HEAT-EXCHANGER

PIPE LOSSES (Outdoor pipes)	
Collector Inlet : Heat loss coefficient (W/K)	0.0
Collector Outlet : Heat loss coefficient (W/K)	0.0

HEAT EXCHANGER	Type : External	
Effectiveness	---	0.930
Secondary flow-rate(between heat-exchanger and storage tank):		
Heat capacitance rate(per unit area of Collector)	(W/K/m2)	69.8

Page 16 STORAGE TANKS

MAIN STORAGE TANK

Capacity (water content)	(kg)	1000
Heat loss coefficient	(W/K)	5.0

DHW SOLAR STORAGE TANK

Capacity (water content)	(kg)	100
Heat loss coefficient	(W/K)	2.0
Storage tank environment temperature	(°C)	20.0
Indoor pipe heat loss coefficient	(W/K)	0.0

ESM1.4 (Version 05/90)   ACTIVE SOLAR SPACE HEATING   C ARMINES-B.BOURGES
Weather Data File : CARPENTRAS

Page 17   DISTRIBUTION SYSTEM

DHW DISTRIBUTION

Effectiveness of heat transfer between
the main storage tank and the DHW distribution   1.000

SPACE HEATING :   Heat Distribution Sub-System

1 - Heat emitters defined by heat capacitance rate and effectiveness
2 - Heat emit. defined by Flow & return temperatures for nominal power rate
3 - Air distribution loop

CHOICE NUMBER 1

Page 18   HEAT EMITTERS AND AUXILIARY

HEAT EMITTERS OR WATER/AIR COIL

Water Heat capacitance rate	(W/K)	1150
Effectiveness	---	0.312

AUXILIARY CONTROL AND COUPLING

1 - Series connected auxiliary, modulated control
2 - Parallel connected auxiliary, modulated control
3 - Series connected auxiliary, on/off control
4 - Parallel connected auxiliary, on/off control

CHOICE NUMBER 1

ESM1.4 (Version 05/90)   ACTIVE SOLAR SPACE HEATING   C ARMINES-B.BOURGES
Step : System   Weather Data File : CARPENTRAS

Page 20   SUMMARY OF DATA
LOADS (MJ) Space Heating : 41288   Dom. Hot Water : 11869   Total : 53157

COLLECTOR   Area : 20.0 m2
Orientation : 0   Tilt angle : 45.0
Optical Efficiency : 0.800
Heat loss (cst term) : 7.00 W/m2/K
(prop.term) : 0.030 W/m2/K2
Fluid Heat Capacit. : 51.6 W/m2/K
Ctrl diff.(K) Start : 5.0   Stop : 1.0
HEAT EXCHANGER   Type : External
Heat transfer (W/K/m2 of col.): ***.*
Effectiveness : 0.930
Second. Heat Capa. : 69.8 W/m2/K
PIPE LOSSES   Heat loss coeff. (W/K)
Outdoor Inlet : 0.0   Outlet : 0.0
Indoor : 0.0

DHW TANK   Capacity : 100 liters
Heat loss : 2.0 W/K
DHW distribution effectiveness : 1.000
HEAT EMITTERS   effectiveness : 0.312
Heat capacitance rate : 1150 W/K
Nominal power : *****
Flow temp.: **.*   Ret.temp. :**.*
AUXILIARY COUPLING AND CONTROL :
Series Auxiliary: temp. modul. control

MAIN STORAGE TANK Capacity : 1000 liters
Heat loss : 5.0 W/K   Env.Temp. : 20.0 °C

ESM1.4 (Version 05/90) ACTIVE SOLAR SPACE HEATING C ARMINES-B.BOURGES
Weather Data File : CARPENTRAS

Page 22 SYSTEM PERFORMANCE SUMMARY

Month	Solar Irrad. (MJ/m2)	Total Load (MJ)	Auxil. Energy (MJ)	Solar Energy (MJ)	Solar Frac -
Jan.	12.4	8914	6429	2485	0.279
Feb.	15.6	7041	4126	2914	0.414
Mar.	17.7	6652	3136	3516	0.529
Apr.	19.1	4879	1545	3334	0.683
May	20.9	2406	63	2342	0.974
June	23.7	976	0	976	1.000
July	25.0	1008	0	1008	1.000
Aug.	22.7	1008	0	1008	1.000
Sep.	22.4	976	0	976	1.000
Oct.	17.9	3843	665	3178	0.827
Nov.	14.3	6808	3827	2982	0.438
Dec.	9.4	8648	6914	1735	0.201
Year	18.4	53157	26705	26452	0.498

Page 23 SYSTEM PERFORMANCE SUMMARY (END)

Total load	53157 MJ	(14766 kWh)
Auxiliary energy	26705 MJ	(7418 kWh)
Useful solar energy	26452 MJ	(7348 kWh)
Useful solar energy per unit area of collector	1323 MJ/m2	(367 kWh/m2)
Solar fraction	0.498	

Table 9.3.1: Example of ESM 2.1 print-out.

ESM2.1 (Vers.09/88)   ACTIVE SOLAR WATER HEATING   C ARMINES
Step : Load   Weather Data File : CARPENTRAS

Page 15   HOT WATER LOAD SUMMARY

Met. file : CARPENTRAS

DISTRIBUTION LOSSES		
Distribution pipes heat loss coefficient	(W/K)	5.0
Environment temperature of distribution circuit, recirculation loop and auxiliary tank	(°C)	20.0
TYPE OF RECIRCULATION LOOP		
Recirculation loop connected to auxiliary tank and solar tank		
Recirculation loop heat loss coefficient	(W/K)	20.0
Recirculation loop daily operating time	(hour)	16.0
AUXILIARY COUPLING AND CONTROL		
Series auxiliary with controlled by pass - Off/peak energy		
Auxiliary tank heat loss coefficient	(W/K)	5.0
Capacity of auxiliary tank	(kg)	800
Auxiliary stopping set-point temperature	(°C)	64.6
Off-peak energy begining time	(hours)	22.0
Off-peak energy ending time	(hours)	6.0

Page 16   HOT WATER LOAD SUMMARY

Month	DHW Dem. (l/day)	Cold Temp. (°C)	Hot Temp. (°C)	Hot Water Load (MJ)
Jan.	800	10.0	50.0	4152
Feb.	800	10.0	50.0	3750
Mar.	800	10.0	50.0	4152
Apr.	800	10.0	50.0	4018
May	800	10.0	50.0	4152
June	800	10.0	50.0	4018
July	800	10.0	50.0	4152
Aug	800	10.0	50.0	4152
Sep.	800	10.0	50.0	4018
Oct.	800	10.0	50.0	4152
Nov.	800	10.0	50.0	4018
Dec.	800	10.0	50.0	4152
Year	800	10.0	50.0	48881

ESM2.1 (Vers.09/88)   ACTIVE SOLAR WATER HEATING   C ARMINES
Weather Data File : CARPENTRAS

Page 17   PROJECT REFERENCES

Project Name :   Example Handbook

Location :   Carpentras

Date :

Author :

Comments :

ESM2.1 (Vers.09/88) ACTIVE SOLAR WATER HEATING C ARMINES
Step : System Weather Data File : CARPENTRAS

Page 22

SUMMARY OF DATA

LOADS

Hot Water Load (MJ) : 48881

COLLECTOR

Surface Area (m2) : 20.0 Orientation : 0 Tilt angle : 44.1
Optical Efficiency : 0.800
Overall Heat loss coef. : Constant term (W/K/m2) : 7.00
Overall Heat loss coef. : Proport. term (W/K2/m2) : 0.030
Collector fluid mass flow-rate (kg/h) : 45.9
Collector fluid specific heat (kJ/kg/K) : 3.92

HEAT EXCHANGER

Type : External
Effectiveness : 0.730 Heat transfer coefficient (W/K/m2) : 75.0
Storage side mass flow-rate (kg/h/m2) : 43.0

PIPE LOSSES

Heat losses for external piping (collector inlet) (W/K) : 0.0
Heat losses for external piping (collector outlet) (W/K) : 0.0
Indoor pipe heat loss coefficient (W/K) : 6.0

STORAGE TANK

Capacity (kg): 800 Heat loss coeff. (W/K): 8.0 Envt Temp. (°C): 20.0

ESM2.1 (Vers.09/88) ACTIVE SOLAR WATER HEATING C ARMINES
Weather Data File : CARPENTRAS

Page 24

SYSTEM PERFORMANCE SUMMARY

Month	Solar Irrad. (MJ/m2)	Ambient Temp. (°C)	Total Load (MJ)	Auxiliary Energy (MJ)	Solar Energy (MJ)	Solar Fraction –	Recirc. Loop Losses (MJ)
Jan.	12.4	5.7	5812	3203	2609	0.449	1071
Feb.	15.6	7.4	5281	2421	2860	0.542	968
Mar.	17.7	9.1	5862	2455	3406	0.581	1071
Apr.	19.2	11.5	5681	2246	3435	0.605	1037
May	21.0	15.9	5894	1957	3937	0.668	1071
June	23.9	19.6	5735	1378	4357	0.760	1037
July	25.1	22.3	5942	1064	4878	0.821	1071
Aug.	22.8	22.1	5933	1286	4647	0.783	1071
Sep.	22.4	18.4	5741	1259	4482	0.781	1037
Oct.	17.9	13.4	5886	2078	3808	0.647	1071
Nov.	14.3	8.5	5652	2691	2961	0.524	1037
Dec.	9.3	6.1	5776	3725	2051	0.355	1071
Year	18.5	13.4	69194	25762	43432	0.628	12614

Page 25

SYSTEM PERFORMANCE SUMMARY (END)

Total load : effective solar load	63072 MJ	(17520 kWh)
and residual load	6122 MJ	(1700 kWh)
Auxiliary energy	25762 MJ	(7156 kWh)
Useful solar energy	43432 MJ	(12064 kWh)
Useful solar energy per unit area of collector	2172 MJ/m2	(603 kWh/m2)

Solar fraction 0.628

Table 9.4.1: List of available weather data files for ESM software.
TRY: Test Reference Years

Location	Country	Lat.	Long.	Alt.	Period
Trappes	France	48°46'	02°01'E	168 m	TRY
Macon	France	46°18'	04°48'E	217 m	TRY
Carpentras	France	44°05'	05°03'E	105 m	TRY
Kew	U.K.	51°28'	00°19'W	5 m	TRY
Aberporth	U.K.	52°08'	04°34'W	133 m	TRY
Eskdalemuir	U.K.	55°19'	03°12'W	242 m	TRY
Uccle	Belgium	50°48'	04°21'E	105 m	TRY
DeBilt	Netherlands	52°06'	05°11'E	40 m	TRY
Copenhagen	Denmark	55°40'	12°19'E	30 m	TRY
Hamburg	F.R.Germany	53°38'	10°00'E	14 m	TRY
Wurzburg	F.R.Germany	49°48'	09°54'E	259 m	TRY
Hohenpeissenberg	F.R.Germany	47°48'	11°01'E	975 m	TRY
Copenhagen	Denmark	55°40'	12°19'E	30 m	1959-73
Dublin	Ireland	53°26'	06°15'W	80 m	1976-86
Birr	Ireland	53°05'	07°54'W	72 m	1979-86
Valentia	Ireland	51°56'	10°15'W	20 m	1964-86
Bracknell	U.K.	51°23'	00°47'W	73 m	1967-76
Uccle	Belgium	50°48'	04°21'E	105 m	1966-75
Trappes	France	48°46'	02°01'E	168 m	1974-81
Nancy	France	48°41'	06°13'E	217 m	1974-81
Rennes	France	48°04'	01°44'W	37 m	1974-81
La Rochelle	France	46°09'	01°09'W	10 m	1974-81
Macon	France	46°18'	04°48'E	217 m	1974-81
Limoges	France	45°52'	01°11'E	402 m	1974-81
Embrun	France	44°34'	06°30'E	876 m	1979-83
Agen	France	44°11'	00°36'E	60 m	1974-81
Carpentras	France	44°05'	05°03'E	105 m	1975-81
Nice	France	43°39'	07°12'E	10 m	1974-81
Carcassonne	France	43°13'	02°19'E	126 m	1979-83
Ajaccio	France	41°55'	08°48'E	9 m	1974-81
Zurich	Switzerland	47°27'	08°33'E	440 m	1963-70
Davos	Switzerland	46°48'	09°49'E	1590 m	1963-70
Locarno	Switzerland	46°10'	08°47'E	380 m	1963-70
Oviedo	Spain	43°21'	05°52'W	335 m	1976-83
Tortosa	Spain	40°49'	00°30'E	44 m	1980-83
Madrid	Spain	40°27'	03°43'W	669 m	1981-83
Palma de Mal.	Spain	39°34'	02°39'E	32 m	1976-83
Badajoz	Spain	38°53'	06°49'W	185 m	1976-82
Murcia	Spain	37°59'	01°07'W	57 m	1976-80
Izana	Spain	28°18'	16°25'W	2353 m	1977-80
Bragança	Portugal	41°49'	06°46'W	725 m	1982-87
Porto	Portugal	41°08'	08°36'W	93 m	1982-87
Coimbra	Portugal	40°12'	08°25'W	141 m	1982-87
Lisbon	Portugal	38°43'	09°09'W	77 m	1982-87
Evora	Portugal	38°34'	07°54'W	309 m	1982-87
Faro	Portugal	37°01'	07°55'W	14 m	1982-87
Athens	Greece	37°58'	23°43'E	107 m	1961-82

APPENDICES

APPENDIX A

SOLAR RADIATION DATA

This appendix contains three sets of tables for twelve representative European locations.

The first set of tables gives data for each month for a South-facing surface at 4 different inclinations (30°, 45°, 60°, and an angle equal to the latitude). The data are:

* Average daily global irradiation H_t (MJ/m2).

* Maximum global irradiance I_{max} (W/m2).

* Reduced integral of the cumulative frequency curve of solar irradiance

  $$v = H_t / (I_{max} \cdot t_d)$$

  where t_d is the time between sunrise and sunset (see section 4.3.2).

* Average incidence angle modifier $\bar{K}$ for single-glazed collectors (b_0 = 0.10, see section 4.4.2).
  For other values of b_0, it can be approximated by

  $$\bar{K} (b_0) = 1 - b_0 * 10 * (1 - \bar{K}).$$

* Values of the products $\bar{K}.H_t$ and $\bar{K}.I_{max}$ (which can be used directly in the calculation sheets, Table B of Appendix D).

The values for H_t, I_{max}, and v are taken from reference /4.13/ in Chapter 4. The values of the incidence angle modifier have been estimated with the help of the computer programme CASAMO /4.22/.

The other sets of tables give solar radiation data for the computation of space heating load, accounting for passive solar gains (see section 5.1.3). They contain:

* Monthly values of the correction factor for glazing transmittance $\bar{s}/s_n$ (this term for the glazings is similar to the incidence angle modifier for collectors, but the latter concerns both transmittance and absorptance).

* Monthly averages of daily global irradiation on vertical surfaces H_t (MJ/m2) for several surface azimuths (South, South-East/South-West, and East/West).

These data are derived from the European Solar Radiation Atlas /4.17/.

AVAILABLE METEOROLOGICAL STATIONS

Number	Station	Country	Latitude	Longitude	Altitude
1	TRAPPES	France	48° 46'	02° 01' E	168 m
2	CARPENTRAS	"	44° 05'	05° 03' E	105 m
3	MACON	"	46° 18'	04° 48' E	217 m
4	KEW	United Kingdom	51° 28'	00° 19' W	5 m
5	ABERPORTH	"	52° 08'	04° 34' W	133 m
6	ESKDALEMUIR	"	55° 19'	03° 12' W	242 m
7	UCCLE	Belgium	50° 48'	04° 21' E	105 m
8	DEBILT	Netherlands	52° 06'	05° 11' E	40 m
9	COPENHAGEN	Denmark	55° 40'	12° 19' E	30 m
10	HAMBURG	Germany	53° 38'	10° 00' E	14 m
11	WURZBURG	"	49° 48'	09° 54' E	259 m
12	HOHENPEISSENBERG	"	47° 48'	11° 01' E	975 m

Solar radiation data for a South-facing surface

Location: TRAPPES   Lat. : 48.77°N   Inc. : 30°

	JAN.	FEB.	MAR.	APR.	MAY	JUNE	JULY	AUG.	SEP.	OCT.	NOV.	DEC.
H_T	4.9	8.2	10.9	18.0	17.5	20.0	19.0	17.1	14.8	9.0	5.7	4.1
I_{MAX}	675	798	924	1025	1029	1032	999	960	931	824	702	763
$\underline{v}$	.24	.29	.28	.36	.32	.34	.34	.35	.35	.28	.25	.18
$\bar{K}$	.93	.94	.93	.93	.92	.93	.93	.93	.93	.94	.94	.93
$\bar{K}.H_T$	4.6	7.7	10.2	16.7	16.1	18.6	17.7	15.9	13.7	8.5	5.4	3.8
$\bar{K}.I_{MAX}$	628	750	859	953	947	960	929	893	866	775	660	710

Location: TRAPPES   Lat. : 48.77°N   Inc. : 45°

	JAN.	FEB.	MAR.	APR.	MAY	JUNE	JULY	AUG.	SEP.	OCT.	NOV.	DEC.
H_T	5.5	8.9	11.3	17.9	16.7	18.7	17.9	16.5	15.1	9.5	6.3	4.6
I_{MAX}	780	901	1001	1053	1013	1000	974	979	969	940	805	905
$\underline{v}$	.23	.28	.27	.35	.30	.33	.33	.33	.35	.26	.24	.17
$\bar{K}$	.95	.95	.94	.93	.92	.92	.92	.93	.94	.95	.95	.95
$\bar{K}.H_T$	5.2	8.5	10.7	16.6	15.3	17.2	16.5	15.4	14.2	9.0	6.0	4.4
$\bar{K}.I_{MAX}$	741	856	941	979	932	920	896	910	911	893	765	860

Location: TRAPPES   Lat. : 48.77°N   Inc. : 60°

	JAN.	FEB.	MAR.	APR.	MAY	JUNE	JULY	AUG.	SEP.	OCT.	NOV.	DEC.
H_T	5.8	9.1	11.2	16.9	15.0	16.5	16.0	15.2	14.6	9.5	6.6	4.9
I_{MAX}	837	947	1024	1005	936	908	887	924	950	981	867	999
$\underline{v}$	.22	.27	.26	.35	.30	.32	.32	.32	.34	.25	.23	.17
$\bar{K}$	.96	.96	.94	.93	.93	.91	.91	.92	.94	.95	.96	.96
$\bar{K}.H_T$	5.5	8.8	10.5	15.7	13.9	15.1	14.6	14.0	13.7	9.0	6.3	4.7
$\bar{K}.I_{MAX}$	804	909	963	935	870	826	807	850	893	932	832	959

Location: TRAPPES   Lat. : 48.77°N   Inc. : 48.77°

	JAN.	FEB.	MAR.	APR.	MAY	JUNE	JULY	AUG.	SEP.	OCT.	NOV.	DEC.
H_T	5.6	9.0	11.3	17.7	16.3	18.2	17.5	16.3	15.0	9.5	6.4	4.7
I_{MAX}	799	918	1012	1048	999	982	958	972	970	958	825	934
$\underline{v}$	.23	.27	.27	.35	.30	.33	.33	.33	.34	.26	.24	.17
$\bar{K}$	.95	.95	.94	.93	.92	.92	.92	.93	.94	.95	.95	.95
$\bar{K}.H_T$	5.3	8.6	10.7	16.5	15.0	16.7	16.1	15.1	14.1	9.0	6.1	4.5
$\bar{K}.I_{MAX}$	762	874	952	975	921	902	879	903	913	911	785	890

Solar radiation data for a South-facing surface

Location: CARPENTRAS   Lat. : 44.08°N   Inc. : 30°

	JAN.	FEB.	MAR.	APR.	MAY	JUNE	JULY	AUG.	SEP.	OCT.	NOV.	DEC.
H_T	10.9	14.4	17.1	19.5	22.2	25.8	26.9	23.5	22.0	16.7	12.6	8.2
I_{MAX}	825	1001	1040	1062	1087	1083	1046	1055	1034	984	840	759
$\underline{v}$	.40	.39	.39	.39	.39	.43	.48	.45	.48	.43	.44	.34
$\bar{K}$	.95	.95	.94	.93	.93	.93	.94	.94	.95	.95	.95	.95
$\bar{K}.H_T$	10.4	13.6	16.1	18.2	20.6	24.0	25.3	22.1	20.9	15.9	12.0	7.8
$K.I_{MAX}$	784	951	978	988	1011	1007	983	992	982	935	798	721

Location: CARPENTRAS   Lat. : 44.08°N   Inc. : 45°

	JAN.	FEB.	MAR.	APR.	MAY	JUNE	JULY	AUG.	SEP.	OCT.	NOV.	DEC.
H_T	12.4	15.6	17.7	19.1	20.9	23.7	25.0	22.7	22.4	17.9	14.3	9.4
I_{MAX}	937	1092	1080	1062	1048	1025	1000	1026	1061	1050	950	881
$\underline{v}$	.40	.39	.39	.38	.38	.42	.46	.44	.47	.44	.44	.34
$\bar{K}$	.97	.96	.95	.93	.93	.93	.93	.93	.95	.96	.97	.96
$\bar{K}.H_T$	12.1	15.0	16.8	17.8	19.4	22.1	23.2	21.1	21.2	17.2	13.9	9.0
$K.I_{MAX}$	909	1048	1026	988	975	953	930	954	1008	1008	922	846

Location: CARPENTRAS   Lat. : 44.08°N   Inc. : 60°

	JAN.	FEB.	MAR.	APR.	MAY	JUNE	JULY	AUG.	SEP.	OCT.	NOV.	DEC.
H_T	13.2	16.0	17.3	17.7	18.5	20.4	21.7	20.6	21.5	18.1	15.2	10.0
I_{MAX}	1001	1116	1068	1001	955	912	899	942	1047	1067	1005	945
$\underline{v}$	.40	.39	.38	.37	.37	.41	.45	.44	.46	.43	.44	.33
$\bar{K}$	.97	.97	.95	.92	.91	.90	.91	.92	.94	.96	.97	.97
$\bar{K}.H_T$	12.8	15.5	16.4	16.3	16.8	18.4	19.7	18.9	20.2	17.3	14.7	9.7
$K.I_{MAX}$	971	1083	1015	921	869	821	818	867	984	1024	975	917

Location: CARPENTRAS   Lat. : 44.08°N   Inc. : 44.08°

	JAN.	FEB.	MAR.	APR.	MAY	JUNE	JULY	AUG.	SEP.	OCT.	NOV.	DEC.
H_T	12.4	15.6	17.7	19.2	21.0	23.9	25.1	22.8	22.4	17.9	14.3	9.3
I_{MAX}	931	1088	1079	1064	1052	1030	1004	1029	1060	1047	945	875
$\underline{v}$	.40	.39	.39	.38	.38	.42	.46	.44	.47	.44	.44	.34
$\bar{K}$	.97	.96	.95	.93	.93	.93	.93	.93	.95	.96	.97	.96
$\bar{K}.H_T$	12.0	15.0	16.8	17.9	19.5	22.3	23.4	21.2	21.3	17.2	13.8	8.9
$K.I_{MAX}$	903	1044	1025	990	979	959	935	958	1008	1005	916	840

Solar radiation data for a South-facing surface

Location: MACON   Lat. : 46.3°N   Inc. : 30°

	JAN.	FEB.	MAR.	APR.	MAY	JUNE	JULY	AUG.	SEP.	OCT.	NOV.	DEC.
H_T	4.7	7.5	12.9	20.1	20.7	19.4	21.3	21.2	18.8	12.0	6.2	3.7
I_{MAX}	610	849	1039	1038	1094	1071	1022	1033	985	920	725	573
$\underline{v}$	.24	.24	.29	.40	.36	.32	.38	.41	.43	.33	.26	.21
$\bar{K}$	.93	.93	.93	.94	.93	.92	.93	.93	.94	.94	.93	.93
$\bar{K}.H_T$	4.4	7.0	12.0	18.9	19.3	17.8	19.8	19.7	17.6	11.3	5.8	3.5
$K.I_{MAX}$	567	790	966	976	1017	985	950	961	926	865	674	533

Location: MACON   Lat. : 46.3°N   Inc. : 45°

	JAN.	FEB.	MAR.	APR.	MAY	JUNE	JULY	AUG.	SEP.	OCT.	NOV.	DEC.
H_T	5.1	7.9	13.3	19.9	19.6	18.0	19.9	20.5	19.2	12.8	6.8	4.1
I_{MAX}	700	926	1115	1046	1071	1025	979	1025	1020	991	837	690
$\underline{v}$	.23	.23	.28	.40	.34	.31	.37	.40	.42	.33	.24	.20
$\bar{K}$	.95	.94	.94	.93	.93	.92	.93	.93	.94	.95	.95	.95
$\bar{K}.H_T$	4.9	7.4	12.5	18.5	18.2	16.6	18.5	19.1	18.1	12.2	6.5	3.9
$K.I_{MAX}$	665	870	1048	973	996	943	910	953	959	941	795	656

Location: MACON   Lat. : 46.3°N   Inc. : 60°

	JAN.	FEB.	MAR.	APR.	MAY	JUNE	JULY	AUG.	SEP.	OCT.	NOV.	DEC.
H_T	5.3	7.9	13.1	18.7	17.4	15.8	17.6	18.8	18.6	13.0	7.1	4.3
I_{MAX}	749	949	1100	986	975	912	886	966	992	1012	900	750
$\underline{v}$	.22	.23	.28	.39	.34	.31	.36	.39	.42	.33	.24	.19
$\bar{K}$	.96	.95	.94	.93	.91	.90	.91	.92	.94	.95	.96	.95
$\bar{K}.H_T$	5.1	7.5	12.3	17.4	15.8	14.2	16.0	17.3	17.5	12.3	6.8	4.1
$K.I_{MAX}$	719	902	1034	917	887	821	806	889	932	961	864	713

Location: MACON   Lat. : 46.3°N   Inc. : 46.3°

	JAN.	FEB.	MAR.	APR.	MAY	JUNE	JULY	AUG.	SEP.	OCT.	NOV.	DEC.
H_T	5.1	7.9	13.3	19.9	19.4	17.9	19.8	20.4	19.2	12.9	6.9	4.2
I_{MAX}	706	930	1117	1043	1066	1018	973	1022	1020	995	845	698
$\underline{v}$	.23	.23	.28	.40	.34	.31	.37	.40	.42	.33	.24	.20
$\bar{K}$	.95	.94	.94	.93	.93	.92	.93	.93	.94	.95	.95	.95
$\bar{K}.H_T$	4.9	7.4	12.6	18.5	18.1	16.4	18.4	19.0	18.1	12.2	6.5	4.0
$K.I_{MAX}$	671	875	1051	970	990	935	904	950	959	945	803	663

Solar radiation data for a South-facing surface

Location: KEW Lat. : 51.52°N Inc. : 30°

	JAN.	FEB.	MAR.	APR.	MAY	JUNE	JULY	AUG.	SEP.	OCT.	NOV.	DEC.
H_T	4.4	5.8	10.0	12.5	17.0	17.5	16.4	15.9	12.0	9.1	4.6	2.7
I_{MAX}	539	648	898	975	948	957	921	926	891	774	581	480
$\underline{v}$	.28	.25	.26	.26	.32	.31	.31	.33	.30	.31	.25	.20
_K	.93	.93	.93	.92	.92	.92	.92	.93	.93	.94	.93	.92
_K.H_T	4.1	5.4	9.3	11.5	15.6	16.1	15.1	14.8	11.2	8.6	4.3	2.5
K.I_{MAX}	501	603	835	897	872	880	847	861	829	728	540	442

Location: KEW Lat. : 51.52°N Inc. : 45°

	JAN.	FEB.	MAR.	APR.	MAY	JUNE	JULY	AUG.	SEP.	OCT.	NOV.	DEC.
H_T	5.1	6.3	10.5	12.4	16.2	16.4	15.6	15.5	12.4	9.9	5.2	3.1
I_{MAX}	649	750	950	1000	950	936	910	946	945	868	700	575
$\underline{v}$	.27	.24	.26	.25	.31	.30	.30	.32	.29	.30	.24	.19
_K	.96	.94	.93	.92	.93	.92	.92	.93	.93	.95	.95	.95
_K.H_T	4.9	6.0	9.7	11.4	15.0	15.1	14.3	14.4	11.6	9.4	5.0	2.9
K.I_{MAX}	623	705	884	920	884	861	837	880	879	825	665	546

Location: KEW Lat. : 51.52°N Inc. : 60°

	JAN.	FEB.	MAR.	APR.	MAY	JUNE	JULY	AUG.	SEP.	OCT.	NOV.	DEC.
H_T	5.5	6.6	10.4	11.6	14.7	14.7	14.0	14.4	12.2	10.2	5.6	3.3
I_{MAX}	713	819	952	960	882	863	850	897	948	903	775	660
$\underline{v}$	.26	.23	.26	.25	.30	.29	.29	.31	.28	.30	.23	.18
_K	.96	.95	.94	.92	.91	.91	.91	.92	.93	.95	.96	.96
_K.H_T	5.3	6.3	9.8	10.7	13.4	13.3	12.8	13.2	11.3	9.7	5.4	3.2
K.I_{MAX}	684	778	895	883	803	785	774	825	882	858	744	634

Location: KEW Lat. : 51.52°N Inc. : 51.52°

	JAN.	FEB.	MAR.	APR.	MAY	JUNE	JULY	AUG.	SEP.	OCT.	NOV.	DEC.
H_T	5.3	6.5	10.5	12.1	15.6	15.7	15.0	15.1	12.4	10.1	5.4	3.2
I_{MAX}	683	785	957	990	929	910	890	933	953	891	738	614
$\underline{v}$	.26	.24	.26	.25	.30	.29	.29	.31	.29	.30	.23	.19
_K	.96	.94	.93	.92	.93	.92	.92	.93	.93	.95	.96	.96
_K.H_T	5.1	6.1	9.8	11.1	14.4	14.4	13.7	14.0	11.5	9.6	5.2	3.1
K.I_{MAX}	657	741	893	911	859	835	816	865	886	847	705	587

Solar radiation data for a South-facing surface

Location: ABERPORTH Lat. : 52.13°N Inc. : 30°

	JAN.	FEB.	MAR.	APR.	MAY	JUNE	JULY	AUG.	SEP.	OCT.	NOV.	DEC.
H_T	3.3	7.0	12.8	15.3	20.7	20.2	20.6	18.5	14.1	8.3	5.4	3.0
I_{MAX}	524	758	925	1007	1041	1042	1054	1021	938	800	746	531
$\underline{v}$	.22	.26	.33	.31	.36	.33	.34	.35	.33	.27	.23	.21
_K	.93	.94	.94	.93	.93	.93	.93	.93	.93	.94	.94	.93
_K.H_T	3.1	6.5	12.1	14.3	19.3	18.8	19.1	17.2	13.1	7.8	5.1	2.8
K.I_{MAX}	487	713	870	937	968	969	980	950	872	752	701	494

Location: ABERPORTH Lat. : 52.13°N Inc. : 45°

	JAN.	FEB.	MAR.	APR.	MAY	JUNE	JULY	AUG.	SEP.	OCT.	NOV.	DEC.
H_T	3.6	7.6	13.5	15.2	19.9	19.0	19.5	18.2	14.5	8.9	6.1	3.4
I_{MAX}	609	866	988	1047	1038	1022	1037	1044	987	893	875	690
$\underline{v}$	.20	.25	.33	.30	.34	.31	.33	.33	.33	.26	.22	.18
_K	.95	.95	.94	.93	.93	.92	.93	.93	.94	.95	.96	.96
_K.H_T	3.5	7.2	12.7	14.1	18.5	17.5	18.2	16.9	13.7	8.5	5.8	3.3
K.I_{MAX}	579	823	929	974	965	940	964	971	928	848	840	662

Location: ABERPORTH Lat. : 52.13°N Inc. : 60°

	JAN.	FEB.	MAR.	APR.	MAY	JUNE	JULY	AUG.	SEP.	OCT.	NOV.	DEC.
H_T	3.8	7.8	13.5	14.3	18.1	16.9	17.7	17.0	14.2	9.1	6.4	3.7
I_{MAX}	713	916	1011	1025	972	934	950	1000	975	931	950	771
$\underline{v}$	.18	.24	.32	.28	.33	.31	.32	.33	.32	.26	.22	.17
_K	.96	.96	.95	.93	.92	.91	.91	.93	.94	.95	.97	.97
_K.H_T	3.6	7.5	12.8	13.3	16.6	15.4	16.1	15.8	13.3	8.6	6.2	3.5
K.I_{MAX}	684	879	960	953	894	850	865	930	917	884	922	748

Location: ABERPORTH Lat. : 52.13°N Inc. : 52.13°

	JAN.	FEB.	MAR.	APR.	MAY	JUNE	JULY	AUG.	SEP.	OCT.	NOV.	DEC.
H_T	3.7	7.7	13.6	14.9	19.1	18.1	18.7	17.7	14.5	9.1	6.3	3.6
I_{MAX}	657	897	1004	1044	1014	988	1004	1031	989	918	918	739
$\underline{v}$	.19	.25	.32	.29	.34	.31	.32	.33	.02	.26	.22	.18
_K	.96	.95	.94	.93	.93	.92	.92	.93	.94	.95	.97	.97
_K.H_T	3.6	7.4	12.8	13.8	17.7	16.6	17.3	16.5	13.6	8.6	6.1	3.4
K.I_{MAX}	627	856	948	972	940	905	927	960	931	873	886	714

Solar radiation data for a South-facing surface

Location: ESKDALEMUIR Lat. : 55.32°N Inc. : 30°

	JAN.	FEB.	MAR.	APR.	MAY	JUNE	JULY	AUG.	SEP.	OCT.	NOV.	DEC.
H_T	3.4	5.7	8.3	12.1	16.0	16.7	15.6	13.7	9.7	7.2	4.4	3.3
I_{MAX}	632	725	847	950	999	1017	974	883	900	755	659	517
$\underline{v}$	.20	.23	.24	.25	.28	.27	.27	.29	.24	.26	.23	.25
$\underline{K}$	.93	.93	.93	.92	.92	.92	.92	.92	.92	.93	.93	.93
$\underline{K}.H_T$	3.2	5.3	7.7	11.1	14.7	15.4	14.4	12.6	9.0	6.7	4.1	3.0
$K.I_{MAX}$	588	674	788	874	919	936	896	812	828	702	613	481

Location: ESKDALEMUIR Lat. : 55.32°N Inc. : 45°

	JAN.	FEB.	MAR.	APR.	MAY	JUNE	JULY	AUG.	SEP.	OCT.	NOV.	DEC.
H_T	4.1	6.3	8.8	12.0	15.4	15.8	14.9	13.5	10.0	7.9	5.2	3.9
I_{MAX}	770	826	968	1000	1000	1005	982	910	1000	874	794	650
$\underline{v}$	.20	.23	.22	.24	.27	.26	.25	.28	.22	.24	.22	.24
$\underline{K}$	.96	.95	.93	.92	.92	.92	.92	.92	.93	.95	.96	.97
$\underline{K}.H_T$	3.9	6.0	8.2	11.1	14.2	14.6	13.7	12.4	9.3	7.5	4.9	3.8
$K.I_{MAX}$	739	785	900	920	920	925	903	837	930	830	762	631

Location: ESKDALEMUIR Lat. : 55.32°N Inc. : 60°

	JAN.	FEB.	MAR.	APR.	MAY	JUNE	JULY	AUG.	SEP.	OCT.	NOV.	DEC.
H_T	4.5	6.6	8.9	11.4	14.2	14.3	13.6	12.6	9.7	8.2	5.6	4.4
I_{MAX}	853	900	1025	975	949	935	931	884	1025	934	875	741
$\underline{v}$	.19	.22	.21	.23	.26	.25	.25	.27	.21	.24	.22	.24
$\underline{K}$	.97	.96	.94	.92	.91	.91	.91	.92	.93	.95	.97	.97
$\underline{K}.H_T$	4.4	6.4	8.3	10.5	12.9	13.0	12.4	11.6	9.0	7.8	5.4	4.2
$K.I_{MAX}$	827	864	964	897	864	851	847	813	953	887	849	719

Location: ESKDALEMUIR Lat. : 55.32°N Inc. : 55.32°

	JAN.	FEB.	MAR.	APR.	MAY	JUNE	JULY	AUG.	SEP.	OCT.	NOV.	DEC.
H_T	4.4	6.6	8.9	11.7	14.6	14.8	14.1	12.9	9.8	8.2	5.5	4.3
I_{MAX}	833	880	1015	991	970	963	953	898	1026	922	856	718
$\underline{v}$	.19	.22	.21	.24	.26	.25	.25	.27	.21	.24	.22	.24
$\underline{K}$	.97	.96	.94	.92	.91	.91	.91	.92	.93	.95	.97	.97
$\underline{K}.H_T$	4.3	6.3	8.3	10.8	13.4	13.6	12.9	11.9	9.1	7.8	5.3	4.1
$K.I_{MAX}$	807	843	950	911	887	880	871	826	955	877	829	698

Solar radiation data for a South-facing surface

Location: UCCLE   Lat. : 50.8°N   Inc. : 30°

	JAN.	FEB.	MAR.	APR.	MAY	JUNE	JULY	AUG.	SEP.	OCT.	NOV.	DEC.
H_T	3.5	6.8	9.8	12.9	15.7	19.3	15.1	15.2	14.9	7.4	3.8	2.5
I_{MAX}	542	827	898	950	1000	1016	975	979	900	794	575	499
$\underline{v}$	.21	.23	.26	.28	.29	.33	.27	.30	.37	.25	.21	.18
$\bar{K}$	.93	.93	.93	.92	.92	.93	.92	.92	.93	.93	.93	.93
$\bar{K}.H_T$	3.2	6.3	9.1	11.9	14.4	17.9	13.9	14.0	13.9	6.9	3.5	2.3
$\bar{K}.I_{MAX}$	504	769	835	874	920	945	897	901	837	738	535	464

Location: UCCLE   Lat. : 50.8°N   Inc. : 45°

	JAN.	FEB.	MAR.	APR.	MAY	JUNE	JULY	AUG.	SEP.	OCT.	NOV.	DEC.
H_T	3.8	7.3	10.2	12.7	15.0	18.1	14.3	14.8	15.4	7.8	4.1	2.7
I_{MAX}	643	927	996	975	986	993	954	989	969	900	675	598
$\underline{v}$	.20	.22	.24	.27	.28	.31	.26	.29	.35	.23	.19	.16
$\bar{K}$	.95	.95	.93	.92	.92	.92	.92	.92	.94	.94	.95	.95
$\bar{K}.H_T$	3.6	6.9	9.5	11.7	13.8	16.7	13.2	13.6	14.5	7.3	3.9	2.6
$\bar{K}.I_{MAX}$	611	881	926	897	907	914	878	910	911	846	641	568

Location: UCCLE   Lat. : 50.8°N   Inc. : 60°

	JAN.	FEB.	MAR.	APR.	MAY	JUNE	JULY	AUG.	SEP.	OCT.	NOV.	DEC.
H_T	3.9	7.4	10.1	11.9	13.5	16.1	12.9	13.7	15.1	7.8	4.2	2.9
I_{MAX}	700	968	1031	934	915	906	879	947	974	938	732	662
$\underline{v}$	.19	.22	.23	.26	.27	.31	.26	.28	.34	.22	.18	.15
$\bar{K}$	.96	.96	.94	.92	.91	.91	.91	.92	.94	.95	.95	.96
$\bar{K}.H_T$	3.8	7.1	9.5	10.9	12.3	14.7	11.7	12.6	14.2	7.4	4.0	2.7
$\bar{K}.I_{MAX}$	672	929	969	859	833	824	800	871	916	891	695	636

Location: UCCLE   Lat. : 50.8°N   Inc. : 50.8°

	JAN.	FEB.	MAR.	APR.	MAY	JUNE	JULY	AUG.	SEP.	OCT.	NOV.	DEC.
H_T	3.9	7.4	10.2	12.4	14.5	17.4	13.8	14.5	15.4	7.8	4.2	2.8
I_{MAX}	671	950	1017	967	965	967	931	979	979	924	703	627
$\underline{v}$	.19	.22	.24	.26	.27	.01	.26	.29	.35	.22	.19	.16
$\bar{K}$	.95	.95	.93	.92	.92	.92	.92	.92	.94	.94	.95	.95
$\bar{K}.H_T$	3.7	7.1	9.6	11.4	13.3	16.0	12.7	13.3	14.5	7.4	4.0	2.7
$\bar{K}.I_{MAX}$	640	907	949	890	886	886	854	901	922	871	669	599

Solar radiation data for a South-facing surface

Location: DEBILT Lat. : 52.1°N Inc. : 30°

	JAN.	FEB.	MAR.	APR.	MAY	JUNE	JULY	AUG.	SEP.	OCT.	NOV.	DEC.
H_T	4.0	6.3	10.9	14.7	17.7	19.2	17.0	18.0	12.0	8.1	4.4	3.5
I_{MAX}	775	726	837	963	1046	983	974	1025	875	812	600	625
$\underline{\nu}$	.18	.25	.31	.31	.30	.33	.30	.34	.30	.26	.23	.21
_K	.93	.93	.93	.93	.92	.93	.92	.93	.93	.93	.93	.93
_K.H_T	3.7	5.8	10.2	13.7	16.3	17.8	15.7	16.8	11.2	7.5	4.1	3.3
K.I_{MAX}	721	675	778	896	962	914	896	953	814	755	558	581

Location: DEBILT Lat. : 52.1°N Inc. : 45°

	JAN.	FEB.	MAR.	APR.	MAY	JUNE	JULY	AUG.	SEP.	OCT.	NOV.	DEC.
H_T	4.6	6.9	11.4	14.7	17.0	18.0	16.2	17.8	12.3	8.7	4.9	4.2
I_{MAX}	925	809	924	995	1046	966	975	1049	942	895	700	782
$\underline{\nu}$	.17	.25	.29	.30	.29	.32	.29	.33	.29	.26	.23	.20
_K	.95	.95	.94	.93	.93	.92	.92	.93	.93	.95	.95	.96
_K.H_T	4.4	6.6	10.7	13.6	15.8	16.5	14.9	16.5	11.4	8.3	4.7	4.0
K.I_{MAX}	879	769	869	925	973	889	897	976	876	850	665	751

Location: DEBILT Lat. : 52.1°N Inc. : 60°

	JAN.	FEB.	MAR.	APR.	MAY	JUNE	JULY	AUG.	SEP.	OCT.	NOV.	DEC.
H_T	5.0	7.2	11.3	13.9	15.4	16.1	14.5	16.6	11.9	8.9	5.2	4.6
I_{MAX}	1003	844	924	966	975	888	900	1000	941	925	775	894
$\underline{\nu}$	.17	.24	.29	.29	.28	.31	.28	.32	.28	.25	.22	.19
_K	.96	.95	.94	.92	.92	.91	.91	.93	.93	.95	.96	.96
_K.H_T	4.8	6.8	10.6	12.8	14.2	14.6	13.2	15.5	11.0	8.5	5.0	4.4
K.I_{MAX}	963	802	869	889	897	808	819	930	875	879	744	858

Location: DEBILT Lat. : 52.1°N Inc. : 52.1°

	JAN.	FEB.	MAR.	APR.	MAY	JUNE	JULY	AUG.	SEP.	OCT.	NOV.	DEC.
H_T	4.8	7.1	11.4	14.4	16.3	17.2	15.5	17.4	12.2	8.9	5.1	4.4
I_{MAX}	970	832	935	989	1021	936	949	1035	950	916	739	841
$\underline{\nu}$	.17	.24	.29	.30	.29	.31	.28	.32	.28	.26	.22	.19
_K	.96	.95	.94	.93	.93	.92	.92	.93	.93	.95	.96	.96
_K.H_T	4.6	6.7	10.7	13.3	15.2	15.7	14.2	16.2	11.3	8.5	4.9	4.2
K.I_{MAX}	927	792	879	916	948	857	870	963	884	872	706	810

Solar radiation data for a South-facing surface

Location: COPENHAGEN Lat. : 55.77°N Inc. : 30°

	JAN.	FEB.	MAR.	APR.	MAY	JUNE	JULY	AUG.	SEP.	OCT.	NOV.	DEC.
H_T	3.3	7.3	9.2	16.9	19.2	22.8	19.0	17.4	13.3	8.1	5.0	3.4
I_{MAX}	684	879	934	1075	1109	1064	1037	1005	924	850	720	621
$\underline{v}$	.18	.25	.24	.31	.30	.35	.31	.32	.32	.26	.24	.22
$\bar{K}$	.92	.94	.93	.93	.93	.93	.93	.93	.93	.93	.93	.92
$\bar{K}.H_T$	3.0	6.8	8.6	15.7	17.9	21.2	17.7	16.2	12.3	7.5	4.6	3.1
$\bar{K}.I_{MAX}$	629	826	869	1000	1031	990	964	935	859	791	670	571

Location: COPENHAGEN Lat. : 55.77°N Inc. : 45°

	JAN.	FEB.	MAR.	APR.	MAY	JUNE	JULY	AUG.	SEP.	OCT.	NOV.	DEC.
H_T	4.0	8.3	9.8	16.9	18.6	21.6	18.1	17.3	13.9	9.0	5.9	4.2
I_{MAX}	850	1025	1032	1100	1111	1057	1041	1025	982	955	883	806
$\underline{v}$	.17	.24	.23	.31	.29	.33	.29	.32	.31	.25	.23	.21
$\bar{K}$	.95	.96	.94	.93	.93	.93	.92	.93	.94	.95	.96	.97
$\bar{K}.H_T$	3.8	8.0	9.2	15.7	17.3	20.1	16.7	16.1	13.0	8.5	5.7	4.1
$\bar{K}.I_{MAX}$	808	984	970	1023	1033	983	958	953	923	907	848	782

Location: COPENHAGEN Lat. : 55.77°N Inc. : 60°

	JAN.	FEB.	MAR.	APR.	MAY	JUNE	JULY	AUG.	SEP.	OCT.	NOV.	DEC.
H_T	4.4	8.9	9.8	16.1	17.0	19.6	16.5	16.2	13.8	9.4	6.5	4.7
I_{MAX}	968	1100	1073	1075	1061	992	986	996	1000	1008	995	936
$\underline{v}$	.17	.24	.22	.30	.28	.32	.28	.30	.30	.25	.22	.20
$\bar{K}$	.96	.96	.94	.93	.92	.91	.91	.93	.94	.96	.97	.97
$\bar{K}.H_T$	4.3	8.5	9.2	15.0	15.7	17.8	15.0	15.1	12.9	9.0	6.3	4.6
$\bar{K}.I_{MAX}$	929	1056	1009	1000	976	903	897	926	940	968	965	908

Location: COPENHAGEN Lat. : 55.77°N Inc. : 55.77°

	JAN.	FEB.	MAR.	APR.	MAY	JUNE	JULY	AUG.	SEP.	OCT.	NOV.	DEC.
H_T	4.3	8.8	9.9	16.4	17.6	20.2	17.0	16.6	13.9	9.3	6.4	4.6
I_{MAX}	940	1086	1068	1087	1081	1016	1008	1009	999	998	969	906
$\underline{v}$	.17	.24	.22	.30	.28	.32	.28	.31	.31	.25	.23	.21
$\bar{K}$	.96	.96	.94	.93	.92	.92	.91	.93	.94	.96	.97	.97
$\bar{K}.H_T$	4.1	8.4	9.3	15.3	16.2	18.6	15.5	15.4	13.0	8.9	6.2	4.5
$\bar{K}.I_{MAX}$	901	1044	1004	1011	999	932	920	939	940	956	939	882

Solar radiation data for a South-facing surface

Location: HAMBURG Lat. : 53.63°N Inc. : 30°

	JAN.	FEB.	MAR.	APR.	MAY	JUNE	JULY	AUG.	SEP.	OCT.	NOV.	DEC.
H_T	2.8	4.6	10.3	14.5	17.0	19.7	16.9	16.9	12.1	6.4	3.4	2.6
I_{MAX}	598	641	860	909	1024	972	925	915	864	756	635	592
$\underline{v}$	.17	.21	.29	.32	.30	.34	.31	.35	.31	.23	.17	.17
$\underline{K}$	.93	.93	.93	.93	.92	.93	.92	.93	.93	.93	.93	.92
$\underline{K}.H_T$	2.6	4.3	9.6	13.5	15.7	18.3	15.6	15.7	11.3	6.0	3.1	2.4
$K.I_{MAX}$	556	596	800	845	942	904	851	851	804	703	591	545

Location: HAMBURG Lat. : 53.63°N Inc. : 45°

	JAN.	FEB.	MAR.	APR.	MAY	JUNE	JULY	AUG.	SEP.	OCT.	NOV.	DEC.
H_T	3.1	4.9	10.7	14.5	16.3	18.6	16.1	16.7	12.5	6.9	3.8	3.0
I_{MAX}	766	752	941	950	1000	950	914	921	925	855	775	762
$\underline{v}$	.14	.19	.27	.31	.29	.33	.30	.34	.30	.21	.16	.15
$\underline{K}$	.96	.94	.94	.93	.93	.92	.92	.93	.94	.94	.95	.96
$\underline{K}.H_T$	3.0	4.6	10.1	13.5	15.2	17.1	14.8	15.5	11.7	6.5	3.6	2.9
$K.I_{MAX}$	735	707	885	884	930	874	841	857	870	804	736	732

Location: HAMBURG Lat. : 53.63°N Inc. : 60°

	JAN.	FEB.	MAR.	APR.	MAY	JUNE	JULY	AUG.	SEP.	OCT.	NOV.	DEC.
H_T	3.2	5.0	10.6	13.8	14.9	16.7	14.5	15.6	12.2	7.0	4.0	3.3
I_{MAX}	894	802	975	950	946	875	847	895	935	923	875	882
$\underline{v}$	.13	.18	.26	.29	.28	.32	.29	.33	.29	.20	.15	.14
$\underline{K}$	.96	.95	.94	.93	.92	.91	.91	.93	.94	.95	.96	.97
$\underline{K}.H_T$	3.1	4.7	10.0	12.8	13.7	15.2	13.2	14.5	11.4	6.7	3.9	3.2
$K.I_{MAX}$	858	762	917	884	870	796	771	832	879	877	840	856

Location: HAMBURG Lat. : 53.63°N Inc. : 53.63°

	JAN.	FEB.	MAR.	APR.	MAY	JUNE	JULY	AUG.	SEP.	OCT.	NOV.	DEC.
H_T	3.2	5.0	10.7	14.2	15.6	17.6	15.3	16.1	12.4	7.0	4.0	3.2
I_{MAX}	846	789	967	956	973	913	882	910	937	898	838	838
$\underline{v}$	.13	.18	.26	.30	.28	.32	.30	.34	.29	.21	.16	.14
$\underline{K}$	.96	.95	.94	.93	.93	.91	.92	.93	.94	.95	.96	.97
$\underline{K}.H_T$	3.0	4.7	10.1	13.2	14.4	16.1	14.0	15.0	11.6	6.6	3.8	3.1
$K.I_{MAX}$	814	746	910	889	902	835	808	847	882	849	802	812

Solar radiation data for a South-facing surface

Location: WURZBURG Lat. : 49.8°N Inc. : 30°

	JAN.	FEB.	MAR.	APR.	MAY	JUNE	JULY	AUG.	SEP.	OCT.	NOV.	DEC.
H_T	4.4	8.5	12.8	15.0	18.5	18.3	18.4	16.6	17.1	10.3	4.9	4.5
I_{MAX}	638	901	873	999	1025	1025	981	962	943	874	700	709
$\underline{v}$	.23	.26	.35	.31	.33	.31	.33	.34	.40	.31	.22	.22
$\bar{K}$	.93	.94	.93	.93	.93	.92	.93	.93	.94	.94	.93	.93
$\bar{K}.H_T$	4.1	8.0	11.9	14.0	17.2	16.9	17.1	15.5	16.1	9.6	4.6	4.2
$\bar{K}.I_{MAX}$	593	847	812	929	953	943	912	895	886	822	651	659

Location: WURZBURG Lat. : 49.8°N Inc. : 45°

	JAN.	FEB.	MAR.	APR.	MAY	JUNE	JULY	AUG.	SEP.	OCT.	NOV.	DEC.
H_T	5.0	9.4	13.3	15.0	17.7	17.2	17.4	16.2	17.7	11.2	5.5	5.3
I_{MAX}	773	1027	933	1034	1003	1000	950	961	1006	970	825	852
$\underline{v}$	.21	.26	.34	.30	.32	.30	.32	.33	.39	.30	.21	.22
$\bar{K}$	.95	.95	.94	.93	.93	.92	.92	.93	.94	.95	.95	.95
$\bar{K}.H_T$	4.7	8.9	12.5	13.9	16.5	15.8	16.0	15.1	16.6	10.6	5.3	5.0
$\bar{K}.I_{MAX}$	734	976	877	962	933	920	874	894	946	922	784	809

Location: WURZBURG Lat. : 49.8°N Inc. : 60°

	JAN.	FEB.	MAR.	APR.	MAY	JUNE	JULY	AUG.	SEP.	OCT.	NOV.	DEC.
H_T	5.3	9.7	13.1	14.1	16.0	15.2	15.5	15.0	17.3	11.4	5.8	5.8
I_{MAX}	850	1076	947	1008	925	925	872	905	1011	1000	919	950
$\underline{v}$	.21	.25	.33	.29	.32	.28	.32	.32	.38	.30	.20	.21
$\bar{K}$	.96	.96	.94	.92	.91	.91	.91	.92	.94	.95	.96	.96
$\bar{K}.H_T$	5.1	9.3	12.3	13.0	14.6	13.8	14.1	13.8	16.2	10.9	5.6	5.6
$\bar{K}.I_{MAX}$	816	1033	890	927	842	842	794	833	950	950	882	912

Location: WURZBURG Lat. : 49.8°N Inc. : 49.8°

	JAN.	FEB.	MAR.	APR.	MAY	JUNE	JULY	AUG.	SEP.	OCT.	NOV.	DEC.
H_T	5.1	9.5	13.3	14.8	17.3	16.6	16.9	15.9	17.7	11.3	5.7	5.5
I_{MAX}	804	1052	943	1032	984	982	930	949	1014	987	859	889
$\underline{v}$	.21	.26	.34	.29	.32	.29	.32	.33	.39	.30	.21	.21
$\bar{K}$	.95	.95	.94	.93	.93	.92	.92	.93	.94	.95	.95	.95
$\bar{K}.H_T$	4.9	9.1	12.5	13.7	16.0	15.3	15.5	14.8	16.6	10.8	5.4	5.2
$\bar{K}.I_{MAX}$	767	1002	887	958	911	901	853	880	953	938	820	847

Solar radiation data for a South-facing surface

Location: HOHENPEISSENBERG Lat. : 47.8°N Inc. : 30°

	JAN.	FEB.	MAR.	APR.	MAY	JUNE	JULY	AUG.	SEP.	OCT.	NOV.	DEC.
H_T	8.4	11.7	13.2	15.9	16.7	16.0	19.0	16.7	17.5	12.2	8.3	7.7
I_{MAX}	875	930	1064	975	1064	1055	1062	976	974	971	866	786
$\underline{v}$	.31	.35	.29	.34	.29	.27	.32	.34	.40	.32	.29	.33
$\underline{K}$	.95	.94	.93	.93	.92	.92	.93	.93	.94	.94	.94	.94
$\underline{K}.H_T$	8.0	11.0	12.3	14.7	15.4	14.7	17.7	15.5	16.4	11.4	7.8	7.3
$K.I_{MAX}$	831	874	990	907	979	971	988	908	916	913	814	739

Location: HOHENPEISSENBERG Lat. : 47.8°N Inc. : 45°

	JAN.	FEB.	MAR.	APR.	MAY	JUNE	JULY	AUG.	SEP.	OCT.	NOV.	DEC.
H_T	9.7	13.0	13.8	15.6	15.9	14.9	17.9	16.2	18.0	13.2	9.4	9.1
I_{MAX}	1023	1052	1123	1000	1049	1020	1025	991	1025	1084	1006	935
$\underline{v}$	.30	.34	.29	.32	.28	.26	.31	.32	.39	.32	.28	.33
$\underline{K}$	.97	.96	.94	.93	.92	.92	.92	.93	.94	.95	.96	.97
$\underline{K}.H_T$	9.4	12.5	12.9	14.5	14.6	13.7	16.5	15.0	16.9	12.6	9.0	8.8
$K.I_{MAX}$	992	1010	1056	930	965	938	943	922	964	1030	966	907

Location: HOHENPEISSENBERG Lat. : 47.8°N Inc. : 60°

	JAN.	FEB.	MAR.	APR.	MAY	JUNE	JULY	AUG.	SEP.	OCT.	NOV.	DEC.
H_T	10.5	13.6	13.6	14.5	14.3	13.1	15.8	14.8	17.5	13.6	10.0	9.9
I_{MAX}	1108	1115	1150	950	974	919	925	928	1018	1132	1075	1029
$\underline{v}$	.30	.34	.28	.32	.27	.25	.31	.31	.38	.31	.28	.32
$\underline{K}$	.97	.96	.94	.92	.91	.90	.91	.92	.94	.96	.97	.97
$\underline{K}.H_T$	10.2	13.0	12.8	13.4	13.0	11.8	14.4	13.6	16.5	13.0	9.7	9.6
$K.I_{MAX}$	1075	1070	1081	874	886	827	842	854	957	1087	1043	998

Location: HOHENPEISSENBERG Lat. : 47.8°N Inc. : 47.8°

	JAN.	FEB.	MAR.	APR.	MAY	JUNE	JULY	AUG.	SEP.	OCT.	NOV.	DEC.
H_T	9.9	13.1	13.8	15.5	15.7	14.6	17.6	16.0	18.0	13.4	9.6	9.3
I_{MAX}	1044	1069	1131	996	1040	1006	1011	985	1028	1098	1025	957
$\underline{v}$	.30	.34	.29	.32	.28	.26	.31	.32	.39	.32	.28	.33
$\underline{K}$	.97	.96	.94	.93	.92	.92	.92	.93	.94	.95	.96	.97
$\underline{K}.H_T$	9.6	12.6	13.0	14.4	14.4	13.4	16.1	14.8	16.9	12.7	9.2	9.0
$K.I_{MAX}$	1013	1027	1064	925	955	923	928	915	966	1045	986	930

TRANSMITTANCE OF GLAZINGS: MONTHLY CORRECTION FACTORS

TRAPPES Lat.: 48.8°N

Azimuth	Glazing	Jan	Feb	Mar	Apr	May	Jun	Jul	Aug	Sep	Oct	Nov	Dec	Mean
South	Single	.97	.96	.92	.89	.87	.86	.86	.88	.91	.94	.96	.97	.90
	Double	.94	.92	.87	.82	.79	.77	.78	.80	.85	.90	.94	.95	.84
SE/SW	Single	.93	.94	.93	.92	.91	.90	.90	.92	.94	.93	.93	.93	.92
	Double	.89	.90	.88	.87	.84	.83	.84	.86	.89	.90	.89	.90	.87
E/W	Single	.88	.90	.92	.92	.92	.93	.93	.93	.92	.92	.91	.88	.92
	Double	.80	.85	.86	.87	.87	.88	.88	.88	.87	.85	.84	.82	.87

CARPENTRAS Lat.: 44.1°N

Azimuth	Glazing	Jan	Feb	Mar	Apr	May	Jun	Jul	Aug	Sep	Oct	Nov	Dec	Mean
South	Single	.97	.95	.91	.87	.84	.82	.82	.85	.90	.95	.97	.98	.90
	Double	.95	.92	.85	.79	.75	.71	.72	.76	.83	.90	.94	.96	.84
SE/SW	Single	.94	.94	.94	.92	.91	.89	.90	.91	.93	.94	.94	.94	.92
	Double	.91	.90	.89	.87	.84	.82	.83	.85	.89	.90	.90	.90	.87
E/W	Single	.90	.91	.93	.93	.93	.93	.93	.93	.92	.91	.90	.89	.92
	Double	.84	.85	.87	.88	.88	.88	.88	.88	.88	.86	.84	.82	.87

MACON Lat.: 46.3°N

Azimuth	Glazing	Jan	Feb	Mar	Apr	May	Jun	Jul	Aug	Sep	Oct	Nov	Dec	Mean
South	Single	.97	.95	.92	.89	.86	.84	.84	.87	.90	.94	.96	.97	.90
	Double	.95	.92	.86	.81	.77	.75	.75	.79	.84	.90	.93	.95	.84
SE/SW	Single	.93	.94	.93	.92	.90	.89	.90	.91	.93	.93	.94	.94	.92
	Double	.89	.90	.89	.87	.84	.82	.83	.86	.89	.89	.89	.89	.86
E/W	Single	.90	.92	.92	.92	.92	.93	.93	.92	.93	.91	.90	.90	.92
	Double	.82	.85	.86	.87	.37	.88	.88	.88	.87	.85	.83	.83	.87

TRANSMITTANCE OF GLAZINGS: MONTHLY CORRECTION FACTORS

KEW Lat.: 51.5°N

Azimuth	Glazing	Jan	Feb	Mar	Apr	May	Jun	Jul	Aug	Sep	Oct	Nov	Dec	Mean
South	Single	.97	.96	.93	.90	.88	.87	.87	.89	.91	.95	.96	.97	.91
	Double	.94	.93	.88	.83	.80	.78	.79	.82	.86	.91	.95	.95	.85
SE/SW	Single	.94	.93	.93	.92	.91	.90	.91	.92	.93	.93	.93	.94	.92
	Double	.90	.90	.89	.87	.84	.83	.84	.86	.89	.89	.89	.91	.86
E/W	Single	.91	.90	.92	.92	.92	.92	.92	.92	.92	.91	.89	.86	.92
	Double	.82	.83	.86	.87	.87	.87	.87	.87	.87	.85	.81	.79	.86

COPENHAGEN Lat.: 55.8°N

Azimuth	Glazing	Jan	Feb	Mar	Apr	May	Jun	Jul	Aug	Sep	Oct	Nov	Dec	Mean
South	Single	.98	.97	.93	.91	.89	.87	.88	.90	.92	.96	.97	.98	.91
	Double	.96	.94	.88	.85	.81	.79	.80	.84	.87	.92	.95	.96	.86
SE/SW	Single	.96	.94	.94	.93	.91	.90	.91	.93	.94	.94	.95	.94	.92
	Double	.89	.90	.90	.88	.86	.84	.84	.87	.89	.90	.90	.91	.87
E/W	Single	.88	.90	.92	.93	.93	.93	.93	.93	.92	.91	.89	.90	.92
	Double	.83	.85	.87	.88	.88	.88	.88	.88	.87	.85	.82	.79	.87

WURZBURG Lat.: 49.8°N

Azimuth	Glazing	Jan	Feb	Mar	Apr	May	Jun	Jul	Aug	Sep	Oct	Nov	Dec	Mean
South	Single	.97	.95	.92	.90	.87	.86	.86	.88	.91	.95	.96	.98	.91
	Double	.94	.92	.87	.83	.79	.77	.78	.81	.85	.90	.94	.95	.84
SE/SW	Single	.94	.94	.93	.92	.91	.90	.90	.92	.93	.94	.93	.94	.92
	Double	.89	.90	.89	.87	.84	.83	.83	.86	.89	.89	.89	.90	.87
E/W	Single	.88	.90	.92	.93	.93	.93	.92	.92	.92	.92	.88	.88	.91
	Double	.82	.84	.86	.87	.88	.87	.87	.87	.87	.86	.82	.82	.87

MONTHLY MEANS OF DAILY GLOBAL RADIATION ON VERTICAL SURFACES (MJ/m2)

TRAPPES Lat.: 48.8°N

Azimuth	Jan.	Feb.	Mar.	Apr.	May	June	July	Aug.	Sep.	Oct.	Nov.	Dec.	Mean
South	4.4	7.4	9.3	10.2	9.7	10.0	10.6	10.7	11.0	9.1	5.7	3.7	8.5
SE/SW	3.5	6.1	8.3	10.1	10.3	11.2	11.7	10.9	10.2	7.7	4.6	3.0	8.1
E/W	2.0	3.9	6.3	8.9	10.1	11.5	11.6	9.8	8.0	5.1	2.7	1.6	6.8

CARPENTRAS Lat.: 44.1°N

Azimuth	Jan.	Feb.	Mar.	Apr.	May	June	July	Aug.	Sep.	Oct.	Nov.	Dec.	Mean
South	10.4	12.7	13.0	12.9	11.5	11.1	12.5	13.8	14.3	14.8	11.3	10.8	12.4
SE/SW	8.2	10.4	11.8	13.4	13.0	13.5	14.9	14.7	13.5	12.3	9.0	8.2	11.9
E/W	4.5	6.4	8.9	11.9	13.0	14.7	15.3	13.4	10.6	7.9	5.1	4.0	9.7

MACON Lat.: 46.3°N

Azimuth	Jan.	Feb.	Mar.	Apr.	May	June	July	Aug.	Sep.	Oct.	Nov.	Dec.	Mean
South	5.5	7.9	10.0	10.9	10.2	10.2	11.3	11.3	12.2	9.5	5.9	4.1	9.1
SE/SW	4.4	6.6	9.1	11.1	11.1	11.7	12.9	11.8	11.4	8.0	4.8	3.2	8.9
E/W	2.4	4.3	7.0	9.8	10.9	12.3	13.1	10.9	9.1	5.5	3.0	1.8	7.5

KEW Lat.: 51.5°N

Azimuth	Jan.	Feb.	Mar.	Apr.	May	June	July	Aug.	Sep.	Oct.	Nov.	Dec.	Mean
South	3.1	5.6	8.0	8.5	9.3	9.8	9.3	9.4	9.4	7.3	5.1	3.3	7.3
SE/SW	2.5	4.6	7.1	8.3	9.6	10.6	9.7	9.3	8.5	6.2	4.0	2.6	6.9
E/W	1.4	3.0	5.2	7.1	9.1	10.6	9.4	8.2	6.6	4.0	2.2	1.4	5.7

COPENHAGEN Lat.: 55.8°N

Azimuth	Jan.	Feb.	Mar.	Apr.	May	June	July	Aug.	Sep.	Oct.	Nov.	Dec.	Mean
South	2.2	5.8	9.4	11.3	11.5	12.4	11.4	12.1	10.2	7.3	4.4	2.7	8.4
SE/SW	1.8	4.7	8.3	11.0	12.0	13.5	12.0	11.9	9.2	6.1	3.4	2.1	8.0
E/W	1.0	2.8	6.0	9.2	11.3	13.8	11.5	10.2	6.9	3.9	1.8	.9	6.6

WURZBURG Lat.: 49.8°N

Azimuth	Jan.	Feb.	Mar.	Apr.	May	June	July	Aug.	Sep.	Oct.	Nov.	Dec.	Mean
South	4.2	7.3	9.1	10.3	10.3	9.9	10.4	10.6	11.2	8.3	4.5	3.7	8.3
SE/SW	3.3	6.0	8.2	10.3	10.9	11.1	11.2	10.7	10.4	7.0	3.6	3.0	8.0
E/W	1.9	3.9	6.3	9.0	10.6	11.3	11.1	9.7	8.2	4.7	2.2	1.6	6.7

APPENDIX B

TEMPERATURE AND DEGREE-DAYS

This appendix contains monthly values of the following parameters for twelve European locations (the same as in Appendix A):

* Ambient temperature T_a (°C).

* Standard deviation of monthly ambient temperatures, S (K).

* Degree-days at base temperatures 20, 19, 18, ..., 15 and 14°C. The calculation method is described in section 5.2.1.

Monthly average ambient temperature and space heating degree-days

TRAPPES

	JAN.	FEB.	MAR.	APR.	MAY	JUNE	JULY	AUG.	SEP.	OCT.	NOV.	DEC.
T_A	2.8	3.8	6.3	9.1	12.6	15.6	17.5	17.3	15.0	11.0	6.3	3.8
S	1.7	1.9	1.8	1.2	1.5	1.2	1.5	1.1	1.4	1.2	1.1	2.1
DD20	533	458	425	327	231	135	94	91	153	279	411	502
DD19	502	430	394	297	201	108	71	66	125	248	381	472
DD18	471	402	363	267	171	82	52	44	99	218	351	441
DD17	440	373	333	237	143	59	37	28	75	187	321	410
DD16	409	345	302	208	115	40	25	16	54	157	291	379
DD15	379	317	272	178	90	26	16	9	37	128	261	349
DD14	348	289	242	149	68	16	11	0	24	101	231	318

CARPENTRAS

	JAN.	FEB.	MAR.	APR.	MAY	JUNE	JULY	AUG.	SEP.	OCT.	NOV.	DEC.
T_A	5.7	7.4	9.1	11.5	15.9	19.6	22.3	22.1	18.4	13.4	8.5	6.1
S	1.5	1.9	1.5	1.1	1.2	1.4	1.1	1.2	1.5	1.2	1.0	1.6
DD20	443	357	338	255	131	45	8	11	71	206	345	431
DD19	412	328	307	225	102	30	0	0	52	175	315	400
DD18	381	300	277	196	77	20	0	0	36	146	285	369
DD17	350	273	246	166	54	12	0	0	24	117	255	338
DD16	320	245	216	137	35	0	0	0	16	90	225	308
DD15	289	218	186	109	22	0	0	0	10	66	195	277
DD14	258	191	157	82	13	0	0	0	0	46	166	246

MACON

	JAN.	FEB.	MAR.	APR.	MAY	JUNE	JULY	AUG.	SEP.	OCT.	NOV.	DEC.
T_A	2.1	3.9	7.1	10.1	14.1	17.5	19.7	19.1	16.3	11.3	6.0	2.8
S	1.8	2.4	1.7	1.2	1.2	1.1	1.3	1.3	1.2	1.5	1.4	2.1
DD20	555	456	400	297	184	82	42	53	116	270	420	533
DD19	524	427	369	267	154	59	27	36	90	240	390	502
DD18	493	399	338	237	126	39	17	23	66	210	360	472
DD17	462	371	308	208	98	24	10	14	45	180	330	441
DD16	431	344	277	178	73	14	0	8	30	151	300	410
DD15	400	316	247	149	51	7	0	0	18	123	270	379
DD14	369	288	217	121	34	0	0	0	11	98	241	349

Monthly average ambient temperature and space heating degree-days

KEW

	JAN.	FEB.	MAR.	APR.	MAY	JUNE	JULY	AUG.	SEP.	OCT.	NOV.	DEC.
T_A	4.2	4.4	6.6	9.3	12.4	15.8	17.6	17.2	14.8	10.8	7.2	5.2
S	1.8	2.3	1.8	1.1	1.2	.9	1.2	1.2	1.1	1.3	1.1	1.8
DD20	490	441	416	321	236	127	85	95	157	285	384	459
DD19	459	413	385	291	205	98	62	71	128	255	354	428
DD18	428	385	354	261	175	71	42	49	101	224	324	397
DD17	397	357	323	231	145	47	27	33	75	194	294	366
DD16	366	329	293	201	116	27	17	20	52	164	264	336
DD15	336	302	262	172	89	14	10	12	34	135	234	305
DD14	305	274	233	143	65	7	0	0	20	107	204	275

ABERPORTH

	JAN.	FEB.	MAR.	APR.	MAY	JUNE	JULY	AUG.	SEP.	OCT.	NOV.	DEC.
T_A	6.2	5.9	7.2	9.1	11.5	14.2	15.6	15.8	14.4	11.6	8.8	7.2
S	.9	1.6	1.2	.7	.9	.8	.9	.8	1.2	1.2	.9	1.4
DD20	428	399	397	327	264	174	137	131	169	261	336	397
DD19	397	370	366	297	233	144	107	101	140	230	306	366
DD18	366	342	335	267	202	115	79	72	112	199	276	335
DD17	335	314	304	237	171	86	53	46	86	169	246	304
DD16	304	286	273	207	140	60	32	25	62	139	216	273
DD15	273	258	242	177	110	37	17	12	42	111	186	243
DD14	242	230	211	147	82	20	8	5	27	84	156	212

ESKDALEMUIR

	JAN.	FEB.	MAR.	APR.	MAY	JUNE	JULY	AUG.	SEP.	OCT.	NOV.	DEC.
T_A	1.4	1.8	3.6	5.8	8.9	11.7	13.3	12.9	10.8	7.6	4.6	2.8
S	1.3	2.1	1.4	1.0	1.2	1.1	1.1	1.1	1.4	1.3	1.3	2.1
DD20	577	515	508	426	344	249	208	220	276	384	462	533
DD19	546	486	477	396	313	219	178	190	247	353	432	502
DD18	515	458	446	366	282	190	148	159	217	323	402	472
DD17	484	430	415	336	251	160	118	129	188	292	372	441
DD16	453	402	384	306	221	131	91	101	159	261	342	410
DD15	422	374	354	276	190	103	66	74	131	230	312	379
DD14	391	346	323	246	160	77	44	51	104	200	282	349

Monthly average ambient temperature and space heating degree-days

UCCLE

	JAN.	FEB.	MAR.	APR.	MAY	JUNE	JULY	AUG.	SEP.	OCT.	NOV.	DEC.
T_A	2.2	2.6	6.0	9.2	13.6	16.0	17.5	17.3	14.7	10.3	6.2	3.3
S	1.8	2.3	1.7	1.2	1.2	1.2	1.2	1.2	1.4	1.4	1.2	2.1
DD20	552	492	434	324	199	123	87	92	162	301	414	518
DD19	521	464	403	294	169	96	62	67	134	270	384	487
DD18	490	436	372	264	139	71	42	46	107	239	354	456
DD17	459	408	342	234	111	50	27	29	82	209	324	425
DD16	428	380	311	205	84	32	16	18	60	179	294	395
DD15	397	352	280	175	60	20	9	10	42	150	264	364
DD14	366	324	250	146	40	11	0	0	28	122	234	333

DEBILT

	JAN.	FEB.	MAR.	APR.	MAY	JUNE	JULY	AUG.	SEP.	OCT.	NOV.	DEC.
T_A	1.7	2.0	5.0	8.5	12.4	15.5	17.0	16.8	14.3	10.0	5.9	3.0
S	1.8	2.3	1.7	1.2	1.2	1.2	1.2	1.2	1.4	1.4	1.2	2.1
DD20	567	509	465	345	236	137	100	105	173	310	423	527
DD19	536	481	434	315	205	110	74	79	145	279	393	496
DD18	505	453	403	285	175	83	52	56	117	249	363	465
DD17	474	425	372	255	145	60	34	37	92	218	333	435
DD16	443	396	342	225	116	40	21	23	68	188	303	404
DD15	413	368	311	196	89	25	12	14	49	158	273	373
DD14	382	341	280	166	65	15	7	0	33	130	243	343

COPENHAGEN

	JAN.	FEB.	MAR.	APR.	MAY	JUNE	JULY	AUG.	SEP.	OCT.	NOV.	DEC.
T_A	.1	-.1	1.9	6.6	11.8	15.6	17.8	17.3	13.9	9.3	5.4	2.5
S	1.8	2.3	1.7	1.2	1.2	1.2	1.2	1.2	1.4	1.4	1.2	2.1
DD20	617	568	561	402	254	135	79	92	185	332	438	543
DD19	586	540	530	372	224	107	56	67	156	301	408	512
DD18	555	512	499	342	193	81	37	46	128	270	378	481
DD17	524	484	468	312	163	58	23	29	102	239	348	450
DD16	493	455	437	282	133	39	14	18	77	209	318	419
DD15	462	427	406	252	105	24	8	10	56	179	288	388
DD14	431	399	375	222	79	14	0	0	39	150	258	358

Monthly average ambient temperature and space heating degree-days

HAMBURG

	JAN.	FEB.	MAR.	APR.	MAY	JUNE	JULY	AUG.	SEP.	OCT.	NOV.	DEC.
T_A	0.0	.4	3.3	7.6	12.2	15.6	17.3	16.8	13.6	9.1	4.9	1.8
S	1.8	2.3	1.7	1.2	1.2	1.2	1.2	1.2	1.4	1.4	1.2	2.1
DD20	620	554	518	372	242	135	92	105	193	338	453	564
DD19	589	526	487	342	211	107	67	79	165	307	423	533
DD18	558	498	456	312	181	81	46	56	136	276	393	502
DD17	527	470	425	282	151	58	29	37	109	246	363	472
DD16	496	441	394	252	122	39	18	23	84	215	333	441
DD15	465	413	363	222	94	24	10	14	62	185	303	410
DD14	434	385	332	193	69	14	0	8	43	156	273	379

WURZBURG

	JAN.	FEB.	MAR.	APR.	MAY	JUNE	JULY	AUG.	SEP.	OCT.	NOV.	DEC.
T_A	-.6	.3	4.5	8.9	13.3	16.7	18.4	17.7	14.2	8.9	4.3	.6
S	1.8	2.7	2.0	1.7	1.2	1.2	1.2	1.2	1.4	1.6	1.5	2.0
DD20	639	557	481	333	209	105	65	81	176	344	471	601
DD19	608	529	450	304	178	80	44	58	148	314	441	570
DD18	577	501	419	274	149	57	28	39	120	283	411	540
DD17	546	473	388	245	120	39	17	24	94	253	381	509
DD16	515	445	358	216	93	24	10	14	71	222	351	478
DD15	484	417	327	187	68	15	0	8	50	193	321	447
DD14	453	389	297	160	48	9	0	0	34	164	291	416

HOHENPEISSENBERG

	JAN.	FEB.	MAR.	APR.	MAY	JUNE	JULY	AUG.	SEP.	OCT.	NOV.	DEC.
T_A	-2.2	-1.0	3.3	7.9	12.5	15.9	17.7	16.9	13.7	8.2	3.1	-.7
S	1.8	2.7	2.0	1.7	1.2	1.2	1.2	1.2	1.4	1.6	1.5	2.0
DD20	688	594	518	363	233	127	81	102	191	366	507	642
DD19	657	566	487	333	202	100	58	77	162	335	477	611
DD18	626	537	456	304	172	75	39	54	134	304	447	580
DD17	595	509	425	274	143	53	24	35	107	274	417	549
DD16	564	481	394	245	114	35	14	22	82	243	387	518
DD15	533	453	364	216	88	22	8	13	60	213	357	487
DD14	502	425	333	187	64	13	0	7	42	184	327	456

APPENDIX C

PROPERTIES OF MATERIALS

WATER (1 Bar)

Freezing point: 0°C
Boiling point: 100°C

Temperature	Density	Thermal conductivity	Specific Heat	Dynamic Viscosity	Prandlt Number
°C	kg/m3	W/m.K	kJ/kg.K	Pa.s ($\times 10^3$)	-
0	999.8	0.569	4.218	1.786	13.20
10	999.8	0.588	4.195	1.303	9.31
20	998.3	0.604	4.183	1.002	6.94
30	995.7	0.619	4.178	0.798	5.39
40	992.3	0.632	4.178	0.654	4.32
50	988.0	0.644	4.181	0.548	3.56
60	983.2	0.654	4.191	0.467	2.99
70	977.7	0.663	4.195	0.405	2.56
80	971.6	0.670	4.199	0.355	2.23
90	965.2	0.676	4.207	0.316	1.96

WATER + PROPYLENE GLYCOL (20% by weight)

Freezing point: -7°C
Boiling point: 102°C

Temperature °C	Density Kg/m3	Thermal conductivity W/m.K	Specific Heat kJ/Kg.K	Dynamic Viscosity Pa.s (x10³)
0	1018	0.492	4.03	4.1
10	1015	0.502	4.02	2.9
20	1012	0.512	4.02	2.0
30	1009	0.522	4.02	1.5
40	1005	0.532	4.02	1.2
50	1000	0.541	4.03	0.90
60	995	0.551	4.04	0.75
70	989	0.561	4.06	0.65
80	983	0.571	4.08	0.55
90	975	0.581	4.10	0.50

WATER + PROPYLENE GLYCOL (40% by weight)

Freezing point: -20°C
Boiling point: 107°C

Temperature °C	Density Kg/m3	Thermal conductivity W/m.K	Specific Heat kJ/kg.K	Dynamic Viscosity Pa.s (x10³)
0	1040	0.417	3.72	11.5
10	1035	0.419	3.74	7.0
20	1030	0.423	3.76	4.4
30	1025	0.424	3.78	3.0
40	1019	0.427	3.81	2.2
50	1013	0.429	3.83	1.7
60	1005	0.432	3.85	1.3
70	998	0.436	3.89	1.1
80	990	0.439	3.91	0.85
90	982	0.441	3.96	0.70

WATER + ETHYLENE GLYCOL (20% by weight)

Freezing point: -8°C
Boiling point: 102°C

Temperature °C	Density Kg/m3	Thermal conductivity W/m.K	Specific Heat kJ/kg.K	Dynamic Viscosity Pa.s (x10³)
0	1029	0.498	3.86	3.1
10	1027	0.507	3.87	2.3
20	1024	0.516	3.89	1.65
30	1020	0.525	3.91	1.25
40	1015	0.534	3.93	1.00
50	1010	0.543	3.95	0.80
60	1004	0.550	3.97	0.67
70	998	0.554	3.98	0.60
80	991	0.557	4.00	0.50
90	984	0.560	4.02	0.44

WATER + ETHYLENE GLYCOL (40% by weight)

Freezing point: -24°C
Boiling point: 105°C

Temperature °C	Density kg/m3	Thermal conductivity W/m.K	Specific Heat kJ/kg.K	Dynamic Viscosity Pa.s (x10³)
0	1059	0.441	3.41	6.3
10	1055	0.444	3.46	4.1
20	1050	0.447	3.51	2.8
30	1045	0.450	3.56	2.1
40	1040	0.453	3.60	1.7
50	1035	0.456	3.64	1.4
60	1029	0.458	3.60	1.1
70	1023	0.456	3.71	0.9
80	1017	0.456	3.74	0.75
90	1009	0.456	3.77	0.65

PROPERTIES OF BUILDING AND INSULATING MATERIALS

Material	Density Kg/m3	Thermal Conductivity W/m.K	Specific Heat KJ/Kg.K
METALS			
Aluminium	2700	208	0.90
Copper	8930	372	0.38
Iron	7870	71	0.45
Steel	7780	52	0.48
Cast iron	7500	50-60	0.54
BUILDING MATERIALS			
Granite	2500-3000	3.5	0.75
Limestone	2300-2600	2.2	0.8
Terracotta	1800-2000	1.15	0.9
Concrete	2200-2400	1.75	0.9
Gypsum plaster	750-1000	0.35	0.8
Hardwoods (oak)	650-800	0.23	2.4
Softwoods (pine)	500-650	0.15	1.2
Glass	2500	1.15	0.84
PVC	1380	0.17	1.0
Polyethylene	940	0.45-0.35	1.5-2.0
INSULATING MATERIALS			
Mineral wool	20-35	0.045	0.9
	35-180	0.040	
Fibreglass	8-12	0.045	0.84
	12-80	0.040	
Expanded Polystyrene	9-15	0.045	1.3
	15-35	0.049	
Extruded Polystyrene	25-40	0.035	1.3
Polyurethane	30-40	0.030	1.4
Foamed glass	110-140	0.050	0.84
Cork board	100-150	0.045	1.4
Bulk Insul. Material	20-100	0.050	0.9
FLUIDS			
Air (25°C)	1.17	0.025	1.00
Water (20°C)	998	0.60	4.183

APPENDIX D

ESM1: A SIMPLIFIED "MANUAL" VERSION

The detailed version of ESM1 (Active Solar Space Heating + Combined Domestic Hot Water) presented in section 6.3 could be used "manually", but this would quickly become tedious, and the method is used better on a microcomputer. In order to retain the possibility of manual calculations, a simplified version has been developed, based on tables, charts, and simple formulae.

INITIAL PARAMETERS AND PRINCIPLES

The initial data are the same as for the detailed version, except that certain secondary parameters have been omitted, e.g., control differentials and storage heat loss coefficient.

The version described in the following table involves the steps for the space heating only calculations, but also deals with combined hot water. Some elements are calculated only once for all cases: a constant collector heat loss coefficient U_L is considered (U_2 = 0) (steps 1 to 4).
In the course of the monthly calculations, certain steps have been removed (the stratification correction, the differential correction, and the correction for storage heat losses). Thus, the monthly calculations lead to an estimate of the load (steps 5 and 6) and of the base temperature (step 7). Subsequent calculations give the reduced threshold y_c (step 8), the utilisability (step 9), the dimensionless ratio input/demand Y (step 10), and the solar contribution, with the help of a storage correction term. An iteration is necessary in the case of parallel auxiliary (step 12). The modulated control only is considered. The hot water load is integrated in the space heating load.

For the summer months, when only hot water is needed, the base temperature is the average of the cold mains water and the hot water temperature. The efficiency of the hot water preheater is no longer taken into account.

Graphs may be used instead of some of the more complicated formulae:

- Figure 2.1.7 in Chapter 2 gives the collector heat removal factor F_r. For $\dot{m}_0 c_0 / AU_L > 5$ this graph is supplemented by the approximate formula

$$F_r = 1 - 0.5.A.U_L / \dot{m}_0 c_0.$$

- Figure 4.3.2 in Chapter 4 (reproduced here as Fig. D.1) gives the utilizability Ø in terms of the dimensionless parameter v and the reduced threshold y_c.

Table D.1: ESM1 (Space Heating + Combined Hot Water): Algorithms of the simplified manual version

---

INITIAL STEPS IN THE CALCULATION

1 - Collector heat removal factor (provisional value)

$$F_r = m_0c_0 \, . \, [\, 1 - EXP \, (\, - \, A \, . \, U_L \, / \, m_0c_0 \,) \,] \, / \, (A.U_L) \, .$$

2 - Correction for piping heat losses

$$U_L{}^* = U_L + (\, A_{in}U_{in} + A_{ou}U_{ou} \,) \, / \, A \, .$$

Collector heat removal factor (final value)

$$F_r{}^* = m_0c_0 \, . \, [\, 1 - EXP \, (\, - \, A \, . \, U_L{}^* \, / \, m_0c_0 \,) \,] \, / \, (A.U_L{}^*) \, .$$

Optical factor

$$\eta_0{}^* = \eta_0 \, . \, F_r \, / \, F_r{}^* \, . \, EXP \, (\, - \, A_{ou}U_{ou} \, / \, m_0c_0 \,) \, .$$

3 - Effectiveness of solar loop

$$C_{min} = MIN \, (\, m_0c_0 \, ; \, m_1c_1 \,) \, .$$

$$F_r{}' = F_r{}^* \, / \, [\, 1 + A \, . \, F_r{}^* \, . \, U_L{}^* \, . \, (m_0c_0/e_1/C_{min} - 1) \, / \, m_0c_0 \,] \, .$$

4 - Storage correction parameters

$$t_r = 75.85 \; 10^3 \; s \, .$$

$$C_s = M \, . \, c \, / \, (\, t_r \, . \, A \, . \, F_r{}' \, . \, U_L{}^* \,) \, .$$

$$C_1 = 1.0205 \, . \, [\, 1 - 0.47 \, . \, EXP \, (\, - \, 1.785 \, . \, C_s \,) \,] \, .$$

$$C_2 = 0.205 + 0.1515 \, . \, LOG \, (\, C_s \,) \, .$$

STEPS IN THE MONTHLY CALCULATION

5 - Space heating load DD(T_{nh}): Degree-days at base temperature T_{nh}

$$T_{nh} = T_c - P_{fg} \, / \, (GV) \, .$$

$$L_0 = t_h \, . \, (GV) \, . \, DD(T_{nh}) \, .$$

---

6 - Hot water load

$$L_1 = N \cdot mc \cdot (T_h - T_m) .$$

$$L = L_0 + L_1 .$$

7 - Base temperature

Winter months (space heating non-zero)

Type of auxiliary		Base temperature
1	Series	$T_b = T_c + L_0 \cdot (1-e_2) / (t_e.m_2c_2.e_2)$
2	Parallel	$T_b = T_c$

Summer months (no space-heating; hot water only)

$$T_b = (T_m + T_h) / 2 .$$

8 - Reduced threshold irradiance

$$y_c = (U_L^* / \eta_0^*) \cdot (T_b - T_a) / (\bar{K} \cdot I_{max}) .$$

9 - Calculation of utilisability Ø (see Figure D.1 or chapter 4).

10 - Ratio Input/demand

$$Y = (A \cdot F_r' \cdot \eta_0^* \cdot \bar{K} \cdot N \cdot Ø \cdot H_t) / L .$$

11 - Solar fraction (final value for series auxiliary and months with hot water only)

$$f = 1 - EXP [- C_1 \cdot Y \cdot (1 + C_2 \cdot Y)] .$$

12 - Iteration for base temperature (parallell auxiliary only; months with space heating)

$$T_b = T_c + f \cdot L_0 \cdot (1-e_2) / (t_e.m_2c_2.e_2) .$$

Repeat steps 8 to 11.

<u>YEARLY BALANCE</u>

The yearly useful solar energy output is the sum of the monthly solar contributions f.L.

- Figure D.2 gives the solar fraction in terms of the dimensionless ratio Y and the storage coefficient C_s (the use of this diagram dispenses with the need to calculate the parameters C_1 and C_2)

CALCULATION TABLES

Blank tables are given below. Tables A and B are used for data preparation: table A for system parameters and table B for climatic data. For a detailed description of the parameters, see section 6.2.

Table C is used for the calculation of heating and domestic hot water requirements, and also for the estimation of the base temperature T_b.
Table D assembles all the initial calculation steps (U_L^*, F_r, F_r', and C_s).
Table E is used only for systems with parallel auxiliary (first iteration in months when space heating is required). Table F is for all the monthly calculations (except the base temperature, which has already been determined in Table C), and hence, the monthly and annual evaluation of solar fraction.

A WORKED EXAMPLE

Consider the system taken as an example in section 6.4. It comprises 20 m2 of ordinary single glazed collectors inclined at 45°, and a tank of 1000 litres of water supplying heating and hot water for a house in Carpentras. An underfloor heating system (designed to operate at between 30 and 35°C at an external air temperature of -4°C) is supplied from both solar heating and an auxiliary source (e.g., a gas boiler). Domestic hot water (150 litres per day at 55°C) is preheated in a 100 litre cistern from a heat exchanger in the main storage tank. The auxiliary for hot water is an additional back-up tank whose heat losses are estimated as 15% of the basic demand and integrated in the hot water load: an effective value of the required hot water temperature equal to 61.8°C is thus assumed.
The system layout is similar to the one in Figure 6.1.2 (Chapter 6). The data are assembled in Table A and the climatic data in Table B.

The degre-days for a base of 17.5°C (allowing for average occupancy gains of 375 W) are calculated by linear interpolation from the data given in Appendix B.
The space heating demand amounts to 41.4 GJ and the total load to 53.3 GJ (Table C).
An average value of 7.4 W/m2.K is assumed for the collector heat loss coefficient: from this the heat removal factor is calculated as F_r = 0.93 and the effectiveness of the solar loop is F_r' = 0.92.
With the total storage capacity of 1100 litres (including the hot water preheating cistern) the storage parameter is C_s = 0.45 (Table D).
The results (Table F) give a solar contribution of 26 GJ (360 kWh per square metre of collector), approximately 49% of the total demand. For the same case, the detailed version of the method, operating as the computer programme ESM1.4, gave a solar contribution of 49.1 %.

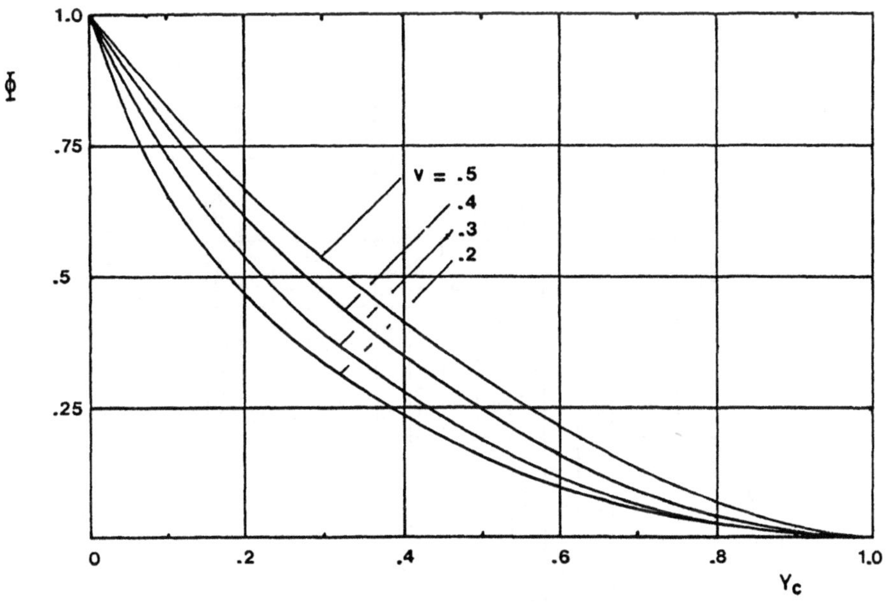

Fig. D.1: Graph giving utilisability Ø as a function of the reduced critical threshold y_c and the dimensionless parameter v.

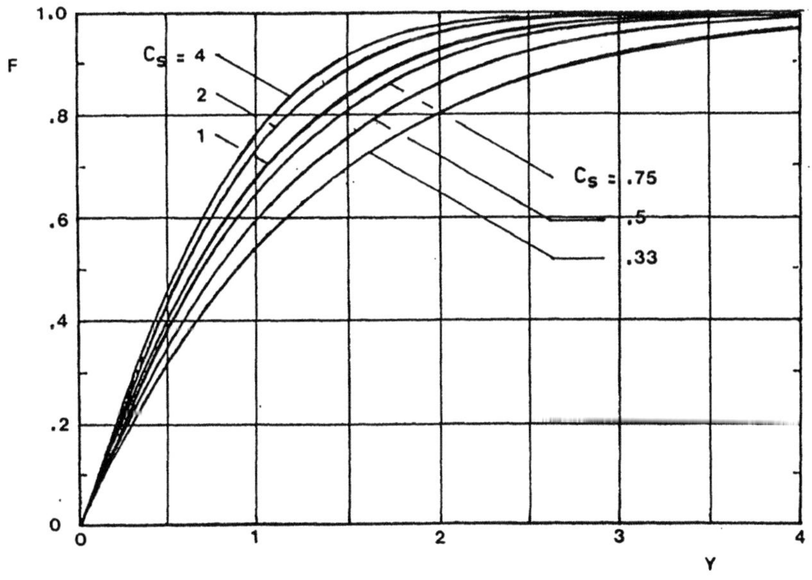

Fig. D.2: Graph showing monthly solar fraction f as a function of Y (available solar energy/demand) and the storage parameter C_s.

ESM1 SYSTEM PARAMETERS TABLE A

Project: EXAMPLE Site: CARPENTRAS
Latitude: 44.1°N Long.:

BUILDING

Heat loss	GV =	250	(W/K)	(A.1)
Demand temperature	T_C =	19	(°C)	(A.2)
Occupancy gains	P_{FG} =	375	(W)	(A.3)

HEATING NETWORK

Capacity rate	$m_Z c_Z$ =	1150	(W/K)	(A.4)
Effectiveness	e_Z =	0.312		(A.5)

Auxiliary type: 1 (Series) [redacted] (A.6)

DOMESTIC HOT WATER

Daily demand	m =	150	(kg/day)	(A.7)
Required temperature	T_H =	61.8	(°C)	(A.8)
Mains temperature	T_M =	10	(°C)	(A.9)

COLLECTORS

Surface area	A =	20	(m2)	(A.10)
Optical efficiency	η_o =	0.80		(A.11)
Heat loss coefficient	U_L =	7.4	(W/m2.K)	(A.12)
Piping losses: inlet	$A_{IN}U_{IN}$ =	0	(W/K)	(A.13)
outlet	$A_{OU}U_{OU}$ =	0	(W/K)	(A.14)
Inclination	β =	45	(°)	(A.15)
Capacity flow rate	$m_C c_O/A$ =	51.6	(W/m2.K)	(A.16)

SOLAR LOOP HEAT-EXCHANGER

Capacity flow rate	$m_I c_I/A$ =	63.8	(W/m2.K)	(A.17)
Effectiveness	e_I =	0.93		(A.18)

STORAGE

Capacity	M =	1100	(kg)	(A.19)
Specific heat	c =	4.18	(kJ/kg.K)	(A.20)
Heat capa. per unit collector area	Mc/A =	230	(kJ/m2.K)	(A.21)

ESM1 METEOROLOGICAL DATA TABLE B

	TEMPERATURE Loc.= CARPENTRAS	SOLAR RADIATION Location = CARPENTRAS Inclination β = 45° (A.15)					
	T_a (°C)	H_T MJ/m2	I_{MAX} W/m2	v -	$\bar{K}$ -	$H_T.\bar{K}$ MJ/m2	$I_{MAX}.\bar{K}$ W/m2
Month	(B.1)	(B.2)	(B.3)	(B.4)	(B.5)	(B.6)	(B.7)
Jan..	5.7	12.4	937	0.40	0.97	12.1	909
Feb..	7.4	15.6	1092	0.39	0.96	15.0	1048
March	9.1	17.7	1080	0.39	0.95	16.8	1026
April	11.5	19.1	1062	0.38	0.93	17.8	988
May	15.9	20.9	1048	0.38	0.93	19.4	975
June	19.6	23.7	1025	0.42	0.93	22.1	953
July	22.3	25.0	1000	0.46	0.93	23.2	930
Aug.	22.1	22.7	1026	0.44	0.93	21.1	954
Sept.	18.4	22.4	1061	0.47	0.95	21.2	1008
Oct.	13.4	17.9	1050	0.43	0.96	17.2	1008
Nov.	8.5	14.3	950	0.44	0.97	13.9	922
Dec.	6.1	9.4	881	0.34	0.96	9.0	846

Data are available in Appendix A (B.2 to B.5) and Appendix B (B.1)

(B.6) = (B.2) . (B.5)
(B.7) = (B.3) . (B.5)

ESM1 TABLE C

CALCULATION OF HEATING LOAD

Project: EXAMPLE

No-heating temperature T_{NH} = 17.5 (°C) (C.1)

(1 - e_2) / ($t_H . m_2 c_2 . e_2$) = 0.0222 (K.day/MJ) (C.2)

$T_H - T_M$ = 51.8 (K) (C.3)

Month	DD	N	L_0 MJ	L_0/N MJ/d	T_B °C	L_i MJ	L MJ
	(C.4)	(C.5)	(C.6)	(C.7)	(C.8)	(C.9)	(C.10)
Jan.	366	31	7906	255	24.7	1007	8913
Feb.	286	28	6178	220.6	23.9	909	7087
March	262	31	5659	182.5	23.1	1007	6666
April	181	30	3910	130.3	21.9	974	4884
May	66	31	1426	46.0	20.0	1007	2433
June	–	30	–	–	36	974	974
July	–	31	–	–	36	1007	1007
Aug.	–	31	–	–	36	1007	1007
Sept.	–	30	–	–	36	974	974
Oct.	132	31	2851	92.0	21.0	1007	3858
Nov.	270	30	5832	194.4	23.2	974	6806
Dec.	354	31	7646	246.6	24.5	1007	8653
TOTAL	1917		41408			11854	53262

(C.1) = (A.2) - (A.3)/(A.1)

(C.2) = 11.574 (1 / (A.5) - 1) / (A.4) (C.3) = (A.8) - (A.9)

(C.4): see Appendix B (Degree-days at base temperature T_{NH})

(C.5): Number of days in the month

(C.6) = 0.0864 (A.1) . (C.4) (C.7) = (C.6) / (C.5)

(C.8) If (C.6) = 0 then (C.8) = 0.5 ((A.8) + (A.9))

Else if (A.6) = 1 then (C.8) = (A.2) + (C.7) . (C.2)

if (A.6) = 2 then (C.8) = (A.2) + (C.7) . (C.2) . (E.7)

(to be filled in after the first iteration)

(C.9) = 4.18 10⁻³ (C.5) . (A.7) . (C.3) (C.10) = (C.6) + (C.9)

ESM1 TABLE D

INITIAL CALCULATIONS

Project: EXAMPLE

COLLECTORS ONLY

$m_o c_o$ / (A . U_L) = 6.973 (D.1)

F_R = 0.93 (D.2)

LOSSES FROM PIPING

U_L^* = 7.4 (W/m2.K) (D.3)

$m_o c_o$ / (A . U_L^*) = 6.973 (D.4)

F_R^* = 0.93 (D.5)

η_o^* = 0.80 (D.6)

EFFECTIVENESS OF SOLAR LOOP

C_{MIN} = MIN ($m_o c_o$/A ; $m_i c_i$/A) = 51.6 (W/m2.K) (D.7)

($m_o c_o$/A) / ($C_{MIN} . e_i$) - 1 = 0.0753 (D.8)

F_R' = 0.92 (D.9)

STORAGE CORRECTION

C_s = 0.446 (D.10)

C_i = 0.804 (D.11)

C_2 = 0.083 (D.12)

FORMULAE

(D.1) = (A.16) / (A.12)

(D.2) = (D.1) . (1 - EXP (- 1 / (D.1)))

(D.3) = (A.12) + ((A.13) + (A.14)) / (A.10)

(D.4) = (A.16) / (D.3)

(D.5) = (D.4) . (1 - EXP (- 1 / (D.4)))

(D.6) = (A.11) . (D.2) / (D.5) . EXP (- (A.14)/(A.16)/(A.10))

If no piping losses (A.14) = (A.13) = 0

then (D.3) = (A.12) (D.5) = (D.2) and (D.6) = (A.11)

(D.7) = Minimum of (A.16) and (A.17)

(D.8) = (A.16) / (A.18) / (D.7) - 1

(D.9) = (D.5) / (1 + (D.5) . (D.8) / (D.4))

(D.10) = 0,0132 . (A.21) / (D.3) / (D.9)

(D.11) = 1.0205 (1 - 0.47 Exp (- 1.785 (D.10)))

(D.12) = 0.205 + 0.1515 Ln (D.10)

SYSTEM CALCULATION

Auxiliary in parallel (First iteration)

Project:

U_L^* / η_0^* = ____ (W/m2.K) (E.1)
A . η_0^* . F_R' = ____ (m2) (E.2)

Month	T_C-T_A	y_C	Ø	Y	f
	(E.3)	(E.4)	(E.5)	(E.6)	(E.7)
Jan .					
Feb.					
March					
April					
May					
June					
July					
Aug .					
Sept.					
Oct.					
Nov.					
Dec.					

(E.1) = (F.1) = (D.3) / (D.6)
(E.2) = (F.2) = (A.10) .(D.9) . (D.6)
(E.3) = (A.2) - (B.1) (F.3) = (C.8) - (B.1)
(E.4) = (E.1) . (E.3) / (B.7) (F.4) = (F.1) . (F.3) / (B.7)
(E.5) and (F.5): Utilisability Ø vs y_C (E.4 or F.4) and v (B.4) (graph D.1)
(E.6) = (E.2) . (B.6) . (E.5) . (C.5) / (C.10)
(F.6) = (F.2) . (B.6) . (F.5) . (C.5) / (C.10)
(E.7) and (F.7): Solar fraction vs Y (E.6 or F.6) and C_S (D.10) (Graph D.2)
(F.8) = (F.7) . (C.10) (see also formula of step 11, Standard Version).

SYSTEM CALCULATION

Series Auxiliary or second iteration for parallel auxiliary

Project: EXAMPLE
U_L^* / η_0^* = 9.25 (W/m2/K) (F.1)
A . η_0^* . F_R' = 14.7 (m2) (F.2)

Month	T_B-T_A	y_C	Ø	Y	f	fL
	(F.3)	(F.4)	(F.5)	(F.6)	(F.7)	(F.8)
Jan .	19.0	0.19	0.63	0.39	0.28	2495
Feb.	16.5	0.15	0.70	0.61	0.40	2835
March	14.0	0.13	0.74	0.85	0.52	3466
April	10.4	0.10	0.78	1.25	0.67	3272
May	4.1	0.04	0.91	3.3	0.96	2336
June	16.4	0.16	0.68	6.8	1.00	974
July	13.7	0.14	0.72	7.6	1.00	1007
Aug .	13.9	0.13	0.72	6.9	1.00	1007
Sept.	17.6	0.16	0.69	6.6	1.00	974
Oct.	7.6	0.07	0.86	1.75	0.80	3086
Nov.	14.7	0.15	0.70	0.63	0.41	2790
Dec.	18.4	0.20	0.57	0.27	0.20	1731
TOTAL						25973 (MJ)

FINAL RESULT

f = 0.49 (F.9) = TOTAL(F.8)/TOTAL(C.10)
fL/A = 1300 (MJ/m2) (F.10) = TOTAL(F.8) / (A.10)

Formulae: see bottom of Table E.

ESM1: "MANUAL" VERSION

BLANK TABLES FOR DESK TOP CALCULATIONS

ESM1

TABLE A

SYSTEM PARAMETERS

Project: ____________________ Site: ________________________
Latitude: ____________________ Long.: ________________________

BUILDING

Heat loss	GV =	_____	(W/K)	(A.1)
Demand temperature	T_C =	_____	(°C)	(A.2)
Occupancy gains	P_{FG} =	_____	(W)	(A.3)

HEATING NETWORK

Capacity rate	m_2c_2 =	_____	(W/K)	(A.4)
Effectiveness	e_2 =	_____		(A.5)

Auxiliary type: 1 (Series) or 2 (Parallel) (A.6)

DOMESTIC HOT WATER

Daily demand	m =	_____	(kg/day)	(A.7)
Required temperature	T_H =	_____	(°C)	(A.8)
Mains temperature	T_M =	_____	(°C)	(A.9)

COLLECTORS

Surface area	A =	_____	(m2)	(A.10)
Optical efficiency	η_0 =	_____		(A.11)
Heat loss coefficient	U_L =	_____	(W/m2.K)	(A.12)
Piping losses: inlet	$A_{IN}U_{IN}$ =	_____	(W/K)	(A.13)
outlet	$A_{OU}U_{OU}$ =	_____	(W/K)	(A.14)
Inclination	β =	_____	(°)	(A.15)
Capacity flow rate	m_0c_0/A =	_____	(W/m2.K)	(A.16)

SOLAR LOOP HEAT-EXCHANGER

Capacity flow rate	m_1c_1/A =	_____	(W/m2.K)	(A.17)
Effectiveness	e_1 =	_____		(A.18)

STORAGE

Capacity	M =	_____	(kg)	(A.19)
Specific heat	c =	_____	(kJ/kg.K)	(A.20)
Heat capa. per unit collector area	Mc/A =	_____	(kJ/m2.K)	(A.21)

METEOROLOGICAL DATA

	TEMPERATURE Loc.=	SOLAR RADIATION Location = Inclination β = (A.15)					
	T_A (°C)	H_T MJ/m2	I_{MAX} W/m2	v -	$\bar{K}$ -	$H_T.\bar{K}$ MJ/m2	$I_{MAX}.\bar{K}$ W/m2
Month	(B.1)	(B.2)	(B.3)	(B.4)	(B.5)	(B.6)	(B.7)
Jan .							
Feb .							
March							
April							
May							
June							
July							
Aug .							
Sept.							
Oct.							
Nov.							
Dec.							

Data are available in Appendix A (B.2 to B.5) and Appendix B (B.1)

(B.6) = (B.2) . (B.5)
(B.7) = (B.3) . (B.5)

CALCULATION OF HEATING LOAD

Project:
No-heating temperature T_{NH} = _____ (°C) (C.1)
(1 - e_2) / ($t_H . m_2 c_2 . e_2$) = _____ (K.day/MJ) (C.2)
$T_H - T_M$ = _____ (K) (C.3)

Month	DD	N	L_0 MJ	L_0/N MJ/d	T_B °C	L_I MJ	L MJ
	(C.4)	(C.5)	(C.6)	(C.7)	(C.8)	(C.9)	(C.10)
Jan .							
Feb .							
March							
April							
May							
June							
July							
Aug .							
Sept.							
Oct.							
Nov.							
Dec.							
TOTAL							

(C.1) = (A.2) - (A.3)/(A.1)
(C.2) = 11.574 (1 / (A.5) - 1) / (A.4) (C.3) = (A.8) - (A.9)
(C.4): see Appendix B (Degree-days at base temperature T_{NH})
(C.5): Number of days in the month
(C.6) = 0.0864 (A.1) . (C.4) (C.7) = (C.6) / (C.5)
(C.8) If (C.6) = 0 then (C.8) = 0.5 ((A.8) + (A.9))
Else if (A.6) = 1 then (C.8) = (A.2) + (C.7) . (C.2)
if (A.6) = 2 then (C.8) = (A.2) + (C.7) . (C.2) . (E.7)
(to be filled in after the first iteration)
(C.9) = 4.18 10^{-3} (C.5) . (A.7) . (C.3) (C.10) = (C.6) + (C.9)

ESM1 TABLE D

INITIAL CALCULATIONS

Project:

COLLECTORS ONLY

$m_o c_o$ / (A . U_L) = _____ (D.1)
F_R = _____ (D.2)

LOSSES FROM PIPING

U_L^* = _____ (W/m2.K) (D.3)
$m_o c_o$ / (A . U_L^*) = _____ (D.4)
F_R^* = _____ (D.5)
n_o^* = _____ (D.6)

EFFECTIVENESS OF SOLAR LOOP

C_{MIN} = MIN ($m_o c_o$/A ; $m_i c_i$/A) = _____ (W/m2.K) (D.7)
($m_o c_o$/A) / (C_{MIN}.e_i) - 1 = _____ (D.8)
F_R' = _____ (D.9)

STORAGE CORRECTION

C_S = _____ (D.10)
C_1 = _____ (D.11)
C_2 = _____ (D.12)

FORMULAE

(D.1) = (A.16) / (A.12)
(D.2) = (D.1) . (1 - EXP (- 1 / (D.1)))
(D.3) = (A.12) + ((A.13) + (A.14)) / (A.10)
(D.4) = (A.16) / (D.3)
(D.5) = (D.4) . (1 - EXP (- 1 / (D.4)))
(D.6) = (A.11) . (D.2) / (D.5) . EXP (- (A.14)/(A.16)/(A.10))
If no piping losses (A.14) = (A.13) = 0
then (D.3) = (A.12) (D.5) = (D.2) and (D.6) = (A.11)
(D.7) = Minimum of (A.16) and (A.17)
(D.8) = (A.16) / (A.18) / (D.7) - 1
(D.9) = (D.5) / (1 + (D.5) . (D.8) / (D.4))
(D.10) = 0,0132 . (A.21) / (D.3) / (D.9)
(D.11) = 1.0205 (1 - 0.47 Exp (- 1.785 (D.10)))
(D.12) = 0.205 + 0.1515 Ln (D.10)

SYSTEM CALCULATION

Auxiliary in parallel (First iteration)

Project:

U_L^* / n_0^* = ______ (W/m2.K) (E.1)
$A \cdot n_0^* \cdot F_R'$ = ______ (m2) (E.2)

Month	$T_C - T_A$	y_C	Ø	Y	f
	(E.3)	(E.4)	(E.5)	(E.6)	(E.7)
Jan.					
Feb.					
March					
April					
May					
June					
July					
Aug.					
Sept.					
Oct.					
Nov.					
Dec.					

(E.1) = (F.1) = (D.3) / (D.6)
(E.2) = (F.2) = (A.10) .(D.9) . (D.6)
(E.3) = (A.2) - (B.1) (F.3) = (C.8) - (B.1)
(E.4) = (E.1) . (E.3) / (B.7) (F.4) = (F.1) . (F.3) / (B.7)
(E.5) and (F.5): Utilisability Ø vs y_C (E.4 or F.4) and v (B.4) (graph D.1)
(E.6) = (E.2) . (B.6) . (E.5) . (C.5) / (C.10)
(F.6) = (F.2) . (B.6) . (F.5) . (C.5) / (C.10)
(E.7) and (F.7): Solar fraction vs Y (E.6 or F.6) and C_S (D.10) (Graph D.2)
(F.8) = (F.7) . (C.10) **(see also formula of step 11, Standard Version).**

ESM1 TABLE F

SYSTEM CALCULATION

Series Auxiliary or second iteration for parallel auxiliary

Project:
U_L^* / n_0^* = ______ (W/m2/K) (F.1)
$A \cdot n_0^* \cdot F_R'$ = ______ (m2) (F.2)

Month	$T_B - T_A$	y_c	Ø	Y	f	fL
	(F.3)	(F.4)	(F.5)	(F.6)	(F.7)	(F.8)
Jan .						
Feb.						
March						
April						
May						
June						
July						
Aug .						
Sept.						
Oct.						
Nov.						
Dec.						
TOTAL						(MJ)

FINAL RESULT

f = ______ (F.9) = TOTAL(F.8)/TOTAL(C.10)

fL/A = ______ (MJ/m2) (F.10) = TOTAL(F.8) / (A.10)

Formulae: see bottom of Table E.

APPENDIX E

ESM 1.4: AIDE-MEMOIRE AND DATA WORKSHEETS.

(SOLAR SPACE HEATING AND COMBINED HOT WATER)

ESM 1.4 is a design software for active solar space heating and combined hot water systems. This "aide-memoire" presents input and output parameters for all ESM 1.4 screen pages, with the corresponding symbol used in the standard version of Chapter 6. This will enable the user to find in the corresponding chapter any additional information required and how to get the data in some difficult cases. Input data can then be prepared with the help of the enclosed blank worksheets, before running the programme. Minimum and maximum values are indicated (input data outside the suitable range will be rejected and a new value will be asked for); default values provided by the programme are also indicated.
Menu options are also given. When a disk data file can be read or updated from a page, this is indicated.

Page 1: MAIN MENU

Origin page: 2, 11 or 23 Destination page: 3

Menu:
1 - Back to DOS (Stopping ESM 1.4)
2 - First step: Weather data and loads
3 - Second step: Solar system characteristics and performance calculation

Page 2: WELCOME

Origin page: 0 Destination page: 1
Page content: Comments

Page 3: WEATHER DATA CHOICE

Origin page: 1 Destination page: 4
Page content: Comments

Page 4: WEATHER DATA

Origin page: 3 Destination page: 5
Available data file: Weather data Type .RA5 (Fixed orientation, all inclinations)
or Type .RA1 (Fixed Orientation, Fixed Inclination)

Menu: Default choice: 2
(F3) - Modify the current weather data file (Reading a new file of current type)
1 - Change the type of file (.RA1 or .RA5)
2 - Keep the current weather data file (indicated in the upper-right corner)

Page 5: SPACE HEATING LOADS: BUILDING CHARACTERISTICS

Origin page: 4 Dest. page: 6 (if option 1), 7 (opt. 2) or 8 (opt. 3)
Available data files: Space heating load (Type .LOD)
Input parameters:

Symbol	Name	Unit	Mini	Maxi	Def.
GV	Building overall heat loss coefficient	W/K	0	9999	250
Tc	Internal set-point temperature	°C	5.0	30.0	20.0
Tnh	No-heating temperature (base for degree-days)	°C	1.0	30.0	18.0

Other input: User's comments

Menu: Monthly Space Heating Loads can be
1 - Provided by the user
2 - Computed by the programme with degree-days from the user
3 - Computed by the programme with degree-days from the programme

Page 6: USER'S SPACE HEATING LOADS

Origin page: 5 (if option 1) Destination page: 8
Available data file: Space heating load (.LOD)
Input parameters:

Symbol	Name	Unit	Mini	Maxi	Def.
LO	Monthly space heating loads (12 values provided by the user)	MJ	0	9999999	0

Other input: User's comments (data source, method used, etc.)

Page 7: USER'S DEGREE-DAYS

Origin page: 5 (option 2) Destination page: 8
Available data file: Degree-days (Type .DDJ)
Output parameter: Tnh No-heating temperature (°C)

Input parameters:

Symbol	Name	Unit	Mini	Maxi	Def.
DD	Monthly degree-days (base Tnh) (12 values provided by the user)		0	999.9	0

Other input: User's comments (data source, method used, etc.)

Page 8: DOMESTIC HOT WATER LOAD

Origin page: 5, 6 or 7 (according to option chosen on page 5)
Destination: 10 (if option 1) or 9 (options 2 or 3)

Input parameters:

Symbol	Name	Unit	Mini	Maxi	Def.
AaUa	DHW back-up tank heat loss coefficient	W/K	0.0	9.9	2.0
Tel	DHW back-up tank environment temperature	°C	10.0	30.0	20.0
m	Average daily hot water demand	Kg/d	0	9999	150
Tm	Average mains cold water temperature	°C	0.0	29.9	15.0
Th	Average required hot water temperature	°C	30.0	90.0	50.0

Other input: User's comments

Menu: Monthly values for m, Tm and Th (Default choice: 1)

1 - To keep these average values for all the months
2 - To modify monthly values initialized with these average values
3 - To provide monthly values without taking account of these average values

<u>Page 9:</u> MONTHLY DHW LOADS

Origin page: 8 Destination page: 10
Available data file: Domestic hot water loads (Type .DHW)
Input parameters:

Symbol	Name	Unit	Mini	Maxi	Def.
m	Average daily hot water demand	Kg/d	0	9999	150
Tm	Average mains cold water temperature	°C	0.0	29.9	15.0
Th	Average required hot water temperature	°C	30.0	90.0	50.0

(12 monthly values for each parameter)

<u>Page 10:</u> SPACE HEATING LOAD CHARACTERISTICS (SUMMARY)

Origin page: 8 or 9 Destination page: 11
Output parameters: (Checking of input parameters from page 5 and 8)
+ Weather data file

<u>Page 11:</u> SPACE HEATING AND DOMESTIC HOT WATER LOAD SUMMARY

Origin page: 10 Destination page: 1
Output parameters: Checking of input parameters (page 9) + load computations
Monthly and yearly values of

DD	Degree-days (base Tnh)	
L0	Space heating load	(MJ)
m	Hot water demand	(kg/day)
Tm	Mains cold water temperature	(°C)
Th	Required hot water temperature	(°C)
L1	Domestic Hot Water load	(MJ)
L	Total load (space heating + hot water)	(MJ)

<u>Page 12:</u> PROJECT REFERENCES

Origin page: 1 (Option 3) Destination page: 13
Available data file: Solar system characteristics (Type .SOL)

Input parameters (User's comments):
Project Name - Location - Date - Author - Comments

Page 13: SOLAR COLLECTOR CHARACTERISTICS

Origin page: 12 Destination page: 14
Output parameter: Collector orientation (degrees)

Input parameters:

Symbol	Name	Unit	Mini	Maxi	Def.
A	Surface area	m2	0.0	999.9	20.0
	Tilt angle	deg.	0.0	90.0	45.0
η_0	Optical efficiency ETA0	--	0.010	1.000	0.800
U1	Overall heat loss coef.: Constant term	(W/K/m2)	0.50	9.99	5.00
U2	Overall heat loss coef.: Proport. term	(W/K2/m2)	0.00	0.10	0.03
m0c0/A	Coll. fluid heat capacitance rate per unit area	(W/K/m2)	5.0	999.9	50.0
don	Control starting differential	(K)	0.0	25.0	5.0
doff	Control stopping differential	(K)	0.0	20.0	1.0

Page 14: TYPE OF SOLAR LOOP HEAT-EXCHANGER

Origin page: 13 Destination page: 15
Menu: (Default choice: 2)

1 - No heat-exchanger (direct loop)
2 - Submerged heat-exchanger defined by its effectiveness
3 - Submerged heat-exchanger defined by its heat transfer coefficient
4 - External heat-exchanger defined by its effectiveness
5 - External heat-exchanger defined by its heat transfer coefficient

Page 15: SOLAR LOOP: HEAT-EXCHANGER AND PIPE LOSSES

Input parameters:

Symbol	Name	Unit	Mini	Maxi	Def.
AinUin	Heat loss for external piping (collector inlet)	W/K	0.0	999.9	0.0
AouUou	Heat loss for external piping (collector outlet)	W/K	0.0	999.9	0.0
Optional input parameters					
	Heat-exchanger: Submerged, defined by its effectiveness (Option 2, page 14)				
e1	Solar loop heat exchanger (collector/storage) effectiveness	--	0.010	1.000	0.500
	Heat-exchanger: Submerged, defined by its heat transfer coefficient (Option 3)				
AeUe/A	Solar loop heat exchanger (collector/storage) heat transfer coefficient (per unit area of collector)	W/K/m2	001.0	999.9	100.0

Symbol	Name	Unit	Mini	Maxi	Def.
	Heat-exchanger: External, defined by its effectiveness (Option 4, page 14)				
e1	Solar loop heat exchanger (collector/storage) effectiveness	--	0.010	1.000	0.500
m1c/A	Heat-exchanger fluid capacitance rate between heat-exch. and storage (per unit area of collector)	W/K/m2	005.0	999.9	050.0
	Heat-exchanger: External, defined by its heat transfer coefficient (Option 5)				
AeUe/A	Solar loop heat exchanger (collector/storage) heat transfer coefficient (per unit area of collector)	W/K/m2	001.0	999.9	100.0
m1c/A	Heat-exchanger fluid capacitance rate between heat-exch. and storage (per unit area of collector)	W/K/m2	005.0	999.9	050.0

Page 16: STORAGE TANKS

Origin page: 15 Destination page: 17

Input parameters:

Symbol	Name	Unit	Mini	Maxi	Def.
M	Main storage tank capacity (water content)	(kg)	00100	99999	01000
AsUs	Heat loss coefficient	W/K	00.0	99.9	05.0
M'	DHW solar storage tank capacity (in the standard version, this parameter is not considered separately; it is included in M)	kg	0000	9999	0150
AsUs'	DHW solar storage tank heat loss coefficient (in the standard version, this parameter is not considered separately; it is included in AsUs)	W/K	00.0	99.9	01.0
Te	Storage tank environment temperature	°C	10.0	30.0	20.0
ApUp	Indoor pipe heat loss coefficient	W/K	00.0	20.0	00.0

Page 17: DISTRIBUTION SYSTEM

Origin page: 16 Destination page: 18

Input parameters:

Symbol	Name	Unit	Mini	Maxi	Def.
e3	Effectiveness of heat transfer between the main storage tank and the DHW distribution	--	0.050	1.000	1.000

Menu: Type of space heating distribution sub-system (default choice: 1)

1 - Heat emitters defined by heat capacitance rate and effectiveness
2 - Heat emit. defined by flow and return temperatures for a nominal power rate
3 - Air distribution loop (defined by heat capacitance rate and effectiveness)

Page 18: HEAT EMITTERS AND AUXILIARY

Origin page: 17 Destination page: 19 (if option 5) or 20
Optional input parameters

Symbol	Name	Unit	Mini	Maxi	Def.
	Heat emitters defined by heat capacitance rate and effectiveness (Opt.1 or 3,p.17)				
m2c2	Water heat capacitance rate	W/K	1	99999	1000
e2	Effectiveness	--	0.010	1.000	0.500
	Heat emitters defined by flow and return temperature (Option 2, page 17)				
Pn	Nominal power	W	1	99999	5000
Tf	Flow temperature (heat emitter inlet)	°C	18.0	99.9	30.0
Tr	Return temperature (heat emitter outlet)	°C	18.0	99.9	35.0

Menu: AUXILIARY CONTROL AND COUPLING (Default choice: 2)
1 - Series connected auxiliary, modulated control
2 - Parallel connected auxiliary, modulated control
3 - Series connected auxiliary, on/off control
4 - Parallel connected auxiliary, on/off control
(Choice has no effect if an air distribution loop has been chosen p.17)

Page 19: AIR DISTRIBUTION LOOP

Origin page: 18 Destination page: 20
Input parameters:

Symbol	Name	Unit	Mini	Maxi	Def.
qf	Fresh air flow-rate	(m3/h)	0	99999	0
qr	Recycled air flow-rate	(m3/h)	0	99999	1000
er	Air recuperator effectiveness	--	0.000	1.000	0.000
Tmax	Maximal air blowing temperature	°C	25.0	50.0	35.0

Page 20: SUMMARY OF DATA

Origin page: 18 or 20 Destination page: 21 or 22
Output parameters: All system parameters (page 12 to 19)
+ Summary of warning messages

Page 21: INPUT PARAMETER WARNINGS

Origin page: 20 Destination page: 22
Output parameters: Detailed warning messages (parameters not in the suitable range)
This page is displayed only if the parameter checking leads to warning messages.

Page 22: SYSTEM PERFORMANCE SUMMARY

Origin page: 20 or 21 Destination page: 23
Output parameters: Monthly and yearly values

Ht	Solar irradiation on the collector surface	MJ/m2/day
L	Total load (space heating + hot water)	MJ
	Auxiliary energy	MJ
	Useful solar energy	MJ
f	Solar fraction	--

System performance calculations are made just before this page is displayed; so, according to your computer type, you may have to wait a little.

Page 23: SYSTEM PERFORMANCE SUMMARY (END)

Origin page: 22 Destination page: 1
Available command: Printing

Output parameters: Yearly values (expressed both in MJ and kWh)

Total load
Auxiliary Energy
Useful solar energy
Useful solar energy per unit area of collector (in MJ/m2 and kWh/m2)
Solar fraction

Summary of warning messages

ESM 1.4: DATA WORKSHEETS

(SOLAR SPACE HEATING + COMBINED HOT WATER)

Input data for ESM 1.4 (SOLAR SPACE HEATING SYSTEMS) can be prepared with the help of these blank worksheets, before running the programme. If you need any information about these parameters or menu options, please consult your AIDE-MEMOIRE ESM 1.4 or Chapters 6 or 9 of this Handbook.

WEATHER DATA CHOICE

Type .RA5 (One orientation, all inclinations) ---> Page 4
or Type .RA1 (One orientation, one inclination) ---> Page 24
Loc.: ________ Orientation: ____ Inclin.: ________
File: ________ Disk: ________ Record: ________

Page 5: SPACE HEATING LOADS: BUILDING CHARACTERISTICS

Available data file: Space heating loads (Type .LOD)
LOAD File ________ from disk ________ Record ________

Page	Symbol	Name	Unit	Value
5	GV	Building overall heat loss coefficient	W/K	____
5	Tc	Internal set-point temperature	°C	____
5	Tnh	No-heating temperature (base for degree-days)	°C	____
5		Menu SPACE HEATING LOAD CALCULATION		____

Comments: ....
....

Page 6: USER'S SPACE HEATING LOADS (Option 1 of menu p. 5)

Available data file: Space heating loads (Type .LOD)
LOAD File ________ from disk ________ Record ________
Comments: ....
....
.... Units: MJ

Month	Jan.	Feb.	Mar.	Apr.	May	June	July	Aug.	Sep.	Oct.	Nov.	Dec.
L0	____	____	____	____	____	____	____	____	____	____	____	____

Page 7: USER'S DEGREE-DAYS (Option 2 of menu p. 5)

Available data file: Degree-days (Type .DDJ)
LOAD File __________ from disk ________ Record ________
Comments: ....
....
....

DEGREE-DAYS:

Month	Jan.	Feb.	Mar.	Apr.	May	June	July	Aug.	Sep.	Oct.	Nov.	Dec.
DD	____	____	____	____	____	____	____	____	____	____	____	____

Page 8: DOMESTIC HOT WATER LOAD: INITIAL VALUES

Page	Symbol	Name	Unit	Value
6		DHW back-up tank heat loss coefficient	W/K	______
6		DHW back-up tank environment temperature	°C	______
6	m	Average daily hot water demand	Kg/d	______
6	Tm	Average mains cold water temperature	°C	______
6	Th	Average required hot water temperature	°C	______
6		Menu monthly DHW values		______

User's comment: ....

Page 9: MONTHLY DHW LOADS

Available data file: Domestic hot water loads (Type .DHW)
LOAD File __________ from disk ________ Record ________

Month	Jan.	Feb.	Mar.	Apr.	May	June	July	Aug.	Sep.	Oct.	Nov.	Dec.
m	____	____	____	____	____	____	____	____	____	____	____	____
Tm	____	____	____	____	____	____	____	____	____	____	____	____
Th	____	____	____	____	____	____	____	____	____	____	____	____

Page 12: SOLAR SYSTEM CHARACTERISTICS

Available data file (p.12): Solar system characteristics (Type .SOL)
LOAD File __________ from disk ________ Record ________
User's comments:
Project name ....
Location ....
Date ....
Author ....
Comments ....
....
....
....
....

Pages 13 to 16:

Page	Symbol	Name	Unit	Value
13	A	Collector surface area	m2	_____
13	β	Collector tilt angle	deg.	_____
13	$\eta 0$	Optical efficiency ETA0	--	_____
13	U1	Overall heat loss coef.: Constant term	(W/K/m2)	_____
13	U2	Overall heat loss coef.: Proportional term	(W/K2/m2)	_____
13	m0c0/A	Collector fluid heat capacitance rate per unit area	(W/K/m2)	_____
13	don	Control starting differential	(K)	_____
13	doff	Control stopping differential	(K)	_____
14		Menu TYPE OF SOLAR LOOP HEAT-EXCHANGER		_____
15	AinUin	Heat loss for external piping (coll. inlet)	W/K	_____
15	AouUou	Heat loss for external piping (coll. outlet)	W/K	_____
		<u>Option 2 of menu p. 14</u> (Submerged heat-exchanger)		
15	e1	Solar loop heat-exch. effectiveness	--	_____
		<u>Option 3 of menu p. 14</u> (Submerged heat-exchanger)		
15	AeUe/A	Solar loop heat-exchanger (collector/storage) heat transfer coeff. (per m2 of coll.)	W/K/m2	_____
		<u>Option 4 of menu p.14</u> (External heat-exchanger)		
15	e1	Solar loop heat-exch. effectiveness	--	_____
15	m1c/A	Solar loop heat-exch. fluid capacitance rate between h.-ex. and storage (per unit area of collector)	W/K/m2	_____
		<u>Option 5 of menu p.14</u> (External heat-exchanger)		
15	AeUe/A	Solar loop heat-exch. (collector/storage) heat transfer coefficient (per m2 of collector)	W/K/m2	_____
15	m1c/A	Solar loop heat-exch. fluid capacitance rate between h.-ex. and storage (per unit area of collector)	W/K/m2	_____
16	M	Main storage tank capacity (water content)	kg	_____
16	AsUs	Heat loss coefficient	W/K	_____
16	M′	DHW solar storage tank capacity	kg	_____
16	AsUs′	DHW solar storage tank heat loss coefficient	W/K	_____
16	Te	Storage tank environment temperature	°C	_____
16	ApUp	Indoor pipe heat loss coefficient	W/K	_____

Pages 17 to 19:

Page	Symbol	Name	Unit	Value
17	e3	Effectiveness of heat transfer between main solar tank and DHW distribution	--	_____
17		Menu SPACE HEATING DISTRIBUTION SUB-SYSTEM		_____
		Option 1 or 3 of menu p.17		
18	m2c2	Heat emitter capacitance rate	W/K	_____
18	e2	Heat emitter effectiveness	--	_____
		Option 2 of menu p.17		
18	Pn	Heat emitter nominal power	W	_____
18	Tf	Flow temperature	°C	_____
18	Tr	Return temperature	°C	_____
18		Menu AUXILIARY CONTROL AND COUPLING		_____
		Option 3 of menu p.17 (Air distribution)		
19	qf	Fresh air flow-rate	m3/h	_____
19	qr	Recycled air flow-rate	m3/h	_____
19	er	Air recuperator effectiveness	--	_____
19	Tmax	Maximal air blowing temperature	°C	_____

SAVING DATA ON DISK FILES (after computation and checking)

Page 6: User's space heating load (Type .LOD)
SAVE File __________ to disk ________ Record ________

Page 7: Degree-days (Type .DDJ)
SAVE File __________ to disk ________ Record ________

Page 9: Domestic hot water loads (Type .DHW)
SAVE File __________ to disk ________ Record ________

Page 12: Solar system characteristics (Type .SOL)
SAVE File __________ to disk ________ Record ________

APPENDIX F

ESM 2.1: AIDE-MEMOIRE AND DATA WORKSHEETS.

(SOLAR HOT WATER SYSTEMS)

ESM 2.1 is a design software for active solar hot water systems. This "aide-memoire" presents input and output parameters for all ESM 2.1 screen pages, with the corresponding symbol used in the standard version of Chapter 7. This will enable the user to find in the corresponding chapter any additional information required and how to get the data in some difficult cases. Input data can then be prepared with the help of the enclosed blank worksheets, before running the programme. Minimum and maximum values are indicated (input data outside the suitable range will be rejected and a new value will be asked for); default values provided by the programme are also indicated. Menu options are also given. When a disk data file can be read or updated from a page, this is indicated.

Page 1: MAIN MENU

Origin page: 2, 7, 13 or 23
Destination page: 3 (Option 2) or 17 (Option 3)

Menu:
1 - Back to DOS (Stopping ESM 1.4)
2 - First step: Weather data and loads. Auxiliary and distribution.
3 - Second step: Solar system characteristics and performance calculation

Page 2: WELCOME

Origin page: 0 Destination page: 1
Page content: Comments

Page 3: WEATHER DATA CHOICE

Origin page: 1 Destination page: 4 (Default) or 5
Page content: Comments

Page 4: WEATHER DATA

Origin page: 3 or 5 (if option 2) Destination page: 6 or 5 (if option 2)
Available data file: Weather data (Type .RA5) (Fixed orientation, all inclinations)

Menu: Default choice: 3
1 - Modify the current weather data file (Reading a new file of type .RA5)
2 - Change the type of file for type .RA1 (fixed orientation, fixed inclin.)
3 - Keep the current weather data file (indicated in the upper-right corner)

<u>Page 5:</u> WEATHER DATA

Origin page: 3 or 4 Destination page: 4 or 6
Available data file: Weather data (Type .RA1) (Fixed orientation, fixed inclination)

Menu: Default choice: 3
1 - Modify the current weather data file (Reading a new file of type .RA1)
2 - Change the type of file for type .RA5 (fixed orientation, all inclin.)
3 - Keep the current weather data file (indicated in the upper-right corner)

<u>Page 6:</u> DOMESTIC HOT WATER LOAD: INITIAL VALUES

Origin page: 4 or 5
Destination: 8 (if option 1) or 7 (options 2 or 3)
Input parameters:

Symbol	Name	Unit	Mini	Maxi	Def.
m	Average daily hot water demand	kg/d	0	9999	800
Tm	Average mains cold water temperature	°C	0.0	29.9	10.0
Th	Average required hot water temperature	°C	30.0	90.0	55.0

Other input: User's comment

Menu: Monthly values for m, Tm and Th (Default choice: 1)
1 - To keep these average values for all the months
2 - To modify monthly values initialized with these average values
3 - To provide monthly values without taking account of these average values

<u>Page 7:</u> MONTHLY DHW LOADS

Origin page: 6 Destination page: 7
Available data file: Domestic hot water loads (Type .DHW)
Input parameters:

Symbol	Name	Unit	Mini	Maxi	Def.
m	Average daily hot water demand	kg/d	0	9999	800
Tm	Average mains cold water temperature	°C	0.0	29.9	10.0
Th	Average required hot water temperature	°C	30.0	90.0	55.0

(12 monthly values for each parameter)

<u>Page 8:</u> DRAW-OFF PROFILE

Origin page: 6 or 7 Destination page: 9
Available data file: Draw-off profile (Type .HW1)
Input parameters:

Symbol	Name	Unit	Mini	Maxi	Def.
qsh	Hourly fraction of daily demand (24 values)	--	0.000	1.000	
tsh	Daily draw-off time	hour	0	24	8

Parameter checking: Sum of qsh equal to 1
Other input parameter: User's comment

Page 9: DISTRIBUTION LOSSES

Origin page: 8 Destination page: 10
Available data file: Distribution circuit and losses (Type .HW2)
Input parameters:

Symbol	Name	Unit	Mini	Maxi	Def.
AdUd	Distribution pipes heat loss coefficient	W/K	0.0	99.9	1.0
Tel	Environment temperature of distribution circuit, recirculation loop and auxiliary tank	°C	10.0	30.0	20.0

Other input parameter: User's comments

Page 10: TYPE OF RECIRCULATION LOOP

Origin page: 9 Destination page: 11 (if option 2 or 3), else 12
Menu: (Default choice: 1)
1 - No recirculation loop
2 - Recirculation loop connected only to auxiliary tank
3 - Recirculation loop connected to auxiliary tank and solar tank (through a controlled three-way valve)

Page 11: RECIRCULATION LOSSES

Origin page: 10 Destination page: 12
Input parameters:

Symbol	Name	Unit	Mini	Maxi	Def.
AcUc	Recirculation loop heat loss coefficient	W/K	0.0	99.9	5.0
Tl	Recirculation loop daily operating time	hours	0.0	24.0	16.0

Page 12: AUXILIARY COUPLING AND CONTROL

Origin page: 10 or 11
Destination page: 13 (Options 1 and 3), 14 (Opt. 2 and 4), or 15 (Opt. 5)

Menu: (Default choice: 3)
1 - Series auxiliary - On/off control
2 - Series auxiliary - Off-peak energy
3 - Series auxiliary with controlled by-pass - On/off control
4 - Series auxiliary with controlled by-pass - Off-peak energy
5 - Instantaneous auxiliary - Perfect control (No losses)

<u>Page 13:</u> AUXILIARY TANK LOSSES (ON/OFF CONTROL)

Origin page: 12 Destination page: 15
Output parameter: Thmin (Minimum required set point temperature)
Input parameters:

Symbol	Name	Unit	Mini	Maxi	Def.
AaUa	Auxiliary tank heat loss coefficient	W/K	0.0	99.9	5.0
Ma	Capacity of auxiliary tank	kg	0	9999	1000
Th1	Auxiliary starting set-point temperature	°C	30.0	120.0	60.0
Th2	Auxiliary stopping set-point temperature	°C	30.0	120.0	60.0

Parameter checking: Th2 > Th1 > Thmin (equality is possible)

<u>Page 14:</u> AUXILIARY TANK LOSSES (OFF-PEAK CONTROL)

Origin page: 12 Destination page: 15

Output parameter: Thmin (Minimum required auxiliary tank temperature)
Thmin' (Minimum auxiliary tank set-point temperature)
Input parameters:

Symbol	Name	Unit	Mini	Maxi	Def.
AaUa	Auxiliary tank heat loss coefficient	W/K	0.0	99.9	5.0
Ma	Capacity of auxiliary tank	kg	0	9999	1000
Th2	Auxiliary stopping set-point temperature	°C	30.0	120.0	60.0
hb	Off-peak energy beginning time	(hours)	00.0	24.0	22.0
He	Off-peak energy ending time	(hours)	00.0	24.0	6.0

Parameter checking: Th2 > Thmin2 (equality is possible)
No draw-off during off-peak energy hours

<u>Page 15:</u> HOT WATER SYSTEM SUMMARY

Output parameters: Weather data file
+ input parameters (pages 9 to 14)

<u>Page 16:</u> HOT WATER LOAD SUMMARY

Output parameters: Monthly and yearly values of
- m daily demand
- Tm mains cold water temperature
- Th required hot water temperature
- Ll Effective solar hot water load (can be met by solar energy) (MJ)

<u>Page 17:</u> PROJECT REFERENCES

Origin page: 1 (Option 3) Destination page: 18
Available data file: Solar system characteristics (Type .HW3)
Input parameters: (User's comments)
Project Name - Location - Date - Author - Comments

Page 18: SOLAR COLLECTOR CHARACTERISTICS

Origin page: 17 Destination page: 19
Output parameter: Collector orientation (degrees)
Input parameters:

Symbol	Name	Unit	Mini	Maxi	Def.
A	Surface area	m2	0.0	999.9	20.0
	Tilt angle	deg.	0.0	90.0	45.0
$\eta 0$	Optical efficiency ETA0	--	0.010	1.000	0.800
U1	Overall heat loss coefficient: Constant term	(W/K/m2)	0.00	25.00	7.00
U2	Overall heat loss coefficient: Proportion.term	(W/K2/m2)	0.00	0.20	0.03
m0/A	Collector fluid mass flow rate per unit area	(Kg/h/m2)	3.0	999.9	40.0
c0	Collector fluid specific heat	(KJ/Kg/K)	2.14	4.19	3.92

Page 19: TYPE OF SOLAR LOOP HEAT-EXCHANGER

Origin page: 18 Destination page: 20
Menu: (Default choice: 2)

1 - No heat-exchanger (direct loop)
2 - Submerged heat-exchanger defined by its effectiveness
3 - Submerged heat-exchanger defined by its heat transfer coefficient
4 - External heat-exchanger defined by its effectiveness
5 - External heat-exchanger defined by its heat transfer coefficient

Page 20: SOLAR LOOP: HEAT-EXCHANGER AND PIPE LOSSES

Origin page: 19 Destination page: 21
Input parameters:

Symbol	Name	Unit	Mini	Maxi	Def.
AinUin	Heat loss for external piping (collector inlet)	W/K	0.0	999.9	0.0
AouUou	Heat loss for external piping (collector outlet)	W/K	0.0	999.9	0.0
Optional input parameters					
	<u>Heat-exchanger: Submerged, defined by its effectiveness</u> (Option 2, page 18)				
e1	Solar loop heat exchanger (collector/storage) effectiveness	--	0.000	1.000	0.500
	<u>Heat-exchanger: Submerged, defined by its heat transfer coefficient</u> (Option 3)				
AeUe/A	Solar loop heat exchanger (collector/storage) heat transfer coefficient <u>(per unit area of collector)</u>	W/K/m2	000.0	999.9	100.0
	<u>Heat-exchanger: External, defined by its effectiveness</u> (Option 4, page 18)				
e1	Solar loop heat exchanger (collector/storage) effectiveness	--	0.000	1.000	0.500
m1/A	Heat-exchanger fluid mass flow rate between heat-exch. and storage (per unit area of collector)	Kg/h/m2	003.0	999.9	040.0

Symbol	Name	Unit	Mini	Maxi	Def.
	Heat-exchanger: External, defined by its heat transfer coefficient (Option 5)				
AeUe/A	Solar loop heat exchanger (collector/storage) heat transfer coefficient (per unit area of collector)	W/K/m2	000.0	999.9	100.0
m1/A	Heat-exchanger fluid mass flow rate between heat-exch. and storage (per unit area of collector)	Kg/h/m2	003.0	999.9	040.0

<u>Page 21:</u> STORAGE TANKS

Origin page: 20 Destination page: 21
Input parameters:

Symbol	Name	Unit	Mini	Maxi	Def.
M	Main storage tank capacity (water content)	(kg)	00100	99999	01000
AsUs	Heat loss coefficient	W/K	00.0	99.9	05.0
Te	Storage tank environment temperature	°C	10.0	30.0	20.0
ApUp	Indoor pipe heat loss coefficient	W/K	00.0	20.0	00.0

<u>Page 22:</u> SUMMARY OF DATA

Origin page: 21 Destination page: 22
Output parameters: All system parameters (page 18 to 21)
+ Summary of warning messages

<u>Page 23:</u> INPUT PARAMETER WARNINGS

Origin page: 22 Destination page: 24
Output parameters: Detailed warning messages (parameters not in the suitable range)
This page is displayed only if the parameter checking leads to warning messages.

<u>Page 24:</u> SYSTEM PERFORMANCE

Origin page: 22 or 23 Destination page: 25
Output parameters: Monthly and yearly values

Ht	Solar irradiation on the collector surface	MJ/m2/day
Ta	Ambient temperature	°C
L	Total hot water load	MJ
	Auxiliary energy (including residual load)	MJ
	Useful solar energy	MJ
f	Solar fraction (with respect to total load)	--
Lc	Recirculation loop thermal losses	MJ

System performance calculations are made just before this page is displayed; hence, according to your computer type, you may have to wait a little.

<u>Page 25:</u> SYSTEM PERFORMANCE SUMMARY (END)

Origin page: 24 Destination page: 1
Available command: Printing

Output parameters: Yearly values (expressed both in MJ and KWh)
Total load split between effective solar load
and residual load
Auxiliary energy
Useful solar energy
Useful solar energy per unit area of collector (in MJ/m2 and KWh/m2)
Solar fraction

Summary of warning messages

ESM 2.1: DATA WORKSHEETS

(SOLAR HOT WATER SYSTEMS)

Input data for ESM 2.1 (SOLAR HOT WATER SYSTEMS) can be prepared with the help of these blank worksheets, before running the programme. If you need any information about these parameters or menu options, please consult your AIDE-MEMOIRE ESM 2.1 or Chapters 7 or 9 of this Handbook.

WEATHER DATA CHOICE

Type .RA5 (One orientation, all inclinations) ---> Page 4
or Type .RA1 (One orientation, one inclination) ---> Page 5

Loc.: ________ Orientation: ____ Inclin.: ________
File: ________ Disk: ________ Record: ________

Page 6: DOMESTIC HOT WATER LOAD: INITIAL VALUES

Page	Symbol	Name	Unit	Value
6	m	Average daily hot water demand	kg/d	______
6	Tm	Average mains cold water temperature	°C	______
6	Th	Average required hot water temperature	°C	______

User's comment: ....
....

Menu: Option number __

Page 7: MONTHLY DHW LOADS

Available data file: Domestic hot water loads (type .DHW)
LOAD File ________ from disk ________ Record ________

Month	Jan.	Feb.	Mar.	Apr.	May	June	July	Aug.	Sep.	Oct.	Nov.	Dec.
m	____	____	____	____	____	____	____	____	____	____	____	____
Tm	____	____	____	____	____	____	____	____	____	____	____	____
Th	____	____	____	____	____	____	____	____	____	____	____	____

Page 8: DRAW-OFF PROFILE

Available data file: Draw-off profile (Type .HW1)
LOAD File __________ from disk ________ Record ________

Hour	0-1	1-2	2-3	3-4	4-5	5-6	6-7	7-8	8-9	9-10	10-11	11-12
qsh	____	____	____	____	____	____	____	____	____	____	____	____
tsh	____	____	____	____	____	____	____	____	____	____	____	____

Hour	12-13	13-14	14-15	15-16	16-17	17-18	18-19	19-20	20-21	21-22	22-23	23-24
qsh	____	____	____	____	____	____	____	____	____	____	____	____
tsh	____	____	____	____	____	____	____	____	____	____	____	____

User's comment: ....
....

Pages 9 to 14 : DISTRIBUTION CIRCUIT, RECIRCULATION LOOP AND AUXILIARY

Available data file (Page 9): Distribution circuit and losses (Type .HW2)
LOAD File __________ from disk ________ Record ________
User's comment: ....
(P.9) ....

Page	Symbol	Name	Unit	Value
9	AdUd	Distribution pipes heat loss coefficient	W/K	______
9	Tel	Environment temperature of distribution circuit, recirculation loop, and auxiliary tank	°C	______
10		Menu TYPE OF RECIRCULATION LOOP		______
11	AcUc	Recirculation loop heat loss coefficient	W/K	______
11	Tl	Recirculation loop daily operating time	hours	______
12		Menu AUXILIARY COUPLING AND CONTROL		______
		Options 1 or 3 of menu p.12 (ON/OFF AUX.)		
13	AaUa	Auxiliary tank heat loss coefficient	W/K	______
13	Ma	Capacity of auxiliary tank	kg	______
13	Th1	Auxiliary starting set-point temperature	°C	______
13	Th2	Auxiliary stopping set-point temperature	°C	______
		Options 2 or 4 of menu p.12 (OFF-PEAK)		
14	AaUa	Auxiliary tank heat loss coefficient	W/K	______
14	Ma	Capacity of auxiliary tank	kg	______
14	Th2	Auxiliary stopping set-point temperature	°C	______
14	hb	Off-peak energy beginning time	(hours)	______
14	He	Off-peak energy ending time	(hours)	______

<u>Pages 17 to 21:</u> SOLAR SYSTEM CHARACTERISTICS

Available data file (p.17): Solar system characteristics (Type .HW3)
LOAD File __________ from disk ________ Record ________
User's comments:
Project Name ....
Location ....
Date ....
Author ....
Comments ....
....
....
....
....

<u>Pages 18 and 19 :</u>

Page	Symbol	Name	Unit	Value
18	A	Collector surface area	m2	____
18	β	Collector tilt angle	deg.	____
18	η0	Optical efficiency ETA0	--	____
18	U1	Overall heat loss coef.: Constant term	(W/K/m2)	____
18	U2	Overall heat loss coef.: Proport. term	(W/K2/m2)	____
18	m0/A	Collector fluid flow-rate per unit area	(kg/h/m2)	____
18	c0	Collector fluid specific heat	(kJ/kg/K)	____
19		Menu TYPE OF SOLAR LOOP HEAT-EXCHANGER		____

Pages 20 and 21:

Page	Symbol	Name	Unit	Value
20	AinUin	Heat loss for external piping (coll. inlet)	W/K	_____
20	AouUou	Heat loss for external piping (coll. outlet)	W/K	_____
		Option 2 of menu p. 19 (Submerged heat-exchanger)		
20	e1	Solar loop heat-exch. effectiveness per unit area	--	_____
		Option 3 of menu p. 19 (Submerged heat-exchanger)		
20	AeUe/A	Solar loop heat-exchanger (collector/ storage) heat transfer coeff. (per m2 of coll.)	W/K/m2	_____
		Option 4 of menu p.19 (External heat-exchanger)		
20	e1	Solar loop heat-exch. effectiveness	--	_____
20	m1/A	Heat-exchanger water mass flow-rate between h.-ex. and storage (per unit area of collector)	kg/h/m2	_____
		Option 5 of menu p.19 (External heat-exchanger)		
20	AeUe/A	Solar loop heat-exch. (collector/storage) heat transfer coefficient (per m2 of collector)	W/K/m2	_____
20	m1/A	Heat-exchanger water mass flow-rate between h.-ex. and storage (per unit area of collector)	kg/h/m2	_____
21	M	Main storage tank capacity (water content)	(kg)	_____
21	AsUs	Heat loss coefficient	W/K	_____
21	Te	Storage tank environment temperature	°C	_____
21	ApUp	Indoor pipe heat loss coefficient	W/K	_____

SAVING DATA ON DISK FILES (after computation and checking)

Page 7: Domestic hot water loads (Type .DHW)
SAVE File __________ to disk ________ Record ________

Page 8: Draw-off profile (Type .HW1)
SAVE File __________ to disk ________ Record ________

Page 9: Distribution circuit and losses (Type .HW2)
SAVE File __________ to disk ________ Record ________

Page 17: Solar System Characteristics (Type .HW3)
SAVE File __________ to disk ________ Record ________

APPENDIX G

FORTRAN SUB-ROUTINE TRIFC

COMPUTATION OF CFC'S

(CUMULATIVE FREQUENCY CURVES OF SOLAR IRRADIANCE)

INPUT:	GLOBAL(J,I)	Hourly values of global irradiation (J/cm2) on the plane of interest, to be sorted as a CFC. Missing values are replaced by -999 J = day (from 1 to 31) I = hour (from 1 to 24) This array is introduced through a labelled COMMON: /RESULT/
	NJOUR	Number of days to be considered (Maximum: 31)
	IHMAX	Number of hours to be considered per day (maximum: 24)
OUTPUT:	MJOUR	Number of days used for the CFC (Days with missing values are eliminated)
	SG	Monthly average of daily irradiation (J/cm2)
	DMIS(J)	Number of missing days (J = day, from 1 to 31)
	NH(K)	Mean daily number of hours during which the average irradiance has exceeded 25*(K-1) W/m2 K = irradiance bin (from 1 to 48)
	FN(K)	Average irradiance (W/m2) of the K-th bin NH and EN are taken out of the sub-routine through a labelled COMMON: /FC/

```
      SUBROUTINE TRIFC(NJOUR,MJOUR,SG,IHMAX,DMIS)
C
C
C     Sorting of a monthly file of solar irradiation
C     GLOBAL for the computation of cumulative frequency
C     curves of solar irradiance. Hourly values of solar
C     irradiation (GLOBAL) are in J/cm2. Bins of 9 J/cm2
C     (25 Wh/m2) are used (i.e. 48 bins).
C     Bin averages are in the array EN (Unit: W/m2).
C
      DIMENSION N(48),DMIS(31)
      COMMON /FC/EN(48),NH(48)
      REAL NH
      COMMON /RESULT/GLOBAL(31,24)
      MJOUR=0
      SG=0.
      DO 1 I=1,48
      NH(I)=0.
      EN(I)=0.
    1 N(I)=0
C     Loop on days
      DO 2 JOUR=1,NJOUR
      IF (DMIS(JOUR).EQ.-1)GOTO 2
      DO 21 IH=1,IHMAX
      IF(GLOBAL(JOUR,IH) .LT. 0.)GOTO 2
      IF(GLOBAL(JOUR,IH) .GT. 423.)GOTO 2
   21 CONTINUE
C     Loop on hours
      DO 20 IH=1,IHMAX
      IF(GLOBAL(JOUR,IH) .EQ. 0.)GOTO 20
      I=INT(GLOBAL(JOUR,IH)/9.)+1
      IF(I .GT. 48)I=48
      EN(I)=EN(I)+GLOBAL(JOUR,IH)
      N(I)=N(I)+1
      SG=SG+GLOBAL(JOUR,IH)
   20 CONTINUE
      MJOUR=MJOUR+1
    2 CONTINUE
      IF (MJOUR.EQ.0)GOTO 999
      SG=SG/MJOUR
C
C     Final loop on bins
      IF (N(48).EQ.0) GOTO 31
      NH(48)=FLOAT(N(48))/FLOAT(MJOUR)
      EN(48)=EN(48)/FLOAT(N(48))*25./9.
31    CONTINUE
      DO 3 I=1,47
      IF (N(I).EQ.0) GOTO 35
      EN(I)=EN(I)/FLOAT(N(I))*25./9.
35    CONTINUE
    3 NH(48-I)=NH(49-I)+FLOAT(N(48-I))/FLOAT(MJOUR)
999   RETURN
      END
```

SUBJECT INDEX

9 780792 312307

GPSR Compliance

The European Union's (EU) General Product Safety Regulation (GPSR) is a set of rules that requires consumer products to be safe and our obligations to ensure this.

If you have any concerns about our products, you can contact us on ProductSafety@springernature.com

In case Publisher is established outside the EU, the EU authorized representative is:

Springer Nature Customer Service Center GmbH
Europaplatz 3
69115 Heidelberg, Germany

Batch number: 10584518

Printed by Printforce, the Netherlands